PILOT JUDGMENT AND CREW RESOURCE MANAGEMENT

Pilot Judgment

and Crew Resource Management

RICHARD S. JENSEN

Routledge
Taylor & Francis Group
LONDON AND NEW YORK

First published 1995 by Ashgate Publishing

2 Park Square, Milton Park, Abingdon, Oxon OX14 4RN
711 Third Avenue, New York, NY 10017, USA

Routledge is an imprint of the Taylor & Francis Group, an informa business

First issued in paperback 2016

British Library Cataloguing-in-Publication Data

Jensen, Richard S.
Pilot Judgement and Crew Resource Management
I. Title
629.13252

Library of Congress Control Number: 95-75938

ISBN 13: 978-0-291-39804-8 (hbk)
ISBN 13: 978-1-138-26310-9 (pbk)

Contents

Foreword
By Allan Diehl

For over half a century human factors problems have been recognized as the major cause of aviation accidents. Methods of reducing 'human error' have evolved over these decades. During the 1940s, because of the Second World War, the focus was upon improved selection tests. The 1950s saw the widespread use of ergonomic principles in cockpit design which improved the human-equipment interface. By the 1960s the focus had moved to instructional systems design. These techniques improved the training of procedural and perceptual-motor skills of the individual aviator, largely through the use of high-fidelity simulators.

In the 1970s several developments occurred which would radically change aviation training. One event was precipitated by the 1973 Middle-East War. The ensuing fuel crisis generated numerous requests for the US Federal Aviation Administration (FAA) to reduce the number of flight hours required to qualify for various ratings. This agency's initial reaction was that pilots needed the specified flight time so they could develop 'judgment'.

But the FAA soon began to question its own assumption, that judgment was simply a bi-product of experience. The agency then took the bold step of launching the first comprehensive study into pilot judgment. R.S. Jensen was selected as the principal investigator for this landmark study (Jensen and Benel, 1977). This study revealed that judgment errors were involved in the majority of US general aviation accidents, and hypothesized that training programs could be developed to enhance pilot judgment.

While Jensen was engaged in this judgment research, another important effort was undertaken by the scientists at the US National Aeronautics and Space Administration (NASA) Ames Research Center. The NASA endeavor, led by Dr John Lauber, was the genesis of the now familiar Crew Resource Management (CRM) training programs (Wiener, Kanki, and Helmreich, 1993). This research cleverly integrated evidence from airline accident investigations, incident reports, aircrew interviews, as well as simulator based studies. Interestingly, while Jensen's work had examined the issues of how the individual general aviation pilot thinks, the NASA focus was on group decision making of airline flight deck crews. The NASA effort was also influenced by other contemporary work such as Line Oriented Flight Training (LOFT). This concept was then being developed by Northwest Airlines to provide realistic human factors training scenarios in flight simulators.

As these watershed efforts were maturing, other important developments were also taking place. During this decade, the US Government initiated airline deregulation, while the National Transportation Safety Board (NTSB) had increased its commitment to examining the underlying causes of human performance related mishaps (Nance 1986). Large numbers of jets were also entering service equipped with cockpit voice and flight data recorders. These remarkable 'black boxes' where effective in revealing the role judgment errors, communication problems and crew coordination failures played in causing accidents.

One such accident occurred on Dec 28, 1978 when an United Airlines DC-8 crew, distracted by a landing gear problem, ran out of fuel and crashed in Portland Oregon. I was assigned as the NTSB human factors group co-chairman for this investigation. After studying this and similar airline accidents, it was obvious that the time had come to apply the CRM concepts. Therefore, in 1979, I drafted the first formal recommendation calling for the operational implementation of CRM training by the US Airlines.

In 1980, shortly after becoming the human performance scientist for the FAA, I was asked to oversee the first operational evaluation of Jensen's theories of judgment training. The experiment was performed at Embry-Riddle Aeronautical University (Berlin, Gruber, Holmes, Jensen, Lau, Mill and O'Kane, 1982). In this demonstration, student pilots who had received such training out-performed their contemporaries by making significantly fewer errors. These results verified the Jensen's earlier prediction that training could improve pilot judgment.

With the approval of the FAA, United Airlines launched the world's most extensive CRM training program in 1981. Several other airlines were carefully watching this pioneering effort. In 1982, I wrote a FAA briefing paper titled 'Developing a Comprehensive Program to Improve Pilot Judgment'. This paper suggested that CRM and judgment training were highly similar concepts. It also recommended the agency examine methods of certifying such programs, as well as develop specialized versions of the these programs tailored to the various categories of airmen.

In 1983 the FAA tasked Jensen with examining methods of certifying CRM training programs. Based on success of this effort, three years later, the agency further tasked Jensen with developing training manuals to enhance decision making for instrument rated private pilots, commercial (air taxi) pilots, and muli-crew operations (CRM). I believe these prototype manuals represent some of the first examples of the 'practical human factors materials' which the International Civil Aviation Organization would later call for in their 1989 resolution (A26-9).

By the mid-1990s, these programs have become an integral part of many airline and military aircrew training programs, particularly the US. Progressive organizations throughout the world have also adopted these concepts to training flight attendants, mechanics, dispatchers, flight test engineers, and air traffic controllers. Some organizations have even extended the techniques to non-aviation occupations, e.g. sailors, industrial plant operators, emergency room physicians. But ironically the programs have made only limited inroads into the general aviation arena; where they are arguably most needed, - given the high accident rates.

R.S. Jensen enjoys a unique position in being instrumental to the launching these truly revolutionary aviation training concepts. His pioneering research, distinguished career as an aviation educator, and extensive flying experience allows him to provide a highly practical treatise of the nexus between these two vital topics: Pilot Judgment and Crew Resource Management.

References

Jensen, R.S. and Benel, R.A. (1977). *Judgment Evaluation and Instruction in Civil Pilot Training*. Springfield, VA: National Technical Information Service, Final Report FAA-RD-78-24.

Berlin, J.I., Gruber, J.M., Holmes, C.W., Jensen, P.K., Lau, J.R., Mills, J.W. and O'Kane, J.M. (1982). *Pilot Judgment Training and Evaluation*. DOT/FAA Report CT-82-56, Vols I, II, III, Washington, DC.

Nance, J.J. (1986). *Blind Trust*. New York: William Morrow and Company, Inc.

Wiener, E.L., Kanki, B.G. and Helmreich, R.L. (1993). *Cockpit Resource Management*. New York: Academic Press.

Acknowledgments

In writing this book, I am standing on the shoulders of many who have gone on the road before me and I dedicate it to them. Anchoring the bottom of the pyramid of shoulders is Dirk Van Dam, the director of Moody Aviation at the time I attended, who provided the first challenge to me in the area of judgment and whose quotations appear frequently in this book. My instructors, Ron Royce and Reed Berry were key judgment examples in my early in flying. My mentor, Stan Roscoe, Director of the University of Illinois Aviation Research Laboratory, provided the freedom and encouragement to me as a young research associate to pursue this line of research. My fellow Illinois graduate student, Russ Benel, worked with me on the first funded aviator judgment contract from the FAA. Dr. Al Diehl, while at the FAA, provided encouragement and funding in the middle years of my judgment and CRM research, much of which is used throughout this book. Capt. Janeen Adrion-Kochan, graduate student and colleague from the middle years to this day, created many of the ideas and words that appear throughout this book. The 'Judgment Junkies', Pete Hwoschinsky, Russ Lawton, Rich Adams, each advanced my own thinking on judgment in the middle years. David O'Hare, a colleague from New Zealand, advanced much of our judgment work and I have quoted his work as well. Larry Kirkbride, and OSU graduate student, provided many ideas and words on soaring in Chapter 13. The 'OSU Research Team', Jerry Chubb, Dave Fournier, Chuck Ventola, John Fisher, Trent Taylor, Steve Caberto, Jim Guilkey and Lori Rakovan, provided many bright ideas and words presented in Chapter's 1 and 14 and represent the future of aviator judgment development. Finally, David Hunter, our FAA contract monitor, has provided ideas, encouragement and funds to advance the field of aviator judgment in recent years, for whose sake I must say that the ideas and data reported in this book are those of the author and do not necessarily reflect those of the sponsoring organization. Other research sponsors in the middle years whose products are used in this book include Sandy Hart, NASA, Brice Hartman, USAF, Clay Foushee, NASA Ames Research Center, Pat Russell, FAA, and Pete Hwoschinsky, FAA.

On a personal note I must thank my first mentor, Marshall Mikesell, who, as my boss from his platform on a painter's scaffold, provided many words of wisdom while I worked my way through undergraduate school. He, together with my brother Paul, provided financial support for my flight training at Moody. This book has taken over four years to complete and therefore the people I most need to thank are my family who have patiently endured it with me, my wife Kathy (who also labored with me through the index), and my children Mark, Rachel, Julie and Angie.

1 Introduction to judgment error

The remarkable thing about people who fly is not that they are so clever; it's that they are so lucky. The clever ones realize it, and even the least perspicacious catch some spark of a hint of their good fortune. It's what keeps them going despite poverty, divorce and bad horoscopes. Sometimes, it keeps them going despite bad weather as well, and that is the dangerous side of the matter: the sirens' song to which few can stop their ears. The most difficult of the pilot's arts is that of knowing when not to fly. It is also the last test of the real pilot; for if flying is cunning and mastery, it is good sense and self-mastery as well. Until a pilot has learned to transcend the faint glory that shines through flight and to subordinate it to his own will, he has still not learned the most essential part of imitating birds: that birds fly because it is natural to do so, because they need to do so, and not because they admire themselves for doing so...Of all the risks that the joy of flight requires, that of falling to the misuse of it is the greatest of all.

Peter Garrison (1972)

The importance of judgment in aviation was made clear to me, when, as a new hire, my boss and I flew a Cessna 205 with a load of passengers from the USA midwest to Colorado for a meeting at the Air Force Academy during the early spring. After landing for fuel in Kansas, we prepared for our final leg to Colorado Springs where the weather was 'Sky obscured, 1/4 mile visibility' and forecast to remain the same for the rest of the day. At my boss's suggestion, we filed for Colorado Springs, with Denver Arapaho (weather, 'clear') as the alternate. During this leg, my boss (flying in the right seat) was on the radio from time to time asking for updates on the weather at Colorado Springs. 'Has anyone landed there today'? 'No', was the reply, 'One Lear tried earlier in the day but missed'. Despite this, he kept asking me to consider going to Colorado Springs.

As we approached Denver (we actually could see Arapaho airport.) he said, 'Let's go to Colorado Springs and give it a try'!

I thought about the situation we would face on a minimums approach into Colorado Springs. I could see him trying to press me (I was flying in the left seat.) to go 'A little lower' and arguing with me about whether or not we could see anything on the ground while trying to maintain glideslope and localizer approaching 200 feet from the ground! He was making me feel 'chicken'. Fortunately, at this point, my better judgment took over and I said, 'No, we're going to Arapaho'! and I took the radio and asked for clearance and landed at Arapaho airport. He did not say anything the rest of the way and on the return trip I was given a seat in the rear of the plane.

That was not my introduction to pilot judgment. My roots in aviation judgment began on my first 'check ride' with Dirk van Dam, the director of the Moody Institute of Aviation, the only civilian flight school during the 1960s that actually taught pilot judgment. As we departed the 300 foot traffic pattern altitude from Moody Woodale Airport (located within the Chicago O'Hare Airport Traffic Area about where Runway 27L now ends, see Figure 1.1) to intercept the outbound corridor, seeing no inbound traffic, I made my climb out of 300 feet to 500 feet before passing through the inbound corridor. At this point, van Dam said, 'Take it back, you failed because you used poor JUDGMENT'! After a very uncertain landing and rollout to a stop, he gave me a lecture on judgment. He said that I had flown right through the inbound corridor at its designated altitude. I explained that my instructor had told me to check for inbound traffic and, if there was none, to go ahead and climb to 500 feet through the inbound corridor because good judgment says flying at 300 feet above the congested Bensenville residential area is not a good idea. Van Dam said that I should have at least announced my intentions to him and furthermore, he said that he had noticed that I had a bad attitude in general which I did not understand at all. I did not argue further but put it into the back of my mind for later study.

Moody taught judgment to their students because its purpose was to train missionary pilots to fly alone in isolated 'bush' locations around the world. Moody wanted its pilots to graduate with judgment sufficient to face the many difficult choices between medical passenger requirements and runway limitations, between missionary needs and weather conditions, etc. A fairly extensive judgment training program was in place at Moody as a part of normal training and testing for Federal Aviation Administration (FAA) certification. After completing my training at Moody, Dirk van Dam generously provided a statement on judgment training that I included in the appendix to my first FAA pilot judgment study (Jensen and Benel, 1977). I have used exerpts from that statement throughout this book.

From these examples you may begin to believe that judgment may be required of general aviation pilots but is rarely used in the highly standardized world of the airline pilot. This is not the case. Consider the following scenario offered by Fahlgren and Hagdahl (1990):

Figure 1.1 The Moody Woodale Airport from the air in the early 1950s

A DC-9 captain is making an approach to Tromso airport in the most mountainous area of northern Norway. The glide path is steep due to mountain peaks. He has a tailwind. The runway is short and braking action is poor due to snow and ice. His new co-pilot is making his first flight into this area, so the experienced and skilled captain demonstrates how it should be done by calculating everything in detail and making a very precise approach. Due to tailwind and a steep glide path his descent is 1,100 feet per minute when fully established on the ILS at the correct speed.

At 1,000 feet, his co-pilot calls out: 'One thousand feet not stabilized' (their SOP says that descent rate shall be maximum 1,000 feet per minute in the stabilized concept). Below clouds, with field in sight, the captain deliberately leaves the steep glide path knowing that an approach angle of 2.5 to 3 degrees will give the shortest landing distance. This action, in combination with an uphill condition before threshold starts the next warning from the ground proximity warning system: 'Glide path -- Glide path -- Glide Path' and then, 'Rate of descent -- Rate of descent -- Rate of descent'.

The captain has made an approach using expert judgment gained during 35 years of flying experience. He has correctly perceived all relevant factors in the situation and expertly used all information. But -- his reward is three slaps in the face from a system without a brain. What will be the reaction of the pilot to a system that requires uniformity and standardized procedures and does not appreciate individual expert performance? A system that prescribes rules, and designs technology to make people behave in the same way regardless of the situation. What happens to the ability of the person to create novel solutions -- still the most important advantage that people have over machines? Does the person fall into blind obedience and complacency rather than use the skills that he or she has gained?

Purposes for this book

My goal in writing this book is not to teach you to make every aviation decision correctly. Instead, it is to make you more aware of and sensitive to the many factors that affect your judgment so that you will improve your chances of making the correct aviation decisions. There is a certain truth to the aviation cliché, 'The superior pilot is one who exercises superior judgment to avoid situations that require superior skill'. Much of what you will learn in this book is aimed at making you sensitive enough to recognize those situations that could be beyond your level of skill and to teach you to have the courage to turn away from those situations in the face of feeling as if you are 'chicken'.

It is my belief that, as pilots are aware of biases that exist in normal human decision making, that awareness is sufficient to reduce those same errors in rational judgment. For example, knowing that our visual/perceptual system will deliver a bias in our perception of distance to the runway depending on the slope of the terrain on the near end, is sufficient to cause a pilot to attempt to control his mental view of the situation and counter the perception given by his or her eyes using other information (e.g. altimeter indications). Accordingly, this book will provide a set of biases in rational cognitive judgment that will enable pilots to compensate for more hazardous biases in difficult decision situations.

During the past 13 years, I have been invited to teach judgment to a variety of aviation audiences in several countries around the world. The audiences have ranged from general aviation pilots, airline pilots, military pilots, Forest Service pilots, balloon pilots, glider pilots and air traffic controllers. Each of these audiences has surprised me with the variety of situations they face involving aviation judgment. The USA FAA for years has required that examiners test pilots for judgment and has suggested that instructors teach it to their flight students. Yet, neither a definition of judgment nor criteria for examining it has been offered in these regulations. More recently, Transport Canada has begun to require judgment training and evaluation for pilots at all levels (Transport Canada, 1992). Other governments are considering such regulations.

Accordingly, another purpose of this book is to provide material that judgment and CRM (crew resource management) instructors can use in such courses.

This book is not meant to provide another basic analysis of human decision making including utility theory and many other theories that have been offered in academic texts. Pilot judgment cannot be explained well using such theories anyway. Instead, this book begins with a descriptive approach to decision making and ends with a prescriptive approach to pilot judgment that can make a difference in how pilots make decisions, hopefully to improve safety. Between these two approaches, ideas and models developed in other fields are presented because they advance our understanding of aviation judgment. My method includes an extensive array of my own experiences as a pilot and those of other pilots.

When I first began aviation judgment research, most aviators did not understand what I meant by judgment. After I explained it, few in the FAA (including Pat Russell, the contract monitor) believed anything could be done about it. Today, I occasionally run into an aviator who believes that judgment cannot be taught or evaluated but this is becoming more and more rare. Through a series of experiments in the USA and Canada and through the experiences of flight programs which have used the training concepts as well as the experiences of air carriers and the military with CRM training we now have sufficient proof that judgment and CRM can be taught and evaluated, and that such teaching can have a major impact on aviation safety (Diehl, 1992). Thus a final purpose of the book is to persuade you that judgment and CRM can and must be taught and evaluated in aviation.

Definitions of judgment and CRM

So that we are clear from the start let us define our most important terms, judgment and CRM. There is a certain truth to the popular belief that judgment is simply 'common sense' applied to making decisions. 'Sense' implies knowledge and intense awareness, realization and understanding of all the factors involved in making a decision. Sense is generally seen as a person's ability to act effectively and positively in any given situation. 'Common' refers to the fact that societal needs and wishes are known and applied to the decision. In Chapter 2 these two concepts come together to formulate a working definition of good pilot judgment that will be used throughout the book.

Another popular conception of judgment is that it is identified as equivalent to the outcome. "If the decision outcome was positive it was 'good judgment', and if it turned out badly, it was 'bad judgment'". This restatement of, 'The end justifies the means', is clearly wrong in my view of aviation judgment. Judgment is not an end, but a *process* for achieving an end.

I would like to define aviation judgment as the *mental process* used to formulate aviation decisions (Jensen, 1982). The decision is the final stage of this mental process. As important as outcome is to aviation, its only part in the judgment process is to provide feedback to help us learn more for the next decision. Much more will be offered on this topic in the next chapter.

CRM is the effective management of all resources available to the crew. CRM is simply the application of aviation judgment to a multi-person crew environment. It is helpful to apply the term 'judgment' when referring to single-pilot operations and CRM when referring to multi-person crews in aviation. CRM adds the social dimension to judgment and requires a strong emphasis on interpersonal communication. In every other way, the two terms mean the same.

One last word about words. Although I use the term 'judgment' in this book, many people, including those in the USA's FAA, have preferred the term 'aeronautical decision making' or ADM to refer to the same concept. Their reasoning is that the term 'judgment' carries with it a negative connotation. Although I mildly disagree with this reasoning, I have used the term 'ADM' in some of my own writing when the work was supported by the FAA. I believe that the term 'judgment' carries the more correct meaning of the concept.

In fact, some revealing pilot insight into the meaning of the two terms (judgment and decision making) was discovered in a recent survey of flight instructors at the University of Newcastle, Institute of Aviation in Australia where instructors said that they thought their commercial pilot graduates had achieved a high level of *decision making ability* but they disagreed with the statement that their students had achieved a high level of *professional judgment* (Ross, 1992). Clearly, a distinction was being made between the two terms by these flight instructors. I believe that the instructors were using the term 'decision making ability' to refer to purely rational information processing. 'Professional judgment' carries a broader meaning to aviation people including the ability and willingness to add societal ethics to the decision making process. I believe that the term 'judgment' encompasses more of the meaning behind the process.

Early aviation judgment research

When I began research on pilot judgment, the focus of attention in pilot training was on technical flying. In such traditional training pilots received a great deal of aeronautical knowledge, skills and experience. It was expected that such training would lead to safe pilots with good judgment. As one FAA person (Bernie Gyer) told me in my visit to the FAA in 1973, pilots learn judgment through hours of experience in the air. The FAA specified that the amount of experience needed for the private pilot certificate was 40 hours (35 in a FAR Part 141 'Approved School') and for the commercial pilot certificate, 250 hours (180 in a FAR Part 141 'Approved School'). In fact, no one really knew if this amount of experience would lead to judgment at all.

Although we still do not know whether these levels of flying experience yield pilots with good judgment, we do know that certain kinds of training lead to better judgment in pilots. We also know that flying requires much more than skilful manipulation of the controls. Some enlightened pilot training programs are now focusing on certain aspects of pilot judgment including pilot attitudes toward risk taking, risk assessment and management techniques, problem diagnosis and resolution and stress management.

Furthermore, traditional flight training was conducted on an individual basis -- one student and one instructor. In this system students were trained and tested as if they would always act alone in an aircraft. They did not work as a cockpit team member until they got into the cockpit of an airliner. In addition, the flying career was sought after by many who preferred to work alone because they thought of the job as an individual rather than as a team effort. After being taught and tested in this way, many did not fit well into the cockpits where they had to work as members of a team. Team performance suffered and over 60 airline accidents between 1960 and 1978 were attributed to poor cockpit resource management (Lauber, 1979).

Today most airlines and many US flying military units are trying to correct this training deficiency through CRM courses. Some innovative flight schools are beginning to teach pilots in teams (instead of individually) from the beginning of their training. The International Civil Aviation Organization (ICAO, 1989) has mandated that their member countries teach 'human factors' courses to pilots. Some countries have begun to require that pilots pass tests on judgment and human factors (Canada and England). Such training has already proven effective in reducing accidents (Diehl, 1992). Furthermore, CRM training serves to elevate the flying career above the level of a 'flying truck driver' to a manager who leads a team of people responsible for large transportation systems, each filled with hundreds of people and equipment worth tens of millions of dollars.

Human error

To understand the problem of aviation judgment it is helpful to look at human error and the human information processing system. Our awareness of the 'human error problem' in aviation began when cockpit voice recordings of pilot communications prior to fatal accidents revealed the enormous factor played by human error in accidents (80 per cent in general aviation, 70 per cent in airline operations). The result was a very large proportion of accidents that were listed as 'pilot error' (see Figure 1.2). Pilots have long been dismayed by the use of this term to describe their failures because it sounds as if the pilot was negligent in some way. In most circumstances that have resulted in pilot error accidents, the pilots involved were putting forth their best efforts to fly safely. To say that an accident was caused by pilot error and go no further is no better than saying that a mechanical failure accident was caused by the airplane. In both cases we are obligated to look further into the reasons for the pilot or mechanical failure that led to the accident. Obviously, mechanical failures are easier to trace because there is usually an abundance of physical evidence to show the way. The reasons for pilot failures are more difficult to discover but they usually can be found as well, if one is willing to use psychological and sociological expertise and probe into the behavior and thought patterns of the crew involved.

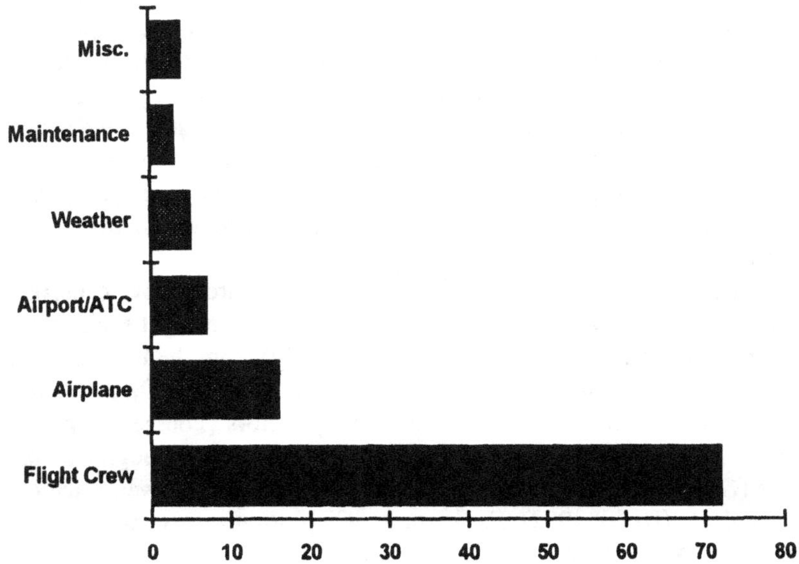

Figure 1.2 Causes of airline accidents as percentages of the total

Pilot error definition

James Reason, in his book, *Human Error*, defines human error as, 'any human activity that fails to accomplish the intended outcome' (Reason, 1991). In continuous controlled flying we are always comparing intended or desired outcome with actual outcome (Roscoe, 1980). As these the actual outcome departs from the intended outcome, we notice it, make a decision, and act by moving the controls to bring the two closer together. In fact, straight and level flight can be defined as a series of *error* correction maneuvers. *To err is human*, to correct errors is what flying is about. Why is it so much easier for pilots to accept that they make these types of control errors and so hard to accept judgmental errors?

Pilots who don't want to admit to any weaknesses, sometimes define pilot error in curious ways. While flying in the jumpseat of an airliner, I once asked the captain about the type of errors that he had observed or made during his flying career. He responded by saying that he had never made an error. He said that every time he had made a mistake, he had recognized and corrected it before it became an error. However, on this particular flight soon after takeoff, I noticed his co-pilot diplomatically point out to him that his altimeter was set at 29.92 in.

Hg., a setting that the captain had not corrected since descending through 18,000 feet on his previous flight, although he had received new settings at three previous points (descending through 18,000 feet, on approach, and on the following takeoff). This captain, whose opinion is not shared by most, prefers not to call a mistake an error unless it results in accident.

Certain types of errors may be difficult to observe because they are obscured in the human information processing system (perceptual and memory errors) or in the machine being controlled. In complex systems with a large amount of inertia, such as nuclear power plants and airplanes, the erroneous movement of a control can be concealed if the error is recognized and corrected early. For example, in a study of advanced cockpit displays for use in curved approaches (Jensen, 1981), very small performance differences (deviation from intended horizontal and vertical course) were observed among the first, second and third-order predictor algorithms as displayed on a CRT in the cockpit. However, an examination of the aileron, elevator and rudder control movements by the pilot revealed highly significant differences among the three levels of this factor indicating that pilots in some conditions were engaging in a great deal of 'trial-and error' control behavior, perhaps to gain perceptual information about the visual scene (Owen *et al.*, 1981). Thus depending on how far back into the pilot's mental model and control manipulation processes one wishes to go in the definition of error, the true human behavioral error rate may be much greater than the observed performance error rate. Certainly it is much larger than the rate observed by the above-mentioned airline captain who waits for the NTSB report to admit to an error.

What types of errors do pilots make?

Paul M. Fitts, a pioneer in aviation human factors, together with Jones, saw that many aircraft accidents in the US military were being caused by the pilot and sought answers to determine why. Believing that at least a part of the reason for the pilot error accidents was in the design and location of the cockpit instruments and controls, he surveyed pilots asking them what types of errors they had made or observed other pilots make during their experience as aviators. In all, pilots reported 270 different types of instrument reading errors (Fitts and Jones, 1947a) and 460 different types of control errors (Fitts and Jones, 1947a).

Control errors were classified as follows (the wording has been edited somewhat into terminology that is used today):

- *Substitution errors*: Confusing one control with another (for example, Confusing flap and gear controls).

  Adjustment errors: Operating a control too rapidly or too slowly, moving a switch to the wrong position, or following the wrong sequence in operating several controls (for example, turning the fuel selector to the wrong tank).

- *Forgetting errors*: Failing to check, unlock, or use a control at the proper time (for example, taking off with the flight controls locked).
- *Reversal errors*: Moving a control in a direction opposite to that necessary to produce the desired results (for example, moving the trim control in the wrong direction).
- *Unintentional activation*: Inadvertently operating a control without being aware of it.

Instrument reading errors were classified as follows:

- *Misreading errors:* Multi-indicator instruments (for example, misreading three-pointer altimeter by more than 1,000 feet).
- *Reversal errors*: Reversing the interpretation of an instrument indication (for example, misinterpretation of gyro compass)
- *Signal interpretation errors*: Misinterpretation of the message conveyed by sound or light warnings.
- *Legibility errors*: Errors resulting from difficulty in seeing the numbers or scales distinctly.
- *Substitution error*: Mistaking one instrument for another (for example, confusing manifold pressure and tachometer).
- *Using an instrument that is inoperative*: Unknowingly accepting as valid the indication of an instrument that is not reliable.
- *Scale interpretation errors*: Errors that result from difficulty in interpolating between numbered gradations of scale.
- *Errors due to illusions*: Misconceptions due to conflict between body sensations and instrument indications and errors due to visual illusions in poor weather.
- *Forgetting errors*: Failing to check instruments prior to takeoff or during flight.

These two classes of pilot errors, identified by Fitts and Jones, were enormously important to the refinement of cockpit instrument panels. One of the results of their study was the 'basic T' arrangement of flight instruments that has been the cornerstone of standardization in the cockpit since the late 1940s. As shown in Figure 1.3, the focal point of the arrangement is the Attitude Indicator which is the instrument on which pilots spend most of their visual attention time. The other three instruments serve as confirmation sources of information and take on more or less importance depending on the maneuver being performed.

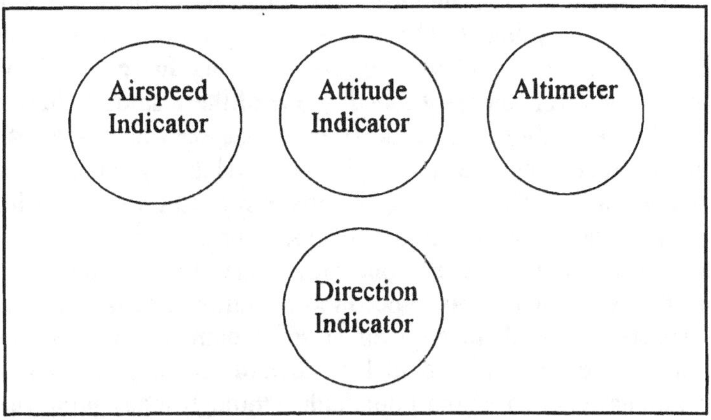

Figure 1.3 'Basic T' arrangement of flight instruments

A second development to come out of the two classes of pilot errors published by Fitts and Jones was a host of human factors research studies to determine how to design instruments and controls for pilots, in particular, but also for people in general who are controlling machines. It is important in human-machine interface design to take advantage of learned stereotypic behavior to minimize the amount of training necessary. For example, we learn through our developmental years that when we see something deviate from where it should be, we move it back in the opposite direction. Naturally, then when people get to the cockpit of an airplane and see a symbol deviating from center, the natural response is to move the controls in the opposite direction to bring it back. This is known as 'control-display compatibility' and is one of the basic tenets of display and control design in the field of human factors.

To see how pervasive are pilot errors, let us look at a more recent flight experiment. In the early days of area navigation (RNAV), the FAA, in an effort to define objective ways to approve airborne systems for certification (FAA, 1969), recognized both continuous control of steering errors (called 'pilotage error') and procedural or computer insertion errors (called 'blunders'). The former were specific values used for purposes of establishing total RNAV system error budgets for certification of airborne avionic equipment. The latter were defined as failure to enter information required for the definition of RNAV routes or entry of uncorrected erroneous values that could have caused the airplane to be steered (perhaps very well) off the desired course. The deviation from the intended course that led to the shooting down of Korean Airlines Flight 007 is a probable example of this type of error.

At the University of Illinois Aviation Research Laboratory number of RNAV simulation and flight experiments were conducted in which reasonably high cockpit workloads were imposed (Jensen, *et al*., 1971; Jensen and Roscoe, 1971; Jensen and Roscoe, 1974; Jensen and Marsh, 1976). In each of these studies the experimenter knew ahead of time the errors that might occur in each operation, he knew the number of required RNAV procedural operations, and during the experiment, he recorded the number of errors occurring for each. Table 1.1 presents the procedural error summary found in one of these studies (Jensen and Roscoe, 1974). The variability in procedural error data seen here suggests that wherever manual procedures are required under relatively high levels of workload, errors will occur. They also suggest that error rates are higher for less frequent than they are for more frequently used operations.

Three conclusions can be drawn from these data, the second and third following from the first. First, every procedural operation that is required of a pilot can be expected to result in an error at some point in time. Second, to reduce the number of errors one should look toward reducing the number of procedural operations required of the pilots either through redesigning the task (as we did) or through automation of the equipment (as many engineers have done since). Third, methods must be found to identify and tell the pilot when he or she has made an error before it becomes catastrophic (design for error tolerance).

Table 1.1
Procedural errors in an RNAV task

Error Type	No. of Operations	No. of Errors	Error Rate
Nav Freq	1620	6	.004
Radial	1620	23	.014
Distance	1620	18	.011
Course	1620	35	.022
Altitude	360	16	.044
Glideslope	360	13	.036
Enter	11860	13	.001

The human must be considered as an irreplaceable but imperfect component in the aircraft system. The human is irreplaceable because there yet exist many tasks that he or she can do better than any machine or computer yet devised (McCormick and Sanders, 1982). However, as we have seen, even those tasks for which the human is better suited than the machine, he or she can be expected to make errors. Thus means must be sought to make these errors tolerable to safe operation in the aviation environment.

Since the Fitts and Jones study, most of the research in the cockpit has focused on the technical aspect of the flying task. Instruments have been provided to 'sensitize' pilots so that they can see these errors easily and quickly and make the corrections before they become catastrophic. Some of the indications provided have become very sophisticated, including a voice that tells you to 'Pull Up' when you are approaching the ground at an unsafe rate without being in the landing configuration. We will see that the new challenge is to devise training and other techniques to sensitize pilots to errors in judgment. Roscoe provided a concise summary of the purpose of human factors '...to design machines to minimize the need for training, and to provide training to complete the job left undone in the design' (Roscoe, 1980).

What pilot errors are causing accidents?

As part of the original FAA judgment study, Russ Benel and I developed a taxonomy of pilot errors designated by the NTSB in accident investigations (Jensen and Benel, 1977). Three broad categories of behavioral activities were identified: procedural, perceptual-motor and decisional activities. Table 1.2 is a summary of fatal and non-fatal procedural, perceptual-motor and decisional error proportions of pilot-caused accidents that occurred during the time period from 1970 through 1974 in US civil aviation.

Procedural activities are those flight activities involving setting switches, selecting frequencies, programming the computer and making required communications. They are usually discrete events that may require careful timing. Therefore, cognitive processes are involved.

Perceptual-motor activities are those flight activities that involve continuous control movements in response, primarily, to what the pilots sees in the environment. The visual information may come from what he or she sees by looking out the window or by looking at the various instruments. The human functions involved are sensory, perceptual and motor processes.

The third type of activity is the most difficult to handle in realistic flight environments because it involves pure cognitive activities and judgments. Cognitive judgments are complex in that they usually involve a large number of relevant pieces of highly probabilistic information, they usually require the specification of and choice from among several alternatives, and they are frequently affected by emotions, values and social pressures. In addition, cognitive judgments usually permit some deliberation before a control response is required.

Since the statistics from Table 1.2 were published, other agencies have examined their own pilot-caused accident rates finding similar results. For example Diehl (1991) found that of 28 major mishaps in the US Air Force, 24 were caused by crew error and in these accidents there were 16 procedural errors,

21 perceptual-motor errors and 48 decisional errors. In an earlier study of major US Air Force accidents from 1987-1989, Diehl (cited in Diehl, 1992) found that 113 of 169 were caused by aircrew error, including 32 procedural, 110 perceptual-motor and 157 decisional errors.

Table 1.2
Fatal and non-fatal accident proportions in per cent of total for US civil aviation pilot caused accidents from 1970-1974

Pilot Behavioral Activity	Fatal	Non-Fatal
Procedural	4.6	8.6
Perceptual-Motor	43.8	56.3
Decisional	51.6	35.1

Source: Jensen and Benel, 1977

Table 1.3 compares three different flying organizations in terms of the proportion of pilot-caused accidents in each of the three categories of pilot behavior. One can see from these data that decisional activities represent a very high proportion of the pilot-caused accidents and thus deserve considerable attention.

Table 1.3
Comparison of aircrew errors in major accidents (including both fatal and non-fatal) for four flying organizations

	Category of Error		
	Procedural	Perceptual-Motor	Decisional
Civil Aviation	9%	56%	35%
Airlines	19%	25%	56%
Military	11%	37%	53%

What types of judgment errors do pilots make?

More recently, a great deal of effort has gone into discovering ways to assist pilots in avoiding, recognizing and correcting errors in judgment. Such errors took longer to recognize as a problem, perhaps because people did not really think anything could be done about them. The Fitts and Jones survey did not reveal much of a problem with judgment perhaps because of the way they phrased the question to pilots. In an effort to bring out judgment related errors, I asked civilian pilots in the Columbus, Ohio area to identify errors that they had made or observed other pilots make in the following areas (Jensen and Chappell, 1983):

- general aircraft handling
- emergency aircraft control and management
- system management including navigation
- communications
- decision making/judgment
- pre-flight operations including crosscountry planning
- IFR operations including approaches
- multiengine procedures

The responses received were far more pervasive than those of the Fitts and Jones study. Out of the 150 pilots who were sent the survey more than half of them (80) responded. Overall, 1,500 errors were reported, of which over 300 were considered unique errors. Table 1.4 is a list of the errors listed as 'Decision Making/Judgment' (see Jensen and Chappell (1983) for the complete taxonomy).

Table 1.4
Decision making/judgment part of pilot error list from survey of Columbus pilots

- continued flight with poor radio
- continued flight with malfunctioning alternator
- didn't stop for fuel because of 'get-homeitis' and landed with critical fuel
- lack of cross check to find precise problem
- lack of desire/ability to gain information
- doing IFR approaches in an aircraft not usually flown
- flying in ill health
- omitting engine instrument scan
- failing to recognize symptoms of carburetor ice in icing conditions
- reluctance to declare emergency due to fear of being penalized
- failure to do 180 degree turn when potentially deeper trouble lies ahead
- failure to ask for help when a severe problem develops
- failure to do an emergency or unscheduled landing when necessary
- failure to follow the published or approved procedure
- lack of knowing own equipment and personal limitations
- presuming that you are a better pilot than your recent experience
- making decisions in haste

- not having alternate plan of action formulated in case of weather problems
- failure to provide margin of safety
- proper decision made too late
- interpreting information incorrectly
- incorrectly determining rate of build up of ice and which way to fly
- failure to recognize/accept 'system' limitations
- failure to properly weigh conditions in the environment
- allowing the desire for instant gratification to bring about poor decision
- getting pressured to fly when job is at stake
- deferred to judgment of more experienced pilot - when he was wrong
- a go around was in order but, trying to save money, landed anyway
- lack of information (not getting the whole picture)
- using out-of-order priorities (i.e., get-homeitis)
- failure to make decisions concerning weather minimums, expecting atc to make your decisions
- optimistic view of a really poor situation
- flight with known non-working equipment
- flying overloaded or with improper c.g.
- continued VFR into IFR weather
- flying while fatigued or while taking medication (e.g., antihistamines)
- flight into icing conditions
- flying at wrong VFR cross-country altitudes
- tried to beat thunderstorm to airport. won - barely.
- entering clouds without clearance
- not complying with aircraft limitations
- unable to detect changing conditions that affect safe flight weather or mechanical
- lack of self-discipline to maintain performance at an adequate level
- inadequate area clearing procedures: head down on the instruments
- failure to use a check list

Source: Jensen and Chappell, 1983

Why do people (and pilots) make errors?

To understand why pilots make errors that result in accidents, let us look at the human information processing system which, though it has many outstanding strengths that cannot be replaced by machines, is fraught with potential error sources. This human information processing system is all we have and by becoming aware of its weaknesses, we can discover ways to better avoid and/or correct our errors before they become catastrophic.

Figure 1.4 gives a broad overview of the human information processing system. From an examination of this overview, one can see both how information is processed and a few reasons why people are prone to make errors. On the left side information enters the body at an enormous rate, perhaps as much as millions of bits per sec through the various sensory organs including the skin, eyes, ears, nose, taste, as well as though internal organs and various combinations of sense organs working in concert.

Because the network of communication channels in the human body as well and the brain is not able to handle information at the rate that the sense organs receive it, mechanisms are provided, called 'filters', to reduce the information down to a manageable size. The first filter occurs at each of the sense organs. The eye, for example, receives all visual information at the lens but, from that point on, reduces it to a fraction of that amount before it reaches the optic nerve on its way to the brain. The light entering is bent and turned upside down (by the lens), colored (by the fluid in the eye ball), and some is cut out entirely (by the wiring to the rods, cones and optic nerve). Filters performing similar functions work at each of the other sensory organs. As a result, information is reduced and lost, leading to possible error in the information getting to the brain for decision making. This is the first source of error in human information processing.

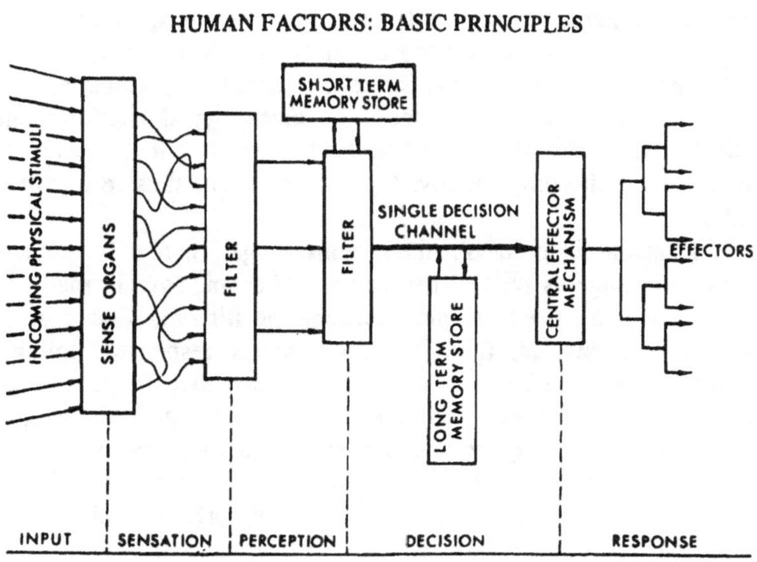

Figure 1.4 Overview of human information processing system (after Allnutt, 1982)

The second source of error occurs when the perceptual system tries to put information that it knows it lost back in. This occurs at the perceptual filter that works with the short term memory store. Short term memory is the memory that we have for events that have occurred during or about the past thirty seconds. In aviation, this is the first stage of 'situational awareness' and it applies equally well to pilots and air traffic controllers. This filter takes what it knows about the current situation and fills in the holes with what it thinks should be there. In the case of the eye, each of the distortions mentioned above are corrected to match the current situation as understood by the person. This is the reason for illusions that we accept so easily as reality (see Figure 1.5).

The third source of error occurs in the decision channel where we are severely limited (to about 4 bits per sec). We have two tasks in this channel. First, we must choose, from among several items competing for attention, what to process first and then how to process it. This decision stage consists of cognitive or thinking activity involving the combination of preprocessed information with information from long-term memory. Between short-term memory and long-term memory there exists working memory. In addition to being a structure used for temporary information storage, perhaps up to several minutes, working memory is also a process that is essential to many other cognitive tasks such as reasoning, comprehension. In aviation, this is the second stage of situational awareness (Tsang, 1992, p 9; Baddeley, 1986).

There exists, in this stage, an enormous source of error because not only does the mind deal with the rational process of making the best decision possible, but from long term memory comes all sorts of motivations, attitudes, emotions, distractions, predispositions (personality traits) and facts, some of which have nothing to do with the rational solution to the immediate problem but everything to do with the 'humanity' of the process. To develop good aviation judgment we must gain control of this stage of the decision process so that we can reduce the errors made here or become sensitive to those errors and thus correct them before they hurt us.

The final stage in the process is the response stage. There exists some potential for error in this stage as well. When activities become 'overlearned', responses can occur without the need for going through the filters and decision processes mentioned above. Instead, these so-called 'motor responses' follow learned (habit) patterns and can be very accurate, provided that the pattern selected is correct. Some flying activities, such as correcting for a slight attitude upset due to turbulence especially during landing, are made in this way by experienced pilots. Some of the best examples of this activity are found in musicians who accurately move their fingers far faster than one could possibly predict if all of the decisions went through the channels mentioned above.

It should also be mentioned that it takes time for information to pass through all of the channels and filters of the human information processing system. This time delay is another source of potential human error. In aviation, decisions sometimes must be made quickly and parts of the process must be short and incomplete.

Classification of human error in aviation

To help us understand judgment error further, let us use the three basic types of human activities in aviation presented earlier, procedural, perceptual-motor or control and judgment. *Procedural errors* are those blunders that pilots make when they move a switch in the wrong direction (such as the gear or flaps) or type the wrong information into the computer. These errors are primarily the result of failures in the response mechanism of our human information processing system. *Control errors* consist of erroneous movements of continuous control devices such as control wheels. In the flight task these errors include altitude, course and airspeed control errors. Control errors find their source in the sensory, perceptual and response mechanisms of the information processing system. *Judgmental errors* are those errors that result from failures in the cognitive process or the decision mechanism of our information processing system. Examples are in-flight decisions to fly into weather conditions that are worse than the pilot's skill level and an air traffic control decision to place a faster aircraft behind a slower aircraft on approach with insufficient distance between them to allow the first to land and get off the runway.

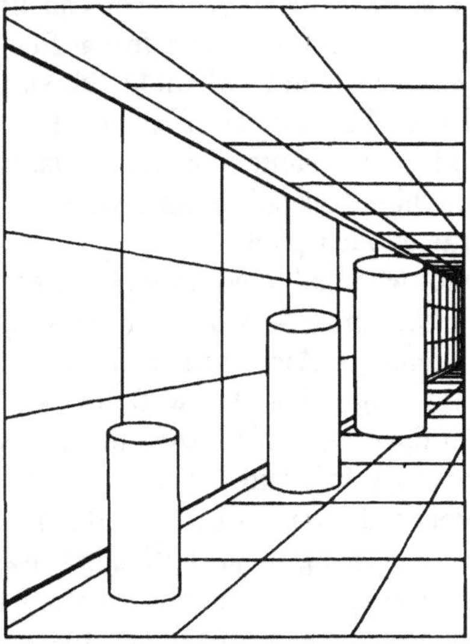

Figure 1.5 The corridor illusion. The cylinders in this perspective drawing of a corridor are depicted as being of increasing size and distance, but on the page they are all identical (From Gibson, 1977)

One of the major weaknesses of the human in the aviation system (compared with computers and other cockpit systems) is that he or she has feelings that can be hurt when their errors are revealed. Pilots do not want to be seen making any of the errors mentioned. However, of the three types of errors, control errors are the least offensive, particularly if they do not get outside of the range of 'acceptable' performance. Procedural errors are next in order of acceptability to pilots. Virtually all of pilot training is focused on minimizing these two types of errors. Judgment errors are the least acceptable to pilots because they are often taken as downgrading the person when they are pointed out or discovered. In some cases there is genuine reason for fear of retribution if judgment errors are discovered. Therefore, pilots try to hide their judgment errors from their peers and especially their passengers and superiors.

An airline captain friend of mine related a story from his experience in which he had to go through an excruciating mental exercise to maintain his mind on the task of flying. He was flying as co-pilot in a B-737 and after making seven approaches to minimums that day, he was tired. This was his leg to fly, and during the approach to the airport, he experienced the worst case of vertigo he had ever experienced in his flying career. He found himself going off course and glideslope as he tried to fight off the strong conflicting feelings arising as a result of sensations from his eyes and his body. To regain control of himself required the greatest effort he had ever used over his mind. He completed the approach successfully and after landing he asked the captain if he had noticed anything unusual about his performance on the approach. The captain said that he noticed what looked like a complacent aircraft control effort at first but that it turned out OK. My friend then revealed the extent to which he felt vertigo on that approach. The captain then reported it to the medical officer for the airline and my friend had another major battle to fight to maintain his flying status. The medical officer thought there must be something basically wrong with a pilot who had such an experience *and* revealed it to another pilot.

Another example of a pilot who had to go through the greatest mind control of his life was told to me by a US Forest Service pilot. This pilot was flying a Twin Otter with a load of smoke jumpers close to the tree tops when he was caught in 'down air'. With the stall warning going off and flying below the tree tops at the end of a clearing, he had to force himself to resist the overwhelming urge to pull back on the control wheel to try to avoid the trees. He said that his whole life past before him as he approached the trees convinced that he was going to crash into them. After successfully clearing them, he landed, walked off the airplane and fell unconscious to the ground. He too told his story to superiors and was reprimanded for telling the story because of the negative effect it might have on US Forest Service aviation.

In fact, each of these types of errors must be treated openly and equally if they are to be avoided and corrected. The stigma that seems to follow judgment errors needs to be dropped in aviation circles if we are to learn to have better judgment. Judgment errors, like control and procedural errors do not make one a poor pilot. They result from weaknesses that we all have. They need to be recognized and corrected before they become accidents.

How do we deal with error in aviation?

Modern cockpit technology has provided several mechanisms for dealing with procedural and control errors in aviation. Less has been done to deal with judgment error. The primary mechanisms for reducing and identifying all forms of error are cockpit design, checklists, training and currency, regulations, self-checking, mnemonics and other crew members.

Cockpit design. It is sometimes surprising to pilots that one of the greatest contributions to reducing pilot error can be made through improvements in cockpit design. It is beyond the scope of this book to go into the many ways that design can improve pilot performance. Such improvements began with the Fitt's 'Basic T' arrangement of flight instruments mentioned above and continues with the addition of computers and automation today. I would refer the reader to a recent book for pilots that deals in more depth with this issue, *Human Factors for General Aviation* (Trollip and Jensen, 1991).

Checklists are used to prevent forgetting errors and to sequence certain procedural operations in the cockpit. They have been a part of aviation since the earliest days of flight because it was found that it is very easy to become distracted in flight and to forget something important. Despite the heavy emphasis on the use of checklists, many pilots do not always use them. We know this because certain kinds of accidents (such as the inadvertent gear up landing which happens on average once every two days in the USA) can only be the result of not using a checklist. Although the purpose of the checklist is to help the pilot avoid procedural errors, not using it is an error in judgment.

Training and currency are meant to provide proficiency in the operation of the aircraft so as to avoid all types of errors. We will see later that another important function of judgment training is to make pilots sensitive to decisional errors so that they can be corrected before they become accidents.

Regulations are an essential guide to preventing errors. If adhered, they can prevent many errors that lead to accidents. However, there is no guarantee of safety even when following the regulations. Furthermore, there are times when regulations must be interpreted for the given circumstances and sometimes it is essential that pilots depart from the regulations for the safety of the operation. A case in point is the Air Florida B-737 accident at Washington National airport in 1982. The pilots were so caught up in company SOPs and regulations that they were not able to function safely. Perhaps the most important to them, a company directive that was drilled into the pilots, was to not over speed the engines because the engines then be required to go through an expensive examination before they could go back into service. The second was an FAA regulation to reduce power for noise abatement in Washington. Both of these regulations needed interpretation by the pilots in these circumstances and should have been disregarded for the safety of the operation. More will be offered about this accident later.

Self-checking is another function of good pilot judgment to discover and prevent error. Good pilots develop elaborate self-checking mechanisms. After completing my landing preparation, on short final, I always say out loud, 'Gear indicates down and locked' whether the aircraft has a retractable or fixed gear. After landing, in retractables, I always say, as I reach for the flaps, 'I have the flaps'. Saying these phrases out loud causes the human response mechanism to trigger the sensory mechanism to check again that the proper action has taken place.

Mnemonics are used by pilots as mental devices to prevent errors when written checklists are not practical. The function of acronyms is the same as the written checklist but require the use of memory. For example the word, 'GUMPS' (Gear, Undercarriage, Mixture and Props) is used in preparation for landing. Mnemonics are not as reliable as written checklists because you must rely on memory (another source of error) both for initiating the sequence of procedures and for interpreting meaning throughout the procedure.

Other crew represent one of the best resources for the identification and prevention of errors of all types. Unfortunately, as we will see later in the CRM chapter, pilots are usually not trained to use other crew members as resources in this way. They are trained and tested individually and many have chosen flight as a profession because they enjoyed working alone. However, to be effective as an error detection resource requires that all members of the crew must be willing to encourage and accept the identification of his or her own errors (including decisional errors) by other crew. This requires a high level of self-esteem and teamwork abilities in all members of the crew.

Why is judgment so important?

Judgment is important in flying because regulatory bodies, such as the FAA, have written the regulations in such a way that pilots are expected to exercise judgment in all phases of flight. In the regulations, you as a pilot are given an enormous amount of responsibility through the application of judgment for your safety and that of others. This responsibility manifests itself in the following ways:

1 *Command of the aircraft:* Regulations clearly state that pilots are to be in command of their aircraft at all times. This means, for example, that you should not give up your authority to make decisions to demanding passengers by letting them unduly influence your judgment.

2 *Adherence to the regulations:* Pilots are expected to adhere to the regulations. This is so basic that it should not need to be mentioned. But, flying by yourself, you are not being monitored by anyone, and it is easy to think you can bypass or stretch the rules. Remember, other pilots fly by the rules expecting that you do too. For example, in VFR conditions, you are expected

to fly at certain altitudes and visually check for traffic. Overtaken traffic expect overtaking traffic to follow this rule, in particular.

3 *Interpretation of the regulations:* The FAA regulations are offered under the expectation that pilots will interpret them based on their own skill and aircraft capabilities. Because the FAA has a dual purpose of promoting aviation and regulating it for safety, it aims its restrictive regulations at the operator with the most skill and equipment capability -- not the one with the least. In the USA, the FAA testing program (for the certification of aircraft equipment and flight procedures) uses the best pilots and equipment to test all operations that it regulates -- not mediocre pilots and equipment. Although pilots do not consider themselves to be mediocre, few have the skill of these test pilots. Therefore, the decision to attempt a certain maneuver (for example, a minimum's approach), is left to you, the pilot. You alone are responsible for assessing your own skill level, state of fatigue and health and the reliability of your equipment for the task you face.

What are the factors that affect our judgment?

The mental factor

The mental factor consists of knowledge, intelligence, information processing skills, computational skills and recall abilities as they affect one's capability for rational judgment. This factor is important to a pilot's ability to diagnose problems, generate alternative solutions and compute risks in flight situations. This factor has been studied extensively in psychology under such topics as subjective probability and mathematical decision theory (Lee, 1971; Edwards and Tversky, 1967). Chapter 3 provides a brief background of this topic as it relates to aviation judgment.

The moral factor

The moral factor refers to the influence of ethical or moral values on pilot judgment. Ethical values are dispositions to behave in certain ways because one believes that he or she ought to behave in those ways. These dispositions are developed early in life but are continually modified by one's environment. Philosophers would say that the moral value of a decision is the moral difference between a choice made by an individual and the choice that society would have made in the same situation. The moral factor tends to bias the decision in the direction believed to be the societal norm.

Parducci (1968) describes a relevant experiment on moral values of college undergraduates. In his experiment he gave one group of subjects a list of mild acts of wrongdoing. A second group of subjects received a much nastier list of acts of wrongdoing. Embedded within both lists was a common set of acts of wrongdoing. Subjects were asked to rate each act on their list independently on his or her own moral value scale from 'not particularly bad' to 'extremely evil'. Results of the experiment showed that in the context of the nastier list, every item common to both lists was rated more leniently than they were rated in the context of the milder list. For example, pocketing a tip left by the previous customer was given a much lower rating in the context of the nasty list than it was in the context of the mild list. This experiment shows how difficult it is to maintain absolute moral values applicable to every situation. It also shows how easily our moral values can be influenced by our environment.

In aviation, these results suggest four implications. First, pilots must be aware that their motivation to choose certain courses of action is influenced by the context in which they are operating. Second, as a significant moral influence to his or her students, the flight instructor has an opportunity and a responsibility to show good judgment to his or her students. Third, the experiment clearly shows that this aspect of judgment can be taught. Fourth, because passengers (and flight students) expect pilots to make decisions based on their (societal) norms (or even more cautiously), it is the responsibility of the pilot to know these norms and to bias decisions in their direction.

The emotional factor

The emotional factor refers to the effects of emotional arousal in the form of fear, worry, stress, anxiety, euphoria, etc. on pilot judgment. Moderate levels of emotional arousal are necessary for optimal performance but high levels have a detrimental effect (Yerkes and Dodson, 1908). Worry and euphoria are especially bad for decisional activities because, even at low levels, they divert attention from the decisional task to the individual himself (Wine, 1971; Liebert and Morris, 1967). Stress may result from concern over the material or social losses one may suffer from whichever course of action is chosen -- including the costs of failing to live up to prior commitments (Janis and Mann, 1977). Stress may also result from loss of self-esteem as a competent decision maker. Surprisingly, stress from fear of physical danger in flight is usually not a factor with experienced pilots until they are very close to an accident.

The emotional factor apparently affects decision making in one of two ways: it may either have a negative affect diverting attention away from the task at hand resulting in a reduced ability to discriminate among factors important to the decision, or it may have a positive affect mobilizing attention, thereby improving discriminability. In either case, it tends to bias the decision in the direction of following through with prior plans and commitments.

The physiological factor

The physiological factor refers to the effects of various physiological factors such as illness, fatigue, alcohol and drugs (including over-the-counter drugs), on pilot judgment. All of these factors are extremely dangerous because they tend to reduce the ability of the pilot to sense and process information. They also have the effect of biasing decisions in the direction of greater risk. For this reason, laws have been established to prevent pilots from flying under their influence. Even a full bladder can bias the decisions of pilots, perhaps to land as quickly as possible.

The social factor

The social factor refers to the effects of pressures from people to follow certain courses of action. Laboratory studies have shown that an individual can be induced to conform to the wishes of others by applying pressures that weaken his confidence in his own beliefs while strengthening his confidence in the correctness of the other's views. In a laboratory study Asch (1951) found that subjects conformed to the incorrect perceptual judgments in a group setting, even in the face of obvious evidence to the contrary. This finding supports what is referred to as social influence. Deutsch and Gerard (1955) divided social influence into normative and informational types which they believed were relevant to Asch's findings. The distinction is that 'normative' applies to an influence to conform to another's expectations, while 'informational' applies to influences to accept information from others as evidence about reality.

It is not surprising that judgment of others (particularly those who are perceived to be competent) should be taken as important evidence to be weighed in making choices. We all know from our past experience that others are frequently reliable sources of information. Once a group is formed (even if loosely structured), 'normative social' influence assumes greater importance as well. Thus, the presence of a group can have subtle effects on the outcome of an individual's judgments.

Sources of social pressure that could affect pilot judgment include fellow pilots, family members, air traffic controllers, passengers, superiors and friends. Effects of social pressure on a pilot who is unsure of his judgment can be great, particularly when his initial belief opposes that of the group. In these circumstances, a very real possibility exists for judgment errors leading to accidents because the pilot's information may be better than that of the group. Errors biased toward the limit of regulations and pilot capabilities are as likely as bias toward over-caution in these circumstances.

The personality factor

The personality factor refers to the effects that certain personality traits have on pilot judgment. Each of us develops strategies over the years to best accomplish our goals of dealing with life and the people around us. Some of these strategies become deeply ingrained and are known as personality traits. These traits are well established by the age of six and are difficult to change thereafter.

There are many aspects of personality that are known to affect judgment. Let us look at three. Two such traits are test anxiety (motive to avoid or fear of failure) and defensiveness (excessive concern with image maintenance). Kogan and Wallach (1964) found that individuals with these two motivational dispositions tend to demonstrate a maximum of conservatism in risk-taking tasks. Individuals with lower levels of these traits tend to evaluate the relevant information more carefully, and they would be more objective in the assignment of likely outcomes of various alternatives.

A third personality trait that has been found to influence judgment is 'locus of control'. A person with high internal locus of control attributes his or her success or failure to his or her own skills and abilities as if he or she is in control. Persons with high external locus of control tend to attribute their own success or failure to luck or forces outside of themselves. Jasper Shealy (1974) found that down-hill skiers with high a external locus of control had a significantly higher injury rate than those with a high internal locus of control.

We can be quite sure that personality factors can influence a person's flying judgment. However, because of the nature of personality traits, it would be futile to attempt to change them in pilot training. Furthermore, most aviation psychologists believe that personality traits are important only in the most extreme cases and can be handled through selection of pilots, at least for airline crew positions.

The attitude factor

Attitudes are personal strategies for dealing with the world that are less deeply ingrained than personality traits, and can be changed, especially under pressure from several sources at the same time. We are constantly bombarded with attempts to change our attitudes by teachers, theologians, advertising people, parents, peers and superiors. Attitudes toward risk taking in aviation are developed within the pilot as a result of flying training and experience. Many of these are developed through watching how our flight instructors deal with decisional issues. Attitudes may be either positive, tending to make us take a safe or cautious approach to flying risk, or negative, causing us to take greater risks than we should.

Attitudes are primarily affected by non-safety factors such as job demands, convenience, monetary gain, self-esteem and commitment. If properly developed, this part of pilot decision-making would minimize information unrelated to the safety of the flight and direct the pilot's decision toward the use of more rational information and processes.

There is no logical, rational reason why pilots should allow hazardous attitudes to influence their decisions. However, these forces are a part of human nature and therefore cannot be avoided entirely. They are, after all, one of the major reasons why we fly at all. Power, speed, adventure, image, are motivations to learn to fly and they stay with us after we become professionals. Good judgment says that we should recognize that hazardous attitudes are present in every human decision and that we must make these attitudinal forces subservient to our more rational thought processes. Unlike personality traits, attitudes can be readily modified through training.

The philosophical factor

The philosophical factor plays a part in most of the factors given above but let us restrict discussion here to the insight it provides into the meaning of judgment. There will always be philosophical differences between aviation people concerning the meaning of judgment. In a nutshell the problem is choosing between absolute criteria that do not vary with circumstance (i.e., an act is judged right or wrong regardless of the circumstances) and relative criteria (i.e., acts are judged in some respect dependent on the conditions in which the judgment occurs). There is by no means a hard and fast distinction between the two because absolute judgments may be deviations from established procedures for specific situations, or, in the *a posteriori* sense, they may be identified strictly with error.

However, if error is the sole criterion of bad judgment, the issue is obscured, for we have merely identified the two phenomena and are now faced with the problem of separating them in our assessment of situations in general. Similarly, if judgments are entirely relative, there is no sense in asking which are good and which are bad since there is literally no general sense of judgment. The correct path may be a happy medium with absolute judgments applied for standard procedures in specific kinds of situations; always modified by the condition that such fixed procedures are based on the relative judgments of society in general.

What is pilot judgment?

As we said earlier, pilot judgment is the mental process that we use in making decisions. To avoid confusion, it is important to distinguish between perceptual and cognitive judgment. Pilots are constantly making decisions based on their visual perceptions. For example, they judge distance, clearance, altitude, closure rate and speed. These perceptual judgments are highly important to the pilot's task of controlling the aircraft. However, they do not require much thought and are relatively easy to learn and to perform consistently, particularly when the pilot is aware of the anomalies and illusions that can occur.

Although perceptual judgment is important to aviation, the type of judgment which we are considering is different. Our present concern is with more complex forms that we refer to as 'cognitive' judgment (Figure 1.6).

PERCEPTUAL JUDGMENT — COGNITIVE JUDGMENT

Increasing Cognitive Complexity ⟶

Figure 1.6 Perceptual/cognitive judgment continuum

In comparison with perceptual judgment, the characteristics of cognitive judgment are:

- the information available is more uncertain
- the pilot usually has more time to think
- there are usually more than two alternatives
- the risks associated with each alternative are harder to assess
- the final decision is more easily influenced by non-flight, background factors such as stress, fatigue, financial pressure, personal commitment and so on.

'Expertise' in aviation

Many decisions that we make in flying, when we have little training or experience are cognitive -- that is, they require a considerable amount of thought. Later, with experience, these decisions become perceptual. For example, when you first learn to land the airplane, you have to think, 'Am I too high, or too low'? Your instructor may tell you, 'You're too high, push the nose down'! Those thoughts and instructions are cognitive. It does not take much training time before you are able to look at the view on final and know, almost automatically and without thought, that you are too high. This behavior is perceptual. A characteristic of highly experienced professional pilots is that most of their decisions are automatic.

Such 'automatic' behavior can occur in procedural activities as well. I recall instructing a student in emergency landings in the flat farming country around Champaign-Urbana, Illinois. After I had 'pulled the engine', the student picked out a field and I let him go farther than usual because I wanted him to see that he would make it. After just reaching the field at about 10 feet in altitude, I called out to him, 'Go around'. Just as he applied throttle, the engine quit! I 'automatically' went for the fuel selector (in the Cherokee, located on the cabin wall on the far side of the student near his feet) and, with my other hand, the fuel pump switch located on the panel. Within a couple of seconds the engine caught and we were climbing away. I had not seen the student turn off these items but, knowing they were on the emergency landing checklist caused me to make the move toward them without thinking -- my experience caused the automatic reaction, saved a great embarrassment and perhaps some damage to the aircraft.

How does one acquire these automatic behaviors or expert reactions or thought patterns? What are the distinguishing characteristics of an expert in any field? In recent years, the study of 'expertise' has led the scientific community away from a 'simplistic' examination of human behavior in which very small subsets of the overall task are isolated and studied with carefully controlled independent and dependent variables. The current trend is to study behavior in more complex and realistic settings where all aspects of the normal environment are free to interact. Nowhere is this trend more apparent than in aviation cognitive research.

In one of the early texts on expertise, Dreyfus and Dreyfus (1986) outline the path from novice to expert with five stages. In the first or 'novice' stage, the learner acquires facts and rules to help associate relationships among those facts. Most of the learning occurs apart from the complete context or situation as a whole. In the second or 'advanced beginner' stage, the learner acquires a marginally acceptable level of performance through applying more sophisticated rules and associations. The learner takes on situational contexts with greater complexities. Experience is more important to performance and learning than verbal description. In the third or 'competence' stage, the learner acquires the proficiency necessary to pass the recognized tests for performance in the field. The learner goes beyond rules to making plans and setting priorities according to preset goals to be accomplished. Thought processes are characterized by information processing and problem solving involving 'detached, deliberative and sometimes agonizing selection from among alternatives' (p.28). In the fourth or 'proficiency' stage, the learner recognizes situations from past experience and reacts without much, if any thought. The complete context is seen as a whole and there is almost total situational awareness. The thought process is intuitive and is described as 'holistic similarity recognition'. The fifth, or expert stage is characterized by automaticity. Experts do not solve problems, they don't even make decisions. They know what to do based on mature and practised understanding. The expert has an immense library of distinguishable situations. In chess, some say a master can recognize 50,000 types of positions.

Dreyfus and Dreyfus (1986) add that interpretation ability (defined as judgment) is an important distinguishing characteristic of the various levels of skill acquisition. In their model, there is no judgment in the novice and advanced beginner stages. The competent performer judges by means of conscious deliberation. The proficient and expert performers make judgments based on prior experiences in ways that cannot be explained. Rationality is the mark of the competent performer. To move on to expert performance requires movement into the area between rationality and irrationality called, 'arationality'. Arationality is defined as 'action without conscious analytic decomposition and recombination'. 'Competent performance is rational; proficiency is transitional; experts act arationally'.

To obtain an understanding of expertise in aviation a new approach to data collection has become necessary. The expertise approach consists of observing pilots operate in the most natural setting that can be arranged (e.g., a flight simulator) conducting a realistic LOS (Line Oriented Simulation) scenario. Data collected consist of audio and video tapes of the behaviors (including communications) of the crew and automatic recordings of control parameters. Retrospective data are collected following the completion of the flight consisting of audio records of crews introspecting about their thought patterns as they are cued with the video records of their performance. These verbal protocol data are proving highly effective in the development of new models of the expert pilot. These data have shown that the major human activity of importance to expert piloting is cognitive rather than psychomotor. Following the Dreyfus and Dreyfus model, acquiring interpretive abilities or pilot judgment is key to becoming an expert in the aviation domain.

In a recent study using the expertise approach, our Ohio State University team of aviation researchers developed a set of nine distinguishing characteristics of the expert general aviation pilot (Jensen *et al.* 1994), based on data from 'semi-structured interviews' with nearly 100 experienced pilots and pilot examiners. The expert in this domain is one who:

- possesses a high level of skill and works constantly to improve it,
- is highly motivated to learn all there is to know about this flight domain,
- has superior ability to focus (or compartmentalize) attention on the flying task at hand and the mental discipline to change his or her focus of attention when new information suggests that a change is necessary,
- is a keen observer of the flight environment, including location of other aircraft, terrain, navigation features, ATC clearances, weather phenomena,
- carefully establishes a baseline for normal instrument indications, aircraft sounds, vibrations and g-forces with respect to control action so that his or her threshold for slight variations is very small,
- is skeptical about 'normal' aircraft functioning and is constantly making contingency plans for those circumstances when things might go wrong,
- possesses superior mental skill and capacity for problem diagnosis, risk assessment and problem resolution,
- has excellent communication skills and can readily adapt them to the audience and situation and

- knows his or her limitations, is motivated to avoid situations that might push his or her skill to those limits, and has the willpower to overcome the pressures of people around his or her to push the limits of his or her skill.

Most decisions in flight do not require an immediate, automatic response. When they do, this is the type of response we hope to achieve in less experienced pilots through the ideas and judgment scenarios presented in this book. When more time is available, we expect pilots to carefully deliberate, using more information.

In the development of technology this new expertise approach is leading the way toward an accelerated knowledge base in humankind. This knowledge base was held back for centuries by a trial and error method from the earliest days of recorded history (Figure 1.7). Around the year 1500, the scientific method was applied to technology development and growth was greatly advanced. Now new methods involving the use of complex systems, unencumbered by strict scientific methods are accelerating technological advancement. In behavioral research it is being called 'cognition in the wild'. This new approach is also changing the way technology is being developed in many other fields.

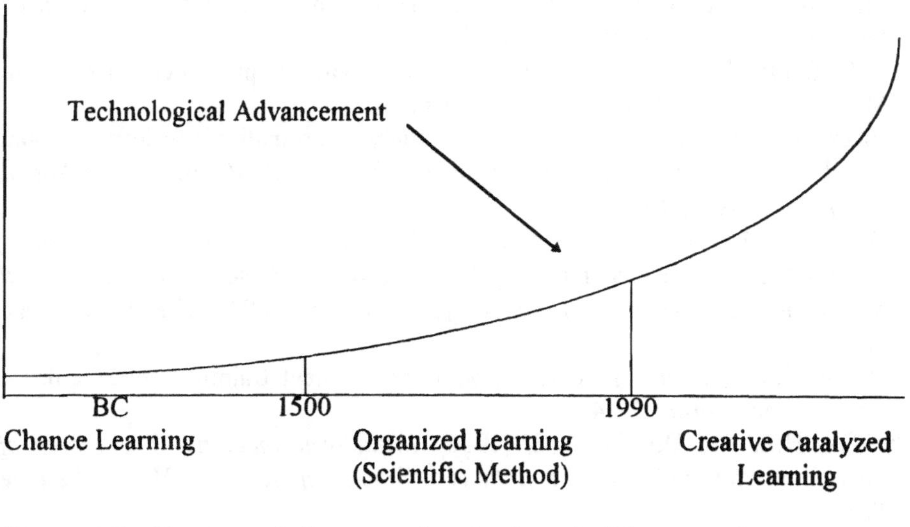

Figure 1.7 Conceptual diagram of technology buildup over human history (adapted from Box, 1994)

AQP in the USA

In aircraft requiring multi-person crews decisions are still made by one person, the captain, but they are made in coordination with other crew members. Thus the decision making process involves teamwork. Until recently, training and evaluation of pilots was done on an individual basis, i.e., teaching each crewmember to function well at his or her task. Now, it has become clear that this narrow approach is not adequate for modern aircraft. In the USA a new FAA program called the Advanced Qualification Program (AQP) has been developed to provide direction to these new efforts to advance team training and evaluation (Birnbach and Longridge, 1993). Keys to the success of AQP in airline operations are a detailed task analyses of each of the aircraft in the fleet and a thorough CRM training program. In Chapter 6, we will discuss CRM in detail and show how this new AQP program will assist airlines in the implementation of team training and evaluation.

References

Allnutt, M. (1982). 'Human factors: Basic principles.' In R. Hurst and L. Hurst, *Pilot Error*. London: Granada Publishing Ltd.

Asch, S.E. (1951). 'Effects of group pressure upon the modification and distortion of judgments.' In H. Gretzkow (ed.) *Groups, Leadership and Men: Research In Human Relations*. Pittsburgh: Carnegie Press, 177-190.

Baddeley, A. (1986). *Working Memory*. New York: Oxford University.

Birnbach, R.A. and Longridge, T.M. (1993). 'The regulatory perspective'. In E.L. Wiener, B.G. Kanki and R.L. Helmreich (eds.) *Cockpit Resource Management*. New York: Academic Press, Inc.

Box, G. E. (1994). 'The scientific basis for total quality improvement'. Dearborn, MI: Ford/OSU CQI Symposium, lecture notes.

Deutsch, M. and Gerard, H.B. (1955). 'A study of normative and informational social influences upon human judgment'. *Journal of Abnormal and Social Psychology*, **51**, 629-636.

Diehl, A.E. (1991). 'The effectiveness of training programs for preventing aircrew "error"'. In R.S. Jensen (ed.) *Proceedings of the Sixth International Symposium on Aviation Psychology*. Columbus, OH: The Ohio State University.

Diehl, A, (1992, January). 'Does cockpit management training reduce aircrew error?' *ISASI Forum*, **24**(4).

Dreyfus, H.L. and Dreyfus, S.E. (1986). *Mind over machine: The power of human intuition and expertise in the era of the computer*. New York: The Free Press.

Edwards, W. and Tversky, A. (Eds) (1967). *Decision Making*. Baltimore: Penguin Books, Inc.

FAA (1969). *Approval of Area Navigation in the US National Airspace System*. Washington, DC: Government Printing Office, FAA Advisory Circular 90-45.

Fahlgren, G. and Hagdahl, R. (1990). 'Complacency'. *Proceedings of the 43rd Annual International Air Safety Seminar*. Rome, Italy: Flight Safety Foundation.

Fitts, P.M. and Jones, R.E. (1947a). 'Psychological aspects of instrument display: Analysis of 270 'pilot error' experiences in reading and interpreting aircraft instruments'. USAF Air Materiel Command Memorandum Report NO. TSEAA-694-12A.

Fitts, P. M. and Jones, R.E. (1947b). 'Analysis of factors contributing to 460 'pilot error' experiences in operating aircraft controls'. USAF Air Materiel Command Memorandum Report NO. TSEAA-694-12.

Garrison, P. (1972). 'Why we fly'. *Flying*, February, p. 38.

Gibson, J.J. (1977). *The Perception of the Visual World.* Boston: Houghton Mifflin Company.

ICAO (1989). *Fundamental Human Factors Concepts.* Montreal: ICAO Circular #216.

Janis, I. and Mann, L. (1977). *Decision Making: A Psychological Analysis of Conflict, Choice and Commitment.* New York: Free Press.

Jensen, R.S. (1981). 'Prediction and quickening in perspective flight displays for curved landing approaches'. *Human Factors*, **23**(3), 355-363.

Jensen, R.S. (1982). 'Pilot judgment: Training and evaluation'. *Human Factors*, **24**(1), 61-73.

Jensen, R.S. and Benel, R.A. (1977). *Judgment Evaluation and Instruction in Civil Pilot Training.* Springfield, VA: National Technical Information Service, Final Report FAA-RD-78-24.

Jensen, R.S. and Chappell, S. (1983). *Pilot Performance and Workload Assessment: An Analysis of Pilot Errors.* Columbus, OH: Contractor Final Report for NASA Contract No. NAG 2-184.

Jensen, R.S. and Marsh, R.W. (1976). *Simulator Tests of Pilot Performance in Terminal Area Navigation Operations: Effects of Various Airborne System Characteristics.* Springfield, VA: National Technical Information Service, Final Report FAA-RD-76-99.

Jensen, R.S. and Roscoe, S.N. (1971). *Performance of Instrument Pilots Using Symbolic Area Navigation Displays.* Savoy, IL: University of Illinois at Urbana-Champaign, Institute of Aviation, Aviation Research Laboratory, Technical Report ARL-7-23/FAA-71-3.

Jensen, R.S. and Roscoe, S.N. (1974). *Pilotage Error in Area Navigation: Effects of Navigational Procedural Complexity.* Springfield, VA: National Technical Information Service, Interim Report FAA-RD-74-143.

Jensen, R.S., Chubb, G.P. Adrion-Kochan, J., Kirkbride, L.A. & Fisher, J. (1994). 'Aeronautical decision making in general aviation: New intervention strategies'. In *Proceedings of the Meeting of the Western Association of Aviation Psychologists*, Dublin, Ireland: European Association of Aviation Psychologists.

Jensen, R.S., Vanderkolk, R.J., FitzHenry, P., Rose, D.J. and Roscoe, S.N. (1971). *Pilotage error in area navigation: Performance of instrument pilots using symbolic displays.* Springfield, VA: National Technical Information Service, Final Report FAA-RD-71-31.

Kogan, N. and Wallach, M. (1964). *Risk Taking: A Study in Cognition and Personality.* New York: Holt, Rinehart and Winston.

Lauber, J.K. (1979). 'Resource management on the flight deck: Background and statement of the problem'. In G.E. Cooper, M.D. White and J.K. Lauber (Eds) *Resource Management in the Cockpit.* Moffett Field, CA: NASA Ames Research Center (NASA Conference Publication 2120).

Lee, W. (1971). *Decision Theory And Human Behavior.* New York: John Wiley and Sons.

Liebert, R. and Morris, L. (1967). Cognitive and emotional components of test anxiety: A destination and some initial data. *Psychological Reports,* **20**, 975-978.

McCormick, E.J. and Sanders, M.S. (1982). *Human Factors in Engineering and Design.* New York: McGraw Hill.

Owen, D.H., Warren, R., Jensen, R.S., Mangold, S.J. and Hettinger, L.J. (1981). 'Optical information for detecting loss in one's own forward speed'. *Acta Psychologica,* **48**, 203-213.

Parducci, A. (1968). 'The relativism of absolute judgments'. *Scientific American,* **219**(6), 84-90.

Reason, J. (1991). *Human Error.* Cambridge: Cambridge University Press.

Roscoe, S.N. (1980). *Aviation Psychology.* Ames, IA: Iowa State University Press.

Ross, M.J. (1992). *The Integrated Commercial Flying School: An Evaluation of Commercial Flight Training and Flight Simulator Effectiveness.* Newcastle, Australia: The University of Newcastle, Department of Education, Unpublished doctoral thesis.

Shealey, J.E. (1974). 'Risk-taking in skilled task performance'. *Proceedings of the Eighteenth Annual Meeting of the Human Factors Society,* Santa Monica, CA: Human Factors Society.

Transport Canada (1992). *Flight test guide: Flight Instructor Rating -- Aeroplane, Helicopter, Aerobatic.* 2nd Edition. Ottawa Canada: Aviation Regulations Directorate.

Trollip, S. and Jensen, R.S. (1991). *Human Factors for General Aviation.* Denver, CO: Jeppesen Sanderson.

Tsang, P.S. 1992). 'A Reappraisal of aging and pilot performance'. *The International Journal of Aviation Psychology,* **2**(3), 193-212.

Wine, J. (1971). 'Test anxiety and direction of attention'. *Psychology Bulletin,* **76**(2), 92-104.

Yerkes, R.M. and Dodson, J.D. (1908). 'The relation of strength of stimulus to rapidity of habit-formation'. *Journal of Comparative Neurology and Psychology,* p. 459.

2 Judgment models

All who are practically concerned with aerial navigation agree that the safety of the operator is more important to successful experimentation than any other point. The history of past investigation demonstrates that greater prudence is needed rather than greater skill. Only a madman would propose taking greater risks than the great constructors of earlier times.

Wilber Wright (1901, quoted in Combs, 1979)

After establishing the need for judgment in aviation, let us now develop a framework for the training and evaluation of aviation judgment. In this chapter we will look at three *normative* models of decision making that tell us how we should make decisions if we were able to optimize all factors known. These models help us to discover the nature of the human judgment process and tell us how good judgment in aviation should be applied. The first is a detailed model which defines the overall framework for the human judgment process. The second called, the Theory of Signal Detectability (TSD) is an old model designed for visual perceptual decisions. In our application this model is generalized to represent cognitive decisions. It is presented because it provides a link between classical decision models and application models in the aviation environment. The third, the Pilot Judgment Model, tells us specifically how we should make decisions in aviation. In chapter's 3 and 4 we will show *descriptively* how people and pilots actually do make decisions. Later we will show *prescriptive* models of how we can teach people to make decisions more closely to the way they should.

At the end of the first chapter the characteristics of an expert pilot were defined. An examination of these characteristics reveals that the primary difference between a competent and an expert pilot is judgment. As the learner acquires expertise, more and more of the complete context or system is learned and becomes a part of his or her mental model. In flying pilots are judged competent when they pass their licensing check flights. It is only after this point that they begin to move out and acquire expertise and judgment.

In this chapter, we explore theories concerning human judgment. At the conclusion we will look at a model of good judgment consisting of two parts: an intellectual part and a motivational part. This model is useful as a way to apply to aviation what scientists have discovered about human judgment. By dividing the task into these two aspects, one can see how each can be taught and evaluated in pilots.

Detailed judgment model

As we indicated in the previous chapter, human judgment is the mental process used in forming a decision. We now look at a detailed model that describes a series of eight steps in this process (Figure 2.1). This is an adaptation of an eight-step judgment model offered in a report for the U.S. Air Force (Jensen *et al.* 1987). This model offers greater detail than the general information processing model (Figure 1.3) and is specifically designed to describe the aviation cognitive judgment process. Understanding each step and how the steps are interrelated will help you to see the ways in which later prescriptive models can be used to develop better aviation judgment.

Problem vigil

The first element of the judgment process is problem vigil. The pilot is in a state of vigilance at all times, looking for changes in the flight environment, including his or her own performance, that could affect the progress and safety of the flight. Most of the human senses are used but by far the greatest amount of information in the cockpit comes to the pilot through his or her eyes and ears. Other useful information comes from the vestibular system (inner ear), the olfactory system (nose), the kinesthetic system (sense of movement or muscle position), tactile system (touch).

Vigilance in the complex aviation environment requires far more than allowing our senses to wait for physical stimulation. In its simplest form, highly developed continuous control decisions (such as small attitude adjustments for light turbulence) can be made without conscious processing. Scientists believe that the information from our senses bypasses the brain going directly to the spine for minimal processing and back out to effectors (hands, feet, etc.) for the response. However, vigilance in aviation involves many other aspects of the brain. To go into short-term, working, or long-term memory, focused attention must be applied to the information being sensed. Such processing subjects the information to the influence of all of the factors that affect it including expectations, fatigue, illness, etc. In most cases, sensation amounts to no more than what the brain thinks it needs. A close approximation to the real world is generally sufficient.

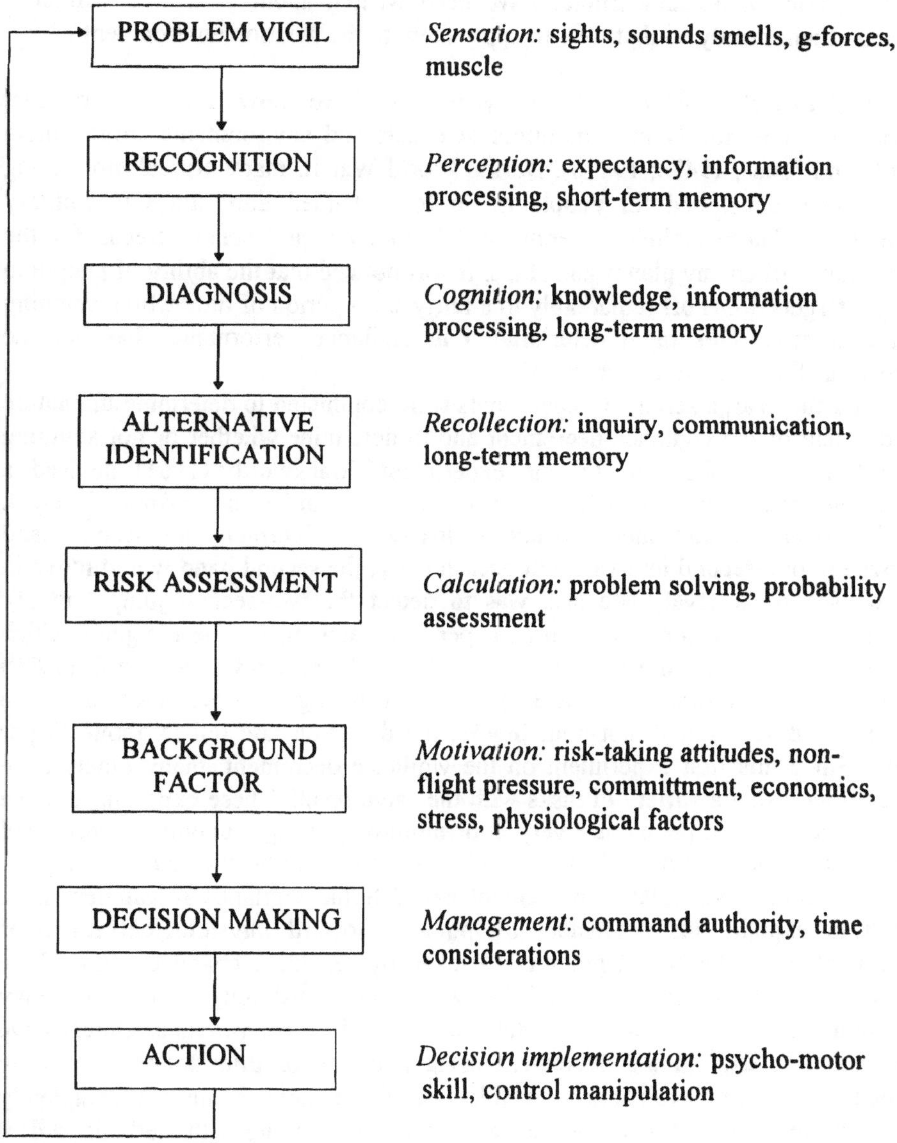

Figure 2.1 Detailed judgment model

There are two factors involved in vigilance that aviation people should know about. Each of these has received a great deal of the attention of psychologists since the 1940s. The first is called the 'vigilance decrement' and refers to the fact that people are not able to maintain vigilant attention for very long. The second is called 'attention' which refers to the mechanism our brain uses to control its focus. Many books and articles have been written about each. We will only discuss them briefly so that we can apply them to the aviation environment.

The vigilance decrement. Aviation psychologists have known for some time that people are notoriously poor monitors of events and environments around them (Fitts and Jones, 1947a; 1947b). During World War II, many critical monitoring tasks were developed for people to do for extended duty times in military operations. These included keeping watch on radar and sonar screens for the appearance of enemy planes and ships. It was noticed that the ability of people to detect targets drops off remarkably in a fairly short period of time while watching these screens. This rapid deterioration in vigilance performance has become known as the 'vigilance decrement'.

After the War, a series of experiments were conducted to determine the nature and extent of the vigilance decrement and to determine whether or not anything could be done about it. In one experiment, Mackworth (1961) devised a 'vigilance task' called the 'Clock Test' in which the subject sat in front of a clock with a sweep second hand that had an irregularity. Normally the second hand moved in one-second intervals, but occasionally, the second hand would move in a two-second interval. The task was to detect the two-second jump (signal). Subjects would sit there for two-hour periods watching for these signals which occurred at random intervals between 0.75 and 10 minutes. Mackworth (1970) found that in the first half-hour 85 per cent of the signals were detected. In the second and subsequent half-hour blocks, the detection rate fell to about 73 per cent. Since this first experiment on the vigilance decrement, many others have been done using a variety of tasks with the same result. These experiments have shown clearly that people are very poor monitors during monotonous tasks and that their vigilance drops off fairly quickly -- too soon to be attributed to fatigue.

Why does this vigilance decrement exist? Some vigilance researchers have tried to explain this decrement by pointing toward physiological functions (habituation of the neural potentials evoked by repetitive characteristics of the task) within the sensory system (Mackworth, 1970). Although habituation may explain some of it, it does not totally explain all of the variance. Others have attempted to explain the decrement using attention or information processing models. The third explanation considers arousal factors. None can completely explain the problem and we are left with trying to come up with methods to deal with the problem without fully knowing its nature. We do know that the problem is basic to human information processing and one cannot automatically attribute negligence to an operator who makes a vigilance error after watching a screen for an extended period of time.

What can we do about the problem that people have of declining vigilance over a relatively short period of time? The first answer is that we must be aware of it because, through such awareness, we will develop individual methods of coping. Second, we should avoid giving tasks to people that are no more complex than simple monitoring of instruments. Later, we will discuss ways in which we can improve our vigilance in judgment training.

Attention. To explain the vigilance decrement, Mackworth says, '...repetition and monotony produce habituation of physiological responses, particularly neural ones and that as a result of such habituation it becomes increasingly difficult to pay attention to the task' (Mackworth, 1970). Thus, the concept of attention, its nature, limits and control are also important to our concept of the problem vigil.

The control of attention is one of the most important elements in aviation. We often see aviation accidents caused by the attention of the pilots being diverted from the flying task. For example, in the Eastern L-1011 accident in the Florida Everglades, the NTSB cited the crew's diversion of attention to a burned out light bulb as a causal factor.

Attention can be defined as the mental faculty that controls the subject matter chosen for conscious information processing. Not much is known about how this mechanism works but it is probably closely associated with motivation which is discussed in the next chapter. One thing we do know about it is that it is severely limited; some say it is limited to a single channel. Others would argue that attention is limited but some parallel processing is possible (Wickens, 1992).

The attention mechanism at work can be seen in what has been called 'the cocktail party phenomenon' (Cherry, 1953). In a large group of people we can listen to one conversation while filtering out other conversations around us. Our attention mechanism chooses the person to whom we listen and we can hear them very well despite other voices. We do not even hear the other conversations going on around us.

However, our listening attention is not so focused that it cannot be interrupted. For example, if our name is mentioned in an unattended conversation at the same sound level or even lower, we may hear it and probably divert our attention to that conversation. This phenomenon seems to indicate that our channel for conscious processing may have a continuous additional capacity for processing certain simple but important information.

In aviation, this additional information processing capacity is both a blessing and a curse. It is a blessing because it can call our attention to items that are important to us while we are thinking about something else, such as our radio call sign. It can be a curse because it can divert our attention from rational, safety-oriented thoughts about a needed decision, to irrational background-oriented thoughts, such as getting home. Through training and experience pilots learn to discipline their attention mechanisms so they attend to items that are safety oriented.

A less obvious example of good pilot vigilance is maintaining a state of awareness of potential problems. For example, if the weather is deteriorating ahead, the pilot must decide whether to continue the flight. Being aware of the negative consequences of such action may prevent an error in judgment. Maintaining vigilance for what may go wrong is just as important as noticing that something has gone wrong. Awareness of potential problems allows the pilot to plan for alternatives ahead of time during low levels of stress, making the correct decision more likely - and bringing about lower levels of overall stress.

In aviation, we can find many examples of accidents that occurred because the pilot did not maintain vigilance. Approximately once every two days, some pilot makes a landing 'forgetting' to extend the landing gear. Frequently, pilots run out of fuel because of a lack of vigilance in their preflight. Sometimes pilots even try to make a takeoff with control locks in place. Weather conditions sometimes change more rapidly than expected trapping the pilot who is not as vigilant as he should be.

Problem recognition

The second step of the judgment process is problem recognition. In this step, a change in the status of the flight is discovered that could affect its safety. Further to this step is the realization that the problem is significant to the safety of the flight. Two psychological factors important to problem recognition are perception and expectancy.

Perception. Hilgard, Atkinson and Atkinson (1975) define perception as 'The process of becoming aware of objects, qualities, or relations by way of sense organs; what is perceived is influenced by prior experience'. Perceptual organization is the process by which particular relationships among potentially separate stimulus elements are determined (i.e., selected from alternative relationships) and in turn guide the interpretation of those elements (Boff, Kaufman and Thomas, 1986). There are two plausible explanations for the way this organization takes place: simplicity and likelihood. Our perceptual system is organized to achieve the simplest or most economical interpretation of the stimulus, and to achieve the interpretation most likely to match the source of distal stimulation (energy striking the sensor). In these ways our perceptual system greatly reduces the information received for our senses to a manageable size usable for cognitive processing and makes possible the rapid recognition of problems in aviation.

On the negative side, our perceptual systems do not always give us the correct interpretation of our environment. Recall, for example, the corridor illusion from the previous chapter (Figure 1.5). The cylinders appear to get larger in size as our perceptual system places them in the context of the perspective. This is the simplest and the most likely interpretation of the stimulus elements and our perceptual system is organized through experiences to choose this interpretation in a most compelling way. To force another interpretation requires intense cognitive control over our minds. Many other examples, such as the Ames Window, the Muller-Lyer Illusion (Weintraub and Walker, 1966) and the Ponzo Illusion (Hilgard, Atkinson and Atkinson, 1975) etc. have been developed by perceptual psychologists to show how we can be mistakened by our perceptual systems.

Expectancy. In each situation with which we become involved, whether on the ground or in flight, we develop an awareness of the predictable elements of the situation. From this awareness, we learn to expect certain things to happen. This becomes like a mind set which reduces the amount of things we have to think about because we can say to ourselves 'This situation has occurred before so I can expect the same result again'. Thus expectancy can be defined as learned thought patterns which tend to bias our interpretation of our environment.

For example, suppose you call air traffic control (ATC) and request 9,000 feet on an east bound IFR flight. ATC returns the call saying 'Cleared to 10,000 feet'. You have in your mind set an expectation of 9,000 feet for two reasons. First, you requested it and you have learned that you usually get what you ask for from ATC, and second it is the correct altitude for the direction you are flying. In this example, if there is anything restricting the transfer of communication, such as static, partial blocking, or your attention focused on something in the cockpit, you may not hear (that is understand), the message as 10,000 feet because you are expecting 9,000 feet.

There are many examples of such expectations that have led to accidents. The collision of the KLM and Pan Am 747s at Tenerife is one such example. The tower cleared the KLM into takeoff position to hold. The captain expected to hear cleared for takeoff and the transmission was partially blocked. He was so sure that he had received the takeoff clearance that he did not question the tower further even in the face of some uncertainty expressed by his co-pilot. This combination led him to make the takeoff when the Pan Am airplane was still on the active runway resulting in the worst air disaster up to that time.

It should be pointed out that some scientists (e.g., Klein, 1989) believe that post hoc analyses of expectancies are not very useful in the heat of the dynamic decision situation such as that faced by the crew of the *Vincennes* (the US Navy ship whose crew mistakenly shot down an Iranian civilian airliner prior to the 1991 Gulf War). He points out that using analytical processes unduly burdens dynamic decision making when time is very short, possibly resulting in decision paralysis. Dynamic decisions such as these are often faced by pilots and air controllers. In such decisions, it becomes important to have a large cadre of knowledge and experience from which to draw so that biases will be properly preset for the correct decisions.

In the first chapter we saw how our senses and perceptual systems are prone to error. Good aviation judgment training can cause these functions to be made less error prone through knowledge, experience and maintenance of a state of alert expectation. For example, some flight instructors tell their multi-engine students that, to be prepared for such an emergency, they should expect to have an engine failure on every takeoff. Expectations are most important when time available is very short.

Problem diagnosis

The third step of the judgment process is problem diagnosis. During this stage, the pilot tries to discover the nature of the problem. Diagnosis is particularly important because the pilot must know the nature of the problem before he or she can develop a solution. An incorrect diagnosis could result in decisions that are very harmful to the safety of the flight. Frequently, problem diagnosis in aviation must be done very quickly.

On a recent VFR cross-country flight in a Cessna 172 from Put-In-Bay to Columbus, I began to experience a communication problem with my radio -- it became difficult to get the approach controllers to understand me. Soon the panel lights on the radios began to flicker. I diagnosed it as an electrical short or severe drain on the battery and shut everything electrical down that I could think of. I did not think to turn off the master switch. Because I could not contact the OSU control tower, I decided to land at a non-controlled airport in Delaware and lined up for the final approach. Just before touching down, I turned the radio on and gave the OSU tower one last try and got them. In the transmission I quickly told them of my potential loss of the radio and they said they would watch for me and give me a light gun signal for landing if I needed it. The radio continued to work so I left it on through the remainder of the flight. When I got back, I discovered that the alternator belt was loose.

Diagnosis requires information processing, knowledge, memory and mathematical problem solving skills. It is often necessary to consider subtle changes to parameters and the effect they may have on the system. It requires an ability to look at the problem from various perspectives, ask 'what if' questions and be able to determine the answers. It requires the ability to follow and/or imagine an extended sequence of events or symptoms and project the result. If time permits, it may require an understanding and application of statistics because prior probabilities for certain events may need to be computed or estimated.

Problem diagnosis is involved in virtually all aviation decisions but it particularly comes into play when there may be a mechanical problem with the aircraft. One of the most stunning recent examples in the airlines was the case of United Flight 232, the DC-10 that made an emergency landing in Sioux City, IA in 1989 (NTSB, 1990). The crew was forced to diagnose a severe mechanical control problem (they lost virtually all hydraulic pressure) and they were required to develop an entirely new control strategy to bring the aircraft down.

Fault diagnosis is a major concern to all mechanics. I recall in aircraft maintenance training it was called 'trouble shooting' and the instructors were fond of 'bugging' the engines with problems that were difficult for us as students to trace. Bill Rouse and his colleagues have performed a series of experiments to understand strategies used by mechanics in fault diagnosis and to develop fault diagnosis training systems to be administered on computer (Johnson, Rouse and Rouse, 1980; Johnson and Rouse, 1982a; Johnson and Rouse, 1982b; Rasmussen and Rouse, 1981; and Rouse, 1979). In the Johnson and Rouse (1982b) study the investigators found that computer simulations with 'moderate fidelity' can provide sufficient problem-solving experience for aircraft maintenance trainees to be competitive with traditional lecture/discussion forms of instruction. The insight developed in these experiments shows that the basic abilities in problem diagnosis can be taught and they are useful to the development of training for in-flight problem diagnosis.

The Ohio State University has been doing research for several years on diagnosis behavior and pilot response to critical in-flight events (Flathers, Giffin and Rockwell, 1981; Giffin and Rockwell, 1984). Most of their experiments have been done on a computer which simulated the cockpit of general aviation aircraft, gave pilot-subjects 'problem' scenarios to 'fly' and asked them to diagnose the problem and make decisions regarding alternatives. Pilot performance and diagnosis strategies were observed through automatic recordings and through visual observation. One scenario that they used went as follows:

> *Scenario:* You are making a day trip from Augusta, ME, to Lebanon, NH....You climb to a cruising altitude of 6,000 feet. After 15 minutes of routine IMC flying, your instruments indicate a straight-and-level attitude, an increase in airspeed, a slow turn to the left and steadily decreasing altitude.
>
> *Diagnosis:* Your vacuum pump has failed as indicated by the low reading on the suction gauge. As the artificial horizon lost its drive, it started to sag to the right and you compensated by turning left, leveling the artificial horizon and putting the plane into a slow, descending left bank. The airspeed increase was due to the slight nose-down attitude.

The experimenters determined diagnosis strategies used by observing 'logic tracks' and inquiries made by the pilot-subjects as they searched for information leading to the cause of the problem. Performance was determined by counting the number of inquiries and/or tracks followed before making the key inquiry (in this case, looking at the suction gage). They found that there is a great variety of information search strategies used by pilots in diagnosis, pilots with greater knowledge of systems did better diagnoses and pilots who had a systematic search pattern used it consistently from one scenario to another (Giffin and Rockwell, 1984).

The role of knowledge and experience is very important to the diagnosis process. It is easy to act in a 'vacuum' with the awareness of limited data and let the 'chips' fall as they may. In flight, this would be very dangerous! The greater our awareness, however, the more data we must assimilate and integrate into our decision making. The more we know, the more complex our decisions become. The more we know, the more it becomes possible to predict just where the chips will fall. However, if we assume the responsibility of attempting to predict accurately just where each chip will fall, we are likely to feel so overwhelmed by the complexity of the task as to sink into inaction.

The quality of the diagnosis can be measured on two dimensions, correctness and the level of certainty that the diagnosis is correct. Often we do not know if our diagnosis is correct until well after the crisis of the decision situation is past. Thus level of certainty in problem diagnosis is a major source of stress in the flying environment.

Alternative identification

The fourth step of the judgment process is alternative identification. The pilot knows the nature of the problem and how serious it may be to the safety of the flight. During this stage, the pilot identifies a set of alternatives that may be available to solve or avoid the problem and get the aircraft down safely. This step requires creativity and knowledge of the aircraft, environment and aviation system.

Flathers, Giffin and Rockwell (1981) identify three factors used by pilots in making a decision to divert from the planned course of action in the flight. First, the results of the diagnosis provide a need for reassessment of the commitment to the flight goal, the capabilities to complete the flight as planned and the limitations of the aircraft. Thus the diagnosis forces the change in plans. Second, the pilot places personal 'worth' or value factors into the decision process that have little or no objective value. These worth factors, discussed in detail later, include economic, social and emotional forces that often bias the judgment process away from the optimum rational decision. Third, the pilot generates a worth function that combines objective and personal worth factors which determine how the decision will be made. In their experiment on diagnosis and alternative generation Flathers *et al.* (1981) gave pilots the following scenario:

> You are at the Bangor International Airport in Bangor, Maine and desire to fly to Glens Falls for a business meeting...You were cleared to the Glens Falls airport 'as filed'. You lifted off from Bangor at 10:00 Am and your departure was routine. At 10:14, you reached your cruising altitude of 8,000 feet, established on V3 northeast of the Augusta VOR (see Figure 2.2)...At 11:21 you cross Grump intersection. One minute later you hear a short burst of static noise over your radio. At the same time you notice your VOR needles flicker unsteadily and then return to normal indications. Curious to know what caused these events you glance over the instrument panel and find a 'zero' reading on the ammeter. You actuate the landing light and notice no change in ammeter indications. From this information, you know that the alternator has failed and you turn off all unnecessary electrical load to conserve the battery.

The battery itself can supply power for a limited time depending on its condition, size, age, etc. Even under ideal conditions battery power is not expected to last longer than 50 minutes. Clearly there is a need to divert from the planned flight and land somewhere else. Glens Falls is beyond the range in terms of battery time. Your task now is to find a suitable place to land.

At this stage, the experimenter placed on the computer screen a map that showed 16 different airports each with five different characteristics: 1) bearing and distance (25 to 70 miles), 2) weather (500-1 to 1000-3), 3) approach aids (NDB, VOR, ILS), 4) ATC service (tower or no tower) and 5) terrain (level, hilly, mountainous). Pilot-subjects were asked to rank order the airports from best to worst alternative for diversion given their present circumstances.

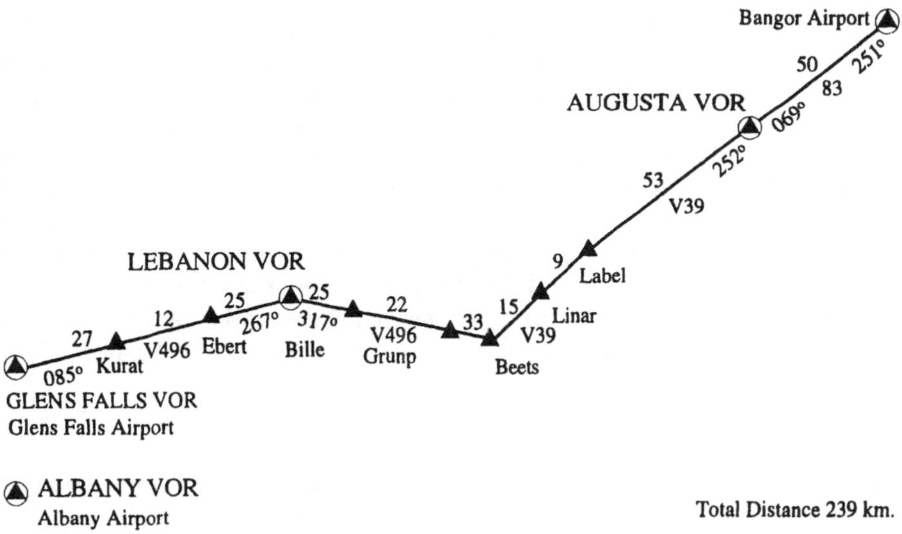

Figure 2.2 IFR Chart given to pilot-subjects in alternative generation experiment (Flathers, Giffin and Rockwell, 1981)

The results indicated that pilots with less experience (instrument pilots with relatively low time) wanted all the help they could get, including the control tower and an ILS approach. More experienced pilots (Air Line Transport rated), picked the closer airports with acceptable landing minima as their first choice. The experienced pilots knew of the uncertain life of the battery (that 50 minutes is no guarantee) and chose to land as quickly as possible to avoid the risk of battery failure. The less experienced pilots did not think of the fact that the battery might not last 50 minutes and set higher priority on the best approach aids with tower assistance.

This study shows two facts important to pilot judgment. First, thought patterns of pilots in the generation of alternatives are different depending on their experience and knowledge. Second, the ability of pilots to generate alternatives can be measured in a relatively simple computer simulation.

Risk analysis

The fifth step of the judgment process is risk analysis. During this stage, the pilot determines the risk involved in taking each of the identified alternatives. The pilot does not actually compute the probabilities of success for each of the alternatives. In most cases, the probability of success for several alternatives is so close to 100 per cent that a person cannot easily discriminate among the alternatives on a purely rational basis. Nevertheless, the better one is able to attach relative probability's of success to alternatives the better one may be at making the right choice.

This step requires computational approximation skills and knowledge of the relative safety in many courses of action that pilots can make. The sorts of factors that may be included in the computation are skill level of the pilot, reliability of mechanical aspects of the aircraft, reliability of the flight and engine instruments, avionics and communication equipment, reliability of the weather forecast, ability of the pilot to determine probable weather conditions to augment the forecast, etc., etc. To cover all of the possibilities, this list could go on indefinitely. The fact is the better the pilot knows each of these factors that are a part of the risk of flying, and the better the pilot becomes at rationally combining these factors, the better will be his or her approximation of the total risk involved in each alternative. This procedure is explained in greater detail in Chapter 11.

The background factor

The sixth step of the judgment process, called The Background Factor, was first introduced by Brecke (1981), who called it the 'background problem'. This factor is far more than the bias that is often considered a part of mathematical decision theory. It represents the motivational forces that keep us from following purely rational decision functions. This factor is more than just a stage in the process. It is like a cloud that covers the whole process but is particularly important just as a decision is about to be made. For this reason it is placed as a step in the judgment process just before the decision is made. It refers to the non-safety related forces that affect every cognitive decision made by the pilot. Examples of the sources of motivational forces included in the background factor are duty, economics, adventure, commitment, ego, social pressures from peers, superiors and passengers and physiological pressures from a full bladder, illness and fatigue.

Some people, looking at human-caused aviation accidents, would say that any influence of the background factor is an act of negligence and pilots using it prior to an accident should be prosecuted. I believe that this factor is a part of human existence and cannot be avoided. Nevertheless, in aviation decisions it is important to minimize the impact of this factor. To combat these forces requires a considerable amount of mental fortitude and discipline which can be learned through training as well as experience. This factor will be the focus of Chapter 4 on motivational judgment.

Decision

The seventh step of the judgment process is the decision. In this stage, the pilot applies his or her decision function and decides what course of action to follow. The pilot is now ready to act. This is the objective or end product of the judgment process. This step requires leadership skills and personal discipline because it consists of an act of the will to initiate action knowing possible negative consequences, especially from the forces represented in the background mentioned above.

Action

The final step of the judgment process is action. During this stage the result of the decision is sent from the brain to the effectors and the pilot acts on the decision. These actions may involve moving the controls or telling the passengers that they will not be able to fly today, asking for confirming information from ATC, or delaying making a response for now. This step requires eye-hand coordination, skills in dealing with people and all of the other skills necessary to respond to personal or group decisions.

Presenting the elements of the judgment model in this way is not meant to imply that they are independent and act in a serial order. This is in fact not the case. Each element is highly influenced by other parts of the model. For example, the first element, Problem Vigil, represents the sensory system standing ready for new stimulation to see, hear, etc. However, it's performance is highly influenced by expectancies, fatigue, boredom, etc. that come from other parts of the model.

The eight steps of the judgment model help us to see the basic human abilities that are needed in the aviation judgment process. They include observation, perception, curiosity, information processing, knowledge, problem solving, creativity, computation, mental fortitude, discipline, leadership and people skills.

The traditional approach to flight training has been to teach student pilots the capabilities and flight characteristics of an aircraft and its systems; knowledge of the national airspace system; general knowledge of meteorology; regulations; emergency procedures and 'stick and rudder' skills. It is generally believed that, if pilots have this kind of information, they will be able to exercise the good judgment required to assure safe flight.

I believe that a more systematic and deliberate approach is needed -- that judgment should be taught as a specific and essential skill in aviation just as we teach meteorology and aerodynamics. In judgment training it is important to place students in situations that require them to grapple with a variety of circumstances requiring difficult decisions. As a pilot considers an action, the consequences of taking, as well as not taking, that action must be carefully considered. Judgment is learned through experience with as many decisions as possible. The place to gain that experience is in the classroom, computer terminal, or flight simulator where the only risk is pride -- not in the air where there is risk of physical injury.

The basis for the aviation judgment has been given in the detailed model. Now we add a popular decision theoretic model that shows us how we can simplify and reduce the process into two factors that will be useful later in developing the prescriptive training model. These two factors are discriminability and bias.

The TSD model

A model that has been used for several years to explain certain aspects of basic human decision making is known as the Theory of Signal Detectability (TSD). The theory arose during the 1950s but, perhaps the best textual source is by David Green and John Swets (1966). The name for this theory comes from its first application, that of explaining the behavior of humans who were asked to watch radar screens and pick out 'targets' from within the 'clutter' ('noise') on the screen. Even though TSD has had a fairly restrictive history of application, I believe it to be the best theoretical model to bridge the gap between mathematical decision theory and aviation judgment.

One application of TSD to aviation decision making was made by Eubanks and Killeen (1983) who used it to investigate the decision processes of F4 pilots deciding to fire either a heat or radar missile during simulated engagements with a computer programmed adversary. Using TSD, these researchers were able to show that pilot-subjects changed their strategies as they gained experience. Although accuracy increased throughout the five-day training program, pilots gradually adopted a more stringent criterion for selecting the heat missile and simultaneously relaxed their criterion for the radar missile. The authors advocated further use of TSD measures in the study of aeronautical decision making.

Let us now examine the TSD model so that we can see how it fits into aviation judgment. Four outcomes or states of the world are possible with the TSD model:

1 *Hit:* correct identification of a target. We take action because we think we see a signal and it is correct.

2 *Miss:* failing to pick a target. There was a target out there and we did not see it and we did not take action.

3 *False alarm:* Identifying an object in the noise as a target when it is actually just noise. We thought there was a target and took action as if there was a target but there was nothing there.

4 *Correct rejection:* Correctly rejecting an object as noise when it is just noise. No action was taken.

In the simplified TSD world, we do not judge whether or not the action taken is correct, only that we made a decision to act because we thought there was sufficient information for us to believe that the signal was there. These states of the world can be represented as shown in Figure 2.4.

STATES OF THE WORLD

	Noise (No reason to take action)	Signal (Reason to take action)
No Action	CORRECT REJECTION	MISS
Action	FALSE ALARM	HIT

Figure 2.4 Payoff (also cost) matrix representing the two possible states of the world (signal and noise) and the two types of action or responses (action taken or no action) in TSD paradigm

In the real world of signals and noise, it is often not clear which we are seeing or hearing. Both come to us as statistical distributions that could be represented as a normal curve as shown in Figure 2.5. Our job as aviation decision makers is to study the information or data coming in and decide whether there is a signal within the noise because that means we must take action.

The difference between the means (averages) of the two distributions is known as the d' or discriminability. If there is a considerable difference (d') between the signal and the noise, it is easy to make the discrimination. This situation is usually the case in the real world. Most aviation decisions are clear cut, either you fly or you don't, for example. However, of greater interest to us here are those decisions that are not so clear cut, those decisions in which the two distributions (signal and signal plus noise) are close together. On the other hand, sometimes there is so much background noise that it is difficult to tell if there is a signal in the noise. At such times, another factor called the response bias (β) may become important. The response bias is made up of *values* or *costs* or, as we said above, personal worth, that we attach to making either the correct or the wrong choice, respectively.

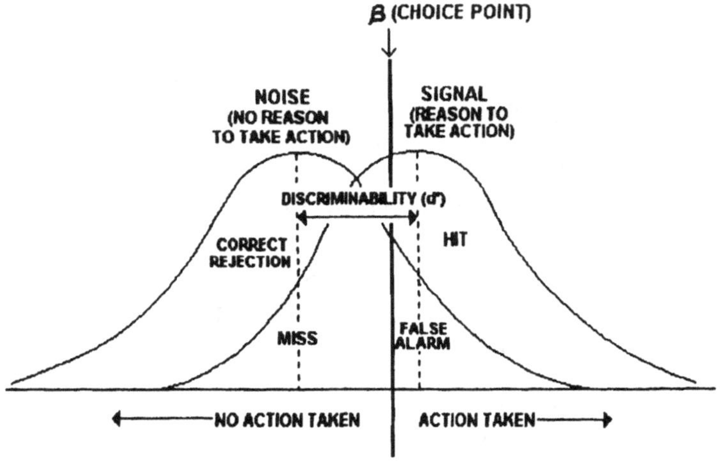

Figure 2.5 Statistical distributions representing noise and signal events in the world according to the TSD model

The sensitivity component depends mainly upon the physical parameters of the stimuli in the situation and the sensory apparatus of the observer. For simplicity, an observer's sensitivity is represented by d' in Figure 2.5. Technically, d' is the difference between the means of the distribution divided by the common standard deviation of the distributions. The more distinct the two distributions become, and hence the greater value of d', the more likely is the observer to detect the presence of the signal. In Figure 2.5 the two distributions overlap. Therefore, it is impossible for the observer to respond with certainty whether or not the signal exists. We can assume that the signal and noise generate central neural effects which, for purposes of decision making, may be summarized by a single quantity. The observer adopts various decision strategies that depend a great deal on non-stimulus parameters.

Despite being independent of sensitivity, the second component of decision behavior also influences discriminative behavior. Because the value of d' is normally considered a constant, variations in signal detections are attributed to such variables as motivation, prior knowledge of the signal's probability of occurrence and the costs and payoffs attendant with a given response. Collectively these affect the bias of the observer labeled β. The detection task is essentially a comparison of the generated central neural activity located along the x-axis (decision axis) of Figure 2.5 with β also located along this axis.

TSD approaches this problem as a statistical decision making task (Wald, 1950). Specifically, the observer responds 'signal' whenever the value of the generated neural event equals or exceeds the value of β and 'no signal' whenever it is less. With overlapping distributions, the observer can vary β and influence the success and failure of his decision-making behavior. Figure 2.5 represents two possible states of the world (also shown in Figure 2.4), noise alone and signal plus noise and the two possible responses, action or no action taken by the observer. The observer has two chances for being correct (correct rejection and hit) and two chances for error (miss and false alarm). Using the observer's personal values for hits and false alarms, the value of β can be calculated. In Figure 2.5, with β set at the indicated position, the observer will correctly identify many true signals and also correctly reject many instances of noise without signals -- hits and correct rejections, respectively. Unfortunately, the portion of the noise distribution to the right of β represents noise identified as signals -- false alarms. The portion of the signal-plus-noise distribution to the left of β represents undetected true signals -- misses.

According to TSD, an observer's decision behavior may be varied by influencing his β. Galenter and Holman (1967) varied the *a priori* (information given prior to the decision) probability of signals, payoff functions and instructions. Although the subject was induced to alter his decision behavior by manipulation of these independent variables, d' (his sensitivity) was unaffected. These results and others strongly support the position that the response criterion affects the outcome of decision behavior.

Birdsall (1955) advanced a definition, The Expected Value Observer, to describe these relationships mathematically. Each of the four possible outcomes has a value to the subject V_h - value of a hit, V_{cr} - value of a correct rejection, V_{fa} - value of a false alarm and V_m - value of a miss. The chosen criterion is that which maximizes the expected value:

$$\beta = \frac{P(N)}{P(SN)} x \frac{V_{cr} - V_{fa}}{V_h - V_m}$$

Optimum performance does not mean errorless performance, and optimality will vary among individuals just as the various values can cause shifts within an individual. β selection maximizes the return for that individual at the time of the selection. Shelley and Bryan (1964, p.9) propose that optimum be considered loosely as the '...selection of an action that is in some sense *best*'. With the personal, subjective nature of the concept, it can be seen that the particular β used by a subject will be entirely consistent with his value structure at the time of selection.

In aviation decision making the situations in which TSD might be applied are numerous. In each aspect of the pilot's task, TSD might be applied from the planning phase, through the preflight inspection, to enroute decision making. A classic TSD application would be the search for 'signals' in the preflighting of the aircraft to determine its airworthiness. The inspection of each item in the preflight provides information that must be compared with an internal representation of the item. Discovery of a discrepancy signal (for example, the engine is down a quart of oil) is an event typical of TSD tasks.

The criterion that a pilot sets for this situation may vary depending on the relative importance of the flight. It would be possible to take vernier caliper readings of a crack and call all those over a certain length 'signals', but even that would not eliminate a human element from the measurement process. When reading engine instruments, we could expect to see similar results, for example, in judging the acceptability of the drop in RPM when checking the magnetos.

Would the criterion of the pilot vary with the costs and payoffs? If a businessman pilot is making a flight to seal a deal that will avert imminent bankruptcy for his company, we might expect his criterion to be somewhat more liberal than if he were making a routine call on an account that someone else was supposed to have done anyway. Implicit within all such situations is the likelihood that 'signal' responses will vary, depending on the personal situation of the respondent. The majority of cognitive decisions arrived at in the aviation environment are based on probabilistic information. Remembering the flexible nature of β and the fact that it can be easily manipulated by background factors, it is easy to see why decisions to continue a flight *against the odds* frequently occur.

Although d' is generally taken to be a fixed value reflecting the individual's sensory apparatus, we know that some improvement in this area is possible through becoming aware of our biases and through training in problem solving. Savage (1954) in his coverage of decision theory touches on this approach to certainty through experience. He indicates that the person typically becomes almost certain of truth as his experience increases indefinitely.

The use of TSD techniques in this way goes quite far beyond its traditional use for auditory signal detection though it is not completely out of its realm. Its major departure is the change in task from the simple detection to a global decision task. However, this less traditional conception of TSD in the pilot's decision task allows investigation of behavior that might previously have been considered too complex or perhaps 'mentalistic'. The unobservable cognitive behavior of deciding does have observable behavior tied to it opening the possibility of actually seeing the influence of decision from the observed behavior.

These examples indicate how the combination of β and d' can account for decision making. The first five factors of the descriptive model given above (problem vigil, recognition, diagnosis, alternative identification and risk assessment) taken together make up the d'. The β contains the whole of the background factor. These two factors have been studied by scientists in many different contexts and we are now able to bring that level of theoretical understanding into the practical world of aviation. In teaching these two parts of aviation judgment, d' in the pilot is primarily improved through knowledge acquisition. Minimizing the negative effect of β can only be learned through experience.

How do we get an individual pilot's aviation experience to reach a level of certainty such that we can claim, 'good judgment'? None of us can live long enough to gain it all from personal flying experience. Instead we must learn from the experiences of others. This is the approach to judgment training that will be advanced in our training chapter. It would be nice if we could expect this level of judgment to be attained through the experience of training for the various certificates -- in conventional aviator training, that is what we have expected.

The pilot judgment model

To help you understand how you can improve your aviation judgment, it is helpful to reduce the eight-step descriptive judgment model into two parts as described by the TSD model. The first part, called *rational judgment*, encompasses the first five steps of descriptive judgment model that relate to intellectual abilities or discriminability. The second, part called *motivational judgment*, encompasses the final three steps, that relate to motivation or bias. In short, good pilot judgment is:

Part I: Rational judgment:

> *The ability to discover and establish the relevance of all available information relating to problems of flight, to diagnose these problems, to specify alternative courses of action and to assess the risk associated with each alternative.*

Part II: Motivational judgment:

> *The motivation to choose and execute a suitable course of action within the available time frame.*
>
> Where:
>
> *a. The choice could be either action or no action and,*
>
> *b. 'Suitable' is a choice consistent with 'societal' norms.*

Rational judgment

The first part of the judgment model is the d' of the TSD model and refers to the mental ability of the pilot to detect, recognize and diagnose problems, to establish available alternatives and to determine the risk associated with each alternative. This part is purely rational, and if it could be used alone (which is not possible), would allow problem solving using mathematical functions in much the same manner as a computer. This does not mean it would be error free; it uses information that is probabilistic and therefore, predicts outcomes that are not certain. In addition, rational judgment depends upon the amount, type and accuracy of the information stored in the pilot's memory as well as his or her learned capabilities to retrieve and process information. Some in aviation circles (see Diehl, 1992) prefer to call this part of pilot judgment 'headwork'. To optimize rational judgment requires high levels of knowledge, experience, organized mental structures and systematic computational and problem solving abilities.

Motivational judgment

The second part of the judgment model is the motivational or bias (β in the TSD model) aspect of judgment. The emphasis is on the directional rather than the aspects of motivation dealing with intensity. This part of judgment says that humans (and pilots) base their decisions, in part, upon bias factors or tendencies to use less than purely rational (as defined by society) information. These factors include immediate gratification such as ego, adventure, commitment, duty, social pressure (Asch, 1951) and emotional arousal in the form of worry, fear, stress, anxiety and euphoria (Wine, 1971; Liebert and Morris, 1967; Steiner, 1972; Janis and Mann, 1977), as well as more long-term biases such as risk-taking attitudes and personality factors, e.g., fear of failure and defensiveness (Kogan and Wallach, 1964). Some in aviation circles (see Diehl, 1992) prefer to call this part of judgment 'attitudes'. To optimize motivational judgment requires both 1) an awareness of biasing factors and 2) a will (motivation) to suppress these error producing factors so that decisions can be made on the basis of relevant safety factors from the physical world.

Neuropsychologists refer to research on the brain showing that the majority of our intellectual rational activity is conducted using the left side of our brain. On the other hand, our creative, artistic, feeling activity is done using the right side of our brain. Although it is an oversimplification, people who are highly rational are said to be 'left brained'. People who are feeling oriented and creative are called 'right brained'. As we consider recommendations for judgment training in aviation, we must not neglect either part. Even though rational thought is highly praised in aviation, it would not be right to try to train everyone to be purely rational because we would then be no better than the computers we are building. We as humans are better than the machines we are building because we can create novel solutions to unusual problems that we encounter. This will always be the advantage humans have over machines. For this reason, it is important to include creativity training in judgment courses for pilots.

A poignant illustration of the two factors of judgment was revealed to me in a recent biographical novel called *The Power of One* (Courtenay, 1990) given to me by one of my students (Flemington, 1990) from a judgment course that I taught in South Africa. On page 124, seven-year-old PeeKay is getting advice about boxing from his new-found hero Hoppie Groenewald who says to him 'First with the head, and then with the heart, that's how a man stays ahead from the start'. This became the secret ingredient in Peekay's phenomenally successful life -- the 'Power of One'. If you really want to see what I mean by these two parts of judgment, read this book. In addition, you will learn a lot about the real turmoil in South Africa.

This now completes our development of the structure upon which aviation judgment will be built. The next two chapters will give the scientific background for the two aspects of judgment. These will be followed by practical ideas for building aviation judgment training programs.

References

Asch, S.E. (1951). 'Effects of group pressure upon the modification and distortion of judgments'. In H. Gretzkow (ed.) *Groups, leadership and men: Research in human relations*, Pittsburg: Carnegie Press, 177-190.

Birdsall, T.G. (1955). 'The theory of signal detectability'. In H. Quastler (ed.), *Information Theory on Psychology: Problems and Methods*, Glencoe, IL: Free Press, 391-402.

Boff, K.R., Kaufman, L. and Thomas, J.P. (1986). *Handbook of Perception and Human Performance*. New York: John Wiley and Sons, 36.

Brecke, F. (1981). 'Instructional design for aircrew judgment training'. In R.S. Jensen, *Proceedings of the First International Symposium on Aviation Psychology*, Columbus, OH: The Ohio State University.

Cherry, E.C. (1953). 'Some experiments on the recognition of speech with one and two ears'. *Journal of the Acoustical Society of America*, **25**, 975-979.

Combs, H. (1979). *Kill Devil Hill*. Boston: Houghton Mifflin Company.

Courtenay, B. (1990). *The Power of One*. London: Mandarin Paperbacks.

Diehl, A, (January, 1992). 'Does cockpit management training reduce aircrew error'? *ISASI Forum*, **24**(4).

Eubanks, J.L. and Killeen, P.R. (1983). 'An application of Signal Detection Theory to air combat training'. *Human Factors*, **25**, 449-456.

Fitts, P.M. and Jones, R.E. (1947a). 'Psychological aspects of instrument display: Analysis of 270 'pilot error' experiences in reading and interpreting aircraft instruments'. USAF Air Materiel Command Memorandum Report NO. TSEAA-694-12A.

Fitts, P. M. and Jones, R.E. (1947b). 'Analysis of factors contributing to 460 'pilot error' experiences in operating aircraft controls'. USAF Air Materiel Command Memorandum Report NO. TSEAA-694-12.

Flathers, G.W., Giffin, W. and Rockwell, T.H. (1981). 'A study of decision-making behavior of aircraft pilots deviating from a planned flight'. In R.S. Jensen (ed.) *Proceedings of the First International Symposium on Aviation Psychology*, Columbus, OH: OSU Department of Aviation.

Flemington, M. (1990). Personal communication.

Galenter, E. and Holman, G.L. (1967). 'Some variations of the intensity function and their implications for the utility of money'. *Journal of Experimental Psychology*, **73**, 333-339.

Giffin, W.C. and Rockwell, T.H. (1984). 'Computer-aided testing of pilot response to critical in-flight events'. *Human Factors*, **26**(5), 573-581.

Green, D.M. and Swets, J.A. (1966). *Signal Detection Theory and Dsychophysics*. New York: Wiley.

Hilgard, E.R., Atkinson, R.C. and Atkinson, R.L. (1975). *Introduction to Psychology*. New York: Harcourt Brace Jovanovich, Inc.

Janis, I. and Mann, L. (1977). *Decision Making: A Psychological Analysis of Conflict, Choice and Commitment*. New York: Free Press.

Jensen, R.S., Adrion, J., Maresh, J., Weinert, C. (1987). *Pilot Judgment Skills Test Battery*. Columbus, OH: Aviation Research Associates/U.S. Air Force Final Report SBIR Contract No. F33615-86-C-4519.

Johnson, W.B., Rouse, S.H. and Rouse, W.B. (1980). *An Annotated Selective Bibliography on Human Performance in Fault Diagnosis Tasks*. Arlington, VA: US Army Research Institute for the Behavioral and Social Sciences, Report No. TR-435, January, ERIC Document Reproduction Service No. ED 192 736.

Johnson, W.B. and Rouse, W.B. (1982a). 'Analysis and classification of human errors in troubleshooting live aircraft powerplants'. *IEEE Transactions on Systems, Man and Cybernetics*.

Johnson, W.B. and Rouse, W.B. (1982b). 'Training maintenance technicians for troubleshooting: Two experiments with computer simulations'. *Human Factors*, **24**(3), 271-276.

Klein, G. (1989). 'Do decision biases explain too much'? *Human Factors Society Bulletin*, **32**(5).

Kogan, N. and Wallach, M. (1964). *Risk Taking: A Study in Cognition and Personality*. New York: Holt, Rinehart and Winston.

Liebert, R. and Morris, L. (1967). 'Cognitive and emotional components of test anxiety: A destination and some initial data'. *Psychological Reports*, **20**, 975-978.

Mackworth, J. F. (1970). *Vigilance and Attention*. Middlesex, England: Penguin Books, Ltd.

Mackworth, N. H. (1961). 'Researches in the measurement of human performance'. In H. A. Sinaiko (ed.), *Selected Papers on Human Factors in Design and Use of Control Systems*, New York: Dover Publications.

NTSB. (1990). *Aircraft Accident Report: United Airlines Flight 232, McDonnell Douglas DC-10-10, Sioux Gateway Airport, Sioux City, Iowa, July 19, 1989*. (NTSB/AAR/90/06). Washington, DC: Author.

Rasmussen, J. and Rouse, W.B. (eds.) (1981). *Human Detection and Diagnosis of System Failures*. New York: Plenum Press.

Rouse, W.B. (1979).'Problem solving performance of maintenance trainees in a fault diagnosis task'. *Human Factors*, **21**, 195-203.

Savage, L.J. (1954). *The Foundations of Statistics*. New York: Wiley.
Shelley, M.W. and Bryan, G.L. (eds.) (1964). *Human Judgments and Optimality*. New York: Wiley.
Steiner, I.D. (1972). *Group Process and Productivity*. New York: Academic Press.
Wald, A. (1950). *Statistical Decision Functions*. New York: Wiley.
Weintraub, D.J. and Walker, E.L. (1966). *Perception*. Belmont CA: Wadsworth Publishing Company, Inc., 45.
Wickens, C.D. (1992). *Engineering Psychology and Human Performance* (second edition). New York: HarperCollins Publishers, Inc.
Wine, J. (1971). 'Test anxiety and direction of attention'. *Psychology Bulletin*, 76(2), 92-104.

3 Rational judgment

In making supposedly rational decisions, we must estimate at least two types of quantity, namely, the relative probabilities of different outcomes of courses of action and the degree of our preference for the different outcomes. The extent to which the fear of failure deters us and the hope of success spurs us on, as well as the degree of confidence in what may happen, all determine our decisions.

Wallach and Kogan (1961)

In this chapter we will look at the rational aspect of pilot judgment. It could be called 'The *science* of aviation decision making'. The concern here is with the facts, probabilities and information processing techniques in mathematical judgment. This part of judgment is considered 'rational' because it could be done on computer. Rational judgment is just as 'human' as motivational judgment, the subject of the next chapter even though we often say to our friends, about certain (irrational?) mistakes, 'that only shows that you are human'.

In fact, there is quite a controversy among scientists over the issue of rational and irrational judgment (Jungermann, 1988). Rational judgment means that there is some norm or set of rules about which one can determine the correctness of a decision. Some would say that these rules lie within the person making the decision. In aviation I would suggest that the norm is set by society, i.e., we are bound to some extent by societal expectations because we can do so much damage to society with wrong decisions. The controversy is over the use of the term 'irrational judgment' to refer to biases and attitudes that all humans sometimes apply in making decisions that defy any set of rules that one could set down. It is interesting to note that psychologists even use the term 'rationalization' to refer to the reasons some people might use to justify what others would call irrational choices. The discussion over this controversy is beyond the scope of this book, but we will look further into motivational judgment in the next chapter. The use of the term 'rational' judgment as the title of this chapter is not meant to convey that there is widespread irrational judgment in aviation -- though I believe that some does exist.

In this chapter, we will make believe that humans (and pilots) can process information and make decisions on a purely intellectual basis, considering just the facts and probabilities almost as a computer would process information. Motivations, emotions, values, social pressures, etc. will be assumed not to be a part of the decision process. Thus we are considering the first five steps in the Detailed Judgment Model from Problem Vigil to Risk Assessment. From the TSD model our subject is the discriminability or d' factor. We consider this part of judgment first because it has been studied widely and applied to many other fields, especially medicine, clinical psychology, stock brokering, business and even fire fighting. Surprisingly, rational judgment has received very little attention in aviation.

Rational judgment is a major part of our everyday life. Kaplan and Schwartz (1977, p.xvi) illustrate the complexity of this judgment process with examples from medicine and public policy. In the field of medicine, a patient presents a physician with a set of complaints. The physician may also have data from a series of tests on the patient. The physician must identify the relevant evidence for diagnostic alternatives and he must weigh each test and the symptoms for their reliability and diagnosticity. In many cases, the results of the tests may point toward different diagnostic choices and thus the strength of the tests must be considered. Finally, the physician must consider the costs, both in terms of time, safety and pain of doing more tests to be sure of the diagnosis.

Consider also the judgment of the city manager who must decide whether to hire more firemen to be more responsive to fire alarms, considering the cost of new hires to the taxpayers. The goal, or value, of being responsive to fires must be weighed against the likelihood that such an alternative will accomplish the desired result.

In aviation the requirements for rational judgment are just as profound as in the previous examples. Actual conditions are never quite the same as those conditions in which pilots are trained and they usually differ from those in which procedures and regulations were tested and adopted. Thus the safety of every flight depends upon a significant amount of interpretation and evaluation of existing conditions by the pilot.

For example, the conditions used to develop flight performance values for a particular type of airplane are usually ideal. They would include a new engine, clean airplane surfaces, a new propeller, an unrestricted air filter and a company test pilot. The results for takeoff and landing performance in these conditions are published in the operating manual. In actual conditions, the pilot must compare these performance values, taken in ideal conditions, with the conditions for a particular flight, where they may include a dirty airplane, a slightly used engine, a few marks on the propeller, a slightly dirty air filter and a less than perfect pilot. The pilot must then look at other variables that could affect the aircraft's performance including gross weight, center of gravity, wind, temperature, humidity, etc., all of which may not be known as well as those in which the book values were derived. Finally, other uncertainties must be considered including present and forecast weather, terrain and expected traffic density and for comparison with self-estimates of the pilot's own capability before making the final decision to make the planned flight. In a recent study, these and other variables were considered risk factors that the pilot must consider in making the takeoff go/no go decision (Jensen, Kirkbride and Chubb, 1994).

In rational judgment we consider what the 'ideal' person should do in the circumstances, expressed as the expected value (*EV*) of the decision and depends on both the value (*v*) of each alternative and the probability (*p*) of achieving the desired goal for each alternative. In risky situations the calculation of the value must also include the amount of loss that is possible and the probability of that loss for each alternative. For a set of alternatives and their associated probabilities, expected value can be expressed as shown in the following equation:

$$EV_i = \sum_{j=1}^{n} p_i v_{ij}$$

This expression could be considered the rule that we all should follow in our decisions. If we were computers (such as 'Data' the Star Trek character 'Android') instead of humans, we would make our decisions in this way. One of the weaknesses in classical rational decision theory is the assumption that humans can think in this way.

Attempts have been made to discover the mental processes that are used by expert judges such as stockbrokers, livestock judges and medical diagnosticians in making their decisions (Shanteau and Phelps, 1977; Slovic, 1969; Anderson, 1969; Hoffman, Slovic and Rorer, 1968; and Hoffman, 1960). The argument is that if models of the mental processes used by expert judges in decision-making were available, these could be used in training people to be expert judges. In each of the areas studied, judgmental training occurs over a fairly long apprenticeship program in which the trainee observes the expert make decisions and learns by observation. However, as in aviation, because of the complexity of the information used to make decisions, observation or even trial and error are very inefficient training methods.

Goldberg (1968) reports an experiment in which various judgmental training methods are used including 'outcome feedback', 'basic formulas' (for integration of information) and 'values' (or coefficients in the formulas, similar to models of mental processes used by expert judges). The task was to learn to make a differential diagnosis of psychosis versus neurosis based on information from a complex personality test. The results revealed that naive subjects (those having no previous psychological training) using this so-called 'value' training method could learn to make judgments as accurately as experienced clinical psychologists and in a much shorter time. Another study reported by Todd and Hammond (1965) found similar results using the Brunswik Lens Model (Brunswik, 1956) as a judgmental training method. These studies indicate that rational judgment can be taught.

The most important factors in rational aviation judgment are: knowledge, risk assessment, creativity and problem solving ability. Knowledge is important to rational decision making in three ways. First, we must have a thorough knowledge of the systems and environment in which we are operating as the basis for rational decision making. Second, because the most important part of the 'system' is the human, we must have a thorough awareness of weaknesses in our own decision making system. Third, we must train ourselves to avoid the pitfalls that these flaws bring about. Risk assessment skills are important to rational judgment because it is expected that the human can compute, in some form, the level of risk associated with each alternative that may be available. Creativity is important to rational aviation judgment because it causes one to step outside of the paradigms (programs and rules) that we have learned as a part of our development into a skeptical attitude toward the status quo that may lead to better analyses of the situation. Problem solving ability is important to rational aviation judgment because it causes us to think clearly, gather the correct information, put together the right analysis of the data and focus on the solution.

Weaknesses in human rational judgment

Because we are human, we have a set of defects (some were discussed in Chapter 1) in our information processing systems that prevent us from making decisions in the ideal way predicted by the classical rational decision theorists. Some of these defects affect our ability to make decisions based solely on the mathematics or rationality. As we will see in the next chapter, other defects, such as our attitudes, actually cause us to make decisions opposed to what we know 'rationally' is right.

The first part of our effort to develop good aviation judgment requires that we become aware of these problems that we humans have with numbers and statistics that can affect our judgment. These weaknesses in our rational information processing system are just as pervasive to flying as are visual illusions that we discussed in Chapter 1. In most of our daily decisions, these weaknesses are used to our advantage. Many of them have been acquired because we have learned that they can be so used. However, when we are placed in an airplane these weaknesses can prevent us from knowing the whole truth about our situation. If we know what they are, we can devise methods for countering their negative effects.

We know that some people are more intelligent than others and some are better problem solvers than others, but these individual differences are not what we are talking about here. Our concern is with inherent and learned weaknesses with rational decision making in most people, regardless of their intelligence. These weaknesses are 'representativeness', 'availability', 'anchoring' and 'primacy and recency' (Tversky and Kahneman, 1987; Kaplan and Schwartz, 1977).

Representativeness

One of the types of decisions that people make is to determine whether an object, event, or process belongs to a class of objects, events, or processes, respectively. In answering such questions people rely on the *representativeness* heuristic. They try to determine to what extent the object, event, or process represents the classes of objects, events, or processes that are known to exist. For example, you know that a person is six feet, six inches tall and plays basketball. You are to determine whether this person is a man or a woman. The person's height and behavior are representative characteristics that you might use to make your choice.

Recently, I was faced with the decision to take either a Cessna 210 or a Cessna 182 on a flight to Put-in-Bay on Lake Erie. The characteristics were two passengers, strong, gusty winds, short, narrow curved runway, with obstacles at both ends. My objective was to pick the safer aircraft for the flight. Using the representativeness heuristic I would choose the aircraft that has the most characteristics represented by a safe outcome. I selected the 182 because I believed that it had better handling characteristics in the gusty winds that were expected. One characteristic not thought of specifically was the interaction of two of the factors, namely strong winds and obstacles at the approach end of the runway. As I approached the trees near Runway 4 at Put-in-Bay, I could see the trees being blown heavily and I was hit by the most severe updraft of my flying career very close to the ground. Fortunately, the wind quickly smoothed out and I made the landing with little effort. I believe that, if I had chosen the 210, this wind gust may have been too much for me to handle. Although successful on this landing, the representativeness heuristic leads to serious errors in general human decision making because it is not influenced by several factors that should affect judgments. The following is a discussion of some of these factors.

Insensitivity to prior probability of outcome. People tend to be unable to compute in their decisions the probability that certain outcomes will occur. Merely looking at representative characteristics of the aircraft, e.g., controllability in gusty winds, and short field characteristic, does not consider the prior safety record that has been established in these conditions for each aircraft.

The Gambler's Fallacy. People tend to believe that random events should be somewhat equally represented throughout a long sequence. For example, on the toss of a coin, people think that heads and tails should be fairly close to alternating one after the other, perhaps because it appears to be more representative of random events. They tend to think that if there has been a series of heads, the chances are better that the next toss will be tails. This is an example of the well known gambler's fallacy. In reality this is not the case at all. The chances of tails appearing is equal on every toss regardless of what happened before. If no accident has occurred to one individual in these conditions, the Gambler's Fallacy might lead a pilot to fear that he or she is 'DUE' for an accident. In reality, in this example, pilots are more likely to believe (irrationally) that they are invulnerable to such an accident than due for it! We will discuss this aspect of decision making in the next chapter.

Conservatism. Theoretically, this thinking weakness consists of the failure to consider the size of the sample. People tend to think that any sample taken from a population should be representative of the population and extremes in a given sample do not occur. For example, if you were to flip a coin 10 times, people tend to expect that most of the time 5 heads and 5 tails should appear. In fact, most of the time this ratio does not appear and sometimes all 10 will land tails. In reality, small samples will vary much more widely from the population than large samples. This is why, the daily stock average moves quite large distances but over a week or year, it moves much less. This thinking weakness leads to the belief in pilots that extremes in circumstances such as forecast peak winds, forecast severe thunderstorms, or stacks of problems occurring at the same time do not or will not occur -- at least not on this flight. This is the cause of the hazardous attitude of invulnerability.

Insensitivity to predictability. People tend to be unable or unwilling to use known predictability of outcomes in their decisions. If they see a situation occurring in the present they tend to predict that the same (representative events) will occur in the future. In one study reported by Kahneman and Tversky (1973, p. 4), students were presented a description of a student teacher in a trial lesson and asked to predict the standing of the teacher five years hence. In every case the students predicted that future performance for that teacher would be very close to the description of the teacher now, unaware of the unpredictability of potential future performance of the teacher based on one lesson. In aviation this thought process could lead to a belief that an existing fair weather pattern will continue or that because I have always landed safely in a crosswind, there is no reason to believe that this landing will be different.

Misconceptions of regression. In real life most repeated events (such as height of males or performance in school) regress toward the mean of those events even though each individual or event taken alone is usually quite different from the average. This means that if a poor performance is observed in one class or even several classes in one term it doesn't mean that that performance will be continued throughout the education program for a given student. Instead the student will perform better and the overall grades will regress toward the mean of that student's ability. As much as these sequences of events occur in everyday life, people do not incorporate it into their thinking. Either they do not expect regression to occur (they are surprised when the student scores much higher on the next test), or they invent rationalizations for the change of fortune. Kahneman and Tversky (1973) relate a story about flight instructors in Israel, the principle of which I can confirm from my own flight instruction experience:

> In a discussion of flight training, experienced flight instructors noted that praise for an exceptionally smooth landing is typically followed by a poor landing on the next try, while harsh criticism after a rough landing is usually followed by an improvement on the next try. The instructors concluded that verbal rewards are detrimental to learning while verbal punishments are beneficial, contrary to accepted psychological doctrine.

In my own experience, this became a pattern that I came to expect as I learned to fly. As Ron Royce, my instructor, introduced me to a new flight maneuver, my first performance (in which I mimicked the instructor's demonstration) was always better than expected and my second, worse. In each case, the instructor was surprised (he would say, 'I thought you had it') and seemed to blame his own instruction. In both my experience and that of the Israeli instructors the student pilots were regressing toward the mean of their performance.

Availability

The second major flaw in human judgment is that people jump to a conclusion about a situation based on characteristics that are most easily brought to mind. They do the first thing that comes to mind. Psychologists refer to this fault in human judgment as availability. Some people are afraid to fly because the most easily retrievable instances about flying are crashes that they have heard about. Some people (and corporations) resist trying new products because they can recall consumer lawsuits resulting from other new products that they or others had introduced. There are two reasons for the availability flaw familiarity and imaginability.

Familiarity. The first factor causing the availability flaw in our thinking is *familiarity*. Characteristics that are familiar tend to be more available. For example, famous people who represent a class (such as males) give that class more emphasis than a class containing only unknowns. In aviation, a pilot/mechanic would give more emphasis to a mechanical explanation for a problem, perhaps forgetting that there may well be another explanation. Another derivative of the availability due to familiarity flaw is that people tend to apply what they know to fix a problem rather than the correct solution. If you go to a surgeon, expect to find that surgery is needed; go to a chiropractor, manipulation. Familiarity also may be caused by the import attached to the characteristic. For example, seeing someone killed in an accident could cause similar circumstances to bring that memory back very easily.

I*maginability.* The second factor causing availability could be called *imaginability* (Tversky and Kahneman, 1987). If we are asked to imagine various results of certain scenarios (for example, losing a friend for telling the truth) those that we think of first and most easily, tend to be considered the most likely. As we gain more and more experience with certain activities and hear of disasters in these activities (such as flying single-engine IFR over mountains), we tend to imagine disaster in these circumstances. I have a friend who has seen quite a number of disasters in his testimony in aircraft accident litigations who refuses to fly single-engine in mountainous terrain.

Anchoring

People evaluate things relative to a reference point or anchor. For example, if you put your hands in cold water, and then put them in a bucket of water that is not quite so cold, you will report that the second bucket of water is warm. The same goes for our evaluation of people. When we are with people who are 'cold' any small bit of friendliness will be considered 'warm'. Given the reference point, any small change means a lot more than it should from a pure rational judgment point of view. In aviation, a ceiling and visibility of 500 and 1 may appear very good after days of 200 and 1/2 but bad after days of CAVU (ceiling and visibility unlimited).

Primacy and recency

The effect of when characteristics of a situation are presented in sequence has been studied extensively with the conclusion that in almost every case the characteristics that are presented first receive the most weight (Asch, 1946; Anderson, 1974, p.259). In other words, first impressions are the most long lasting. Most of the research reported used personal characteristics (traits) of people and had subjects rate whether they were favorably or unfavorably inclined toward the object with the characteristics. If favorable (H) traits and unfavorable (L) traits are given in the following order,

H H H L L L

they will result in a more favorable inclination toward the object than if they are given in this order:

L L L H H H.

These findings tell us that we need to be careful to give equal weight to information about our situation whether it is given to us first or last. We need to be aware of our tendency to give more credence to information heard first and try to overcome this tendency by applying greater weight to information given later in sequence.

The recency effect says that characteristics given most recently have greater effect than those given in some distant past. Again, those things heard recently are more available to us for recall. As we are listening to the weather conditions and forecast, the recency effect could cause us to place greater emphasis on items given last. This effect can be countered by taking careful notes as the weather person is speaking and using all of the information provided equally.

Rational judgment models

There is no need to do an exhaustive review of the models of rational judgment in an applied aviation book such as this because they are extensive and, in my opinion, most have little to contribute to aviation judgment. Nevertheless, we in aviation should be aware of the 'Bayes Formula' because it can increase our understanding of our own judgment processes and therefore, could help us make better decisions.

'Bayes' formula

One of the most important models in decision theory is known as Bayes' formula, named after the English clergyman and mathematician Thomas Bayes (1702-1761), who at least demonstrated that he knew how to use it. This formula is important because it has been widely used in decision theory to explain rational thought and because it encompasses both prior odds and new data in its computation of posterior probabilities. The purpose of the formula is to determine the probability of a hypothesis $p(H_j)$ being true given that new data (D) is now available $p(H_j / D)$ and the fact that we know something about the prior odds before we start. The probability for which we are seeking, $p(H_j / D)$, is called the *posterior* probability. The information we knew before getting new data, $p(D / H_j)$, is called the *likelihood.* For several (n) known hypotheses upon which we are trying to decide, Bayes' formula is written as follows:

$$p(H_j/D) = \frac{p(D/H_j) * p(H_j)}{\sum_{j=1}^{n} p(D/H_j) * p(H_j)}$$

Suppose we have three candidates (Perot, Bush and Clinton) running for the Presidency of the USA and we can establish through opinion polls that each has a certain *prior* probability of winning. On election night, we hear the voting returns from one state. These returns are data (D) that we can input into our equation to determine a new *posterior* probability, $p(H_j / D)$, that our hypotheses are right. After each calculation, the resulting *posterior* probability becomes the *prior* probability for the next calculation.

Similar subjective calculations can be made in aviation judgment. Frequently we are faced with growing evidence that a certain course of action could lead to disaster but we are reluctant to change our minds because we are committed to the action. As new data become available, these can be put into our decision process and we get closer to the correct solution.

The question we might ask is, are humans good 'Bayesian' decision makers? Do they make decisions according to this formula? The answer is clearly no (Lee, 1971, p. 251). In almost every experiment people are found not to give enough impact to the new evidence. They under estimate the impact of the new data. Another way of putting it: when faced with new data, people do not sufficiently adapt to changes in their situation. This phenomenon was stated earlier in the discussion of the impact sequential information. When people are given the state of the world initially, they are very reluctant to decide that the world has changed in the face of new data.

In aviation, we know that adaptability is extremely important. In fact, as a predictor of pilot performance, the Israelis have found that adaptability is the single best test of pilot performance, at least in training (Gopher and Kahneman, 1971). To improve rational pilot judgment requires that pilots be aware of their tendency to give less importance to new data and to force changes more quickly when new data becomes available.

The classic example of this problem is the most common fatal 'pilot error' in general aviation, that of 'continued VFR into known adverse weather'. Unfortunately, sometimes airline pilots make similar errors. For example, consider an Allegheny Airlines Convair 580 accident that occurred in 1971 at New Haven, Connecticut as described by Richard Taylor (1994):

> The pilots were faced with company pressures to hurry which may have led them to decide not to add reserve fuel. The pilots had probably made two landings in weather below IFR approach minimums that day so they knew the weather situation. Earlier, at Groten, Connecticut, with coastal fog limiting visibility to less than a mile, the captain requested a 'contact approach' and made a successful landing in conditions well below IFR minimums. He took off into the weather again with another load of passengers and at New Haven, he was again faced with weather below minimums which were 400 and 1. This time, low on fuel, he told his co-pilot to 'keep a sharp eye out there' while he again descended below minimums over water. Just before the crash, the co-pilot said, 'We're right over the water. Man, we ain't 20 feet off the

> water'! Seconds later the Convair hit three beach houses at 25 feet above sea level killing 28 people, all but two people on board.

Compelled by company pressures (see Chapter 12 for a discussion of the effect of management pressure on pilots), this captain ignored the new data concerning the weather. Undoubtedly, he was also affected by the representativeness factor causing him to think that he had made it through three times earlier, why would he not be able to do it again? This accident shows how important it is to know how easily these weaknesses in our rational thoughts can lead us to ignore new, important data.

Risk assessment

It is well accepted that aviation is an activity that involves risk or the probability of suffering a loss. Based on accident statistics, the risk in aviation varies considerably depending on the type of flying activity. The risk in air carrier flying (in death rate per billion passenger miles) is at least 10 times less than in driving an automobile, whereas the risk in recreational flying is some ten times worse than driving an automobile. However, within general aviation there are large differences as well. The risk in USA corporate jet flying is nearly as low as airline flying but in aerial application it is much higher. The risk of flying with someone in general aviation who has an airline transport pilot certificate is much less than flying with someone who has only a private pilot certificate (O'Hare, 1990). The risk of an engine failure in a twin-engine aircraft is twice that of a single.

One of the most important aspects of aviation judgment is risk assessment. What do we mean by risk in aviation? Our primary concern with risk in aviation is probability of loss in the form of an accident involving property damage, personal injury, death and/or loss of face before our superiors and peers. Closely associated with risk in aviation are hazards which, are the sources of danger. We may be aware of the hazards without knowing the risks associated with each hazard. Our goal in aviation judgment is to be fully aware of both the hazards and the risks involved in each of the decisions that we make.

It is too much to expect pilots to be able actually to compute the risk for each decision in terms of probability of failure or loss for each alternative. What the pilot must do is to be aware of each potential loss connected with each alternative and be able to compute the potential for loss for each alternative *relative* to others that may be available. This computation becomes important because the pilot must then determine what are his or her *outs* for each alternative. This is what pilots would call 'calculated risk'. By computing the relative risk with each decision we are taking calculated risks, but with each risk, we have an out.

For example, suppose a pilot is to make a takeoff in a light-twin engine aircraft with a load of passengers and baggage. He or she computes the density altitude, weight of the aircraft and runway length and determines that the takeoff can be made easily with two operating engines. Then, looking at single engine performance, it is found that, with the load, a successful go-around cannot be made if one engine quits between 50 and 200 feet altitude after takeoff. A subjective estimate of the probability of an engine failure indicates an engine failure at that point to be very low, but if it did occur the pilot would plan to land straight ahead (even if it were into the trees) because the alternative would be a relatively high risk of a stall-spin into the trees. His or her out is not very desirable but considering the low probability that it will be needed and that it is a planned out, he or she makes the takeoff (without going into this amount of detail with his or her passengers) and is highly vigilant for the first sign of engine failure during the takeoff roll through 200 feet. In fact, most warm climate, loaded, light-twin engine takeoffs are made with this type of risk assessment in mind.

In the textbooks on decision making, the term *risk* is used in the very general sense to categorize decision situations (Lee, 1971, p. 114). In this context, a 'risky condition' exists when the state of nature is determine by a set of probabilities. Much has been written about this form of risk which has been taken to be almost synonymous with decision making. Many of the examples given refer to gambling situations.

Attempts have been made in the general decision making literature to determine 'risk-taking tendencies'. Wallach and Kogan (1961) developed a *choice-dilemmas questionnaire* to measure these tendencies. The questionnaire consists of a series of potential employment situations in which the subject is supposed to rate the probability of success in the new job. If the subject rated the probability as low he was considered high in risk-taking. If he rated the probability as high, he was given a low risk-taking score. One of the findings, after giving this test to many people, is the not too surprising conclusion that older people are more conservative (less risk-taking) than younger people. A more interesting finding is that, when discussion among people are permitted, the group decision tends to be more risky than individual decisions. This phenomenon has been called the *risky shift*. President Kennedy noted, 'The trouble is that, when you get a group of senators together, they are always dominated by the man who takes the boldest and strongest line' (quoted in Lee, 1971, p. 118). Could this risky shift occur in multi-person aircraft crews? There are some accidents that seem to support this conclusion. We will see more of this in Chapter 6 on crew resource management.

Conclusion

We must conclude that people and pilots are not good rational decision makers. They depart from rationality both because they are not good information processors and on purpose. We have seen that rational models showing the best course of action toward the result people desire can be constructed, but people (and pilots) often avoid following these models. We have pointed out why people are often unable to follow the rational models. Why do they sometimes purposely avoid rationality? That is the topic of the next chapter. The question we have remaining is, 'What can be done about it'? This will be the topic of our chapter on judgment training. A key to good rational judgment in aviation is knowledge and experience in the field. The more pilots know about the whole of flying, and the more experience they have, the better they are at rational judgment and the more likely they will be to use creative solutions when they are needed.

References

Anderson, N.H. (1969). 'Comment on "analysis of variance model for assignment of configural cue utilization in clinical judgment"'. *Psychological Bulletin*, **72**, 63-65.

Asch, S.E. (1946). 'Forming impressions of personality'. *Journal of Abnormal and Social Psychology*, **41**, 258-290.

Anderson, N.H. (1974). 'Information integration theory: A brief Survey'. In D.H. Krantz, R.C. Atkinson, R.D. Luce and P. Suppes (eds.), *Measurement, Psychophysics and Neural Information Processing*, San Francisco: W. H. Freeman and Company.

Brunswik, E. (1956). *Perception and the Representation Design of Psychological Experiments*. Berkeley, CA: University of California Press.

Goldberg, L.R. (1968). 'Simple models of simple processes? Some research on clinical judgment'. *American Psychologist*, **23**, 483-496.

Gopher, D. and Kahneman, D. (1971). 'Individual differences in attention and the prediction of flight criteria'. *Perceptual and Motor Skills*, **33**, 1335-1342.

Hoffman, P.J., Slovic, P. and Rorer, L.G. (1968). 'An analysis of variance model for the assessment of configural cue utilization in clinical judgment'. *Psychological Bulletin*, **57**, 338-349.

Hoffman, P.J. (1960). 'The paramorphic representation of clinical judgment'. *Psychological Bulletin*, **57**, 116-131.

Jensen, R.S., Kirkbride, L.A. and Chubb, G.P. (1994). *Developing the Personal Minimums Tool for Managing Risk During Preflight Go/No-Go Decisions*. Columbus, OH: Ohio Sate University Department of Aviation/FAA Final Report, Contract No. DTFA01-92-10204.

Jungermann, H. (1988). 'The two camps on rationality'. In H.R. Arkes and K.R. Hammon (eds.), *Judgment and Decision Making: An Interdisciplinary Reader*, Cambridge: Cambridge University Press.

Kahneman, D. and Tversky, A. (1973). 'On the psychology of prediction'. *Psychological Review*, **80**, 237-251.

Kaplan, M.F. and Schwartz, S. (1977). *Human Judgment and Decision Processes in Applied Settings*. New York: Academic Press, Inc.

Lee, W. (1971). *Decision Theory and Human Behavior*. New York: John Wiley and Sons, Inc.

O'Hare, D. (1990). 'Pilot's perception of risks and hazards in general aviation'. *Aviation, Space and Environmental Medicine,* **61**, 599-603.

Shanteau, J. and Phelps, R.H. (1977). 'Judgment and swine: Approaches and issues in applied judgment analysis'. In M.F. Kaplan and S. Swartz (Eds), *Human Judgment and Decision Processes: Applications in Problem Settings*, New York: Academic Press.

Slovic, P. (1969). 'Analyzing the expert judge: A descriptive study of a stockbroker's decision processes'. *Journal of Applied Psychology*, **53**, 255-263).

Taylor, R.L. (1994). 'Hurrying to disaster'. *IFR Refresher*, **8**(10), 1-4.

Todd, F.J. and Hammon, K.R. (1965). 'Differential feedback in two multiple-cue probability learning tasks'. *Behavioral Science*, **10**, 429-435.

Tversky, A. and Kahneman, D. (1987). 'Judgment under uncertainty: Heuristics and biases'. In D. Kahneman, P. Slovic and A. Tversky (eds.), *Judgment Under Uncertainty: Heuristics and Biases*, New York: Cambridge University Press.

Wallach, M.A. and Kogan, N. (1961). 'Aspect of judgment and decision making: Interrelationships and changes with age'. *Behavioral Science,* **6**, 23-36.

4 Motivational judgment

'I still feel just as convinced as ever that as soon as I start playing on my own account I can't fail to win'.

'But what makes you so sure'?

'Well, since you ask me -- I don't know. All I know is that I must win, that it is the only way out for me. Well, maybe that's why I feel so sure'.

The Gambler by Dostoevsky (Lee, 1971)

How can we account for the fact that many pilots feel just as invulnerable to the risks of flying as the gambler in the above quote? How can we account for the fact that many of the decisions people make are far worse than one would predict even considering that they can't compute probabilities very well as we saw in the previous chapter? What are the forces that cause us to make decisions 'against our better judgment'? Such decisions may be considered 'stupid' or irrational afterward, but are they really? In this chapter we will look at motivational judgment in aviation. We will see that people may be just as defective in assigning proper values to the judgment equation as they are in computing probabilities. If rational judgment is the *science* of flying, motivational judgment is 'the *art* of flying'.

Recall the expected value equation given in the previous chapter:

$$EV_i = \sum_{j=1}^{n} p_i v_{ij}$$

Whereas in the previous chapter we discussed the probability (p_i) part of the equation, in this chapter we will focus on the value (v_{ij}) part of the equation referred to as motivational judgment in our two-part model pilot judgment model

(Chapter 2). From the TSD model motivational judgment is called the bias (β). In the Detailed Judgment model, it is called the *background factor*. This discussion will move into what many would consider irrational pilot behavior. Many people who have not actually faced these motivational forces, find it difficult to believe they are as strong or as important to aviator decision making as they are. Very little research has been done on this factor and we must rely on anecdotal data, both to show the extent of the problem in aviation and also to show how the problem can be corrected.

Motivation definition and theory

A motive is 'any condition of the organism that affects its readiness to start upon or continue in a sequence of behavior' (Hilgard, Atkinson and Atkinson, 1975). The word motive is loosely synonymous with 'drive'. Deprivation of 'basic' human physical needs such as food, water, and sex, brings on drives to fulfill these needs. There are numerous theories of motivation. Freud thought that our actions are determined by instinctive inner forces, often operating below the conscious level. It seems more likely that the type of motivation for which we are concerned, can be better explained by social learning theory (Bandura, 1971) which says that the behaviors we exhibit partly determine the rewards or punishments we receive, which in turn influence our behavior. This is quite different from behavior theory in that the primary emphasis is on cognitive processes, not stimulus-response associations. Our actions are governed primarily by the anticipated consequences of those actions.

A popular approach for the representation of human motivation is offered by Maslow (1954) called the 'Needs Hierarchy'. He places levels of human needs in the form of a pyramid (Figure 4.1) starting at the bottom with basic 'Physiological Needs' and ending at the top with 'Self-Actualization'. According to Maslow's theory, the lower items on the hierarchy are met first, followed by those higher on the hierarchy. 'Safety and security' needs are more basic than 'Belongingness', 'Esteem', and 'Cognitive' needs. We have found in aviation that, contrary to this model, sometimes pilots appear to ignore safety and security needs in favor of upper level needs. One of your tasks as a pilot is to be sure your motives emphasize the safety needs, not only of yourself as the pilot, but also of your passengers.

Consider some of the factors that motivate the general aviation pilot to fly rather than to take another form of transportation for a personal trip. Faced with this decision many pilots think about and are motivated by adventure, ego, building flying hours and experience, financial gain, duty, status, peer pressure, family pressure to make a flight. These become the 'values' that are inserted into the equation and often are given higher priority than either the risks or other more 'rational' values that could be calculated such as economics, convenience, and safety in comparison with other modes of transportation.

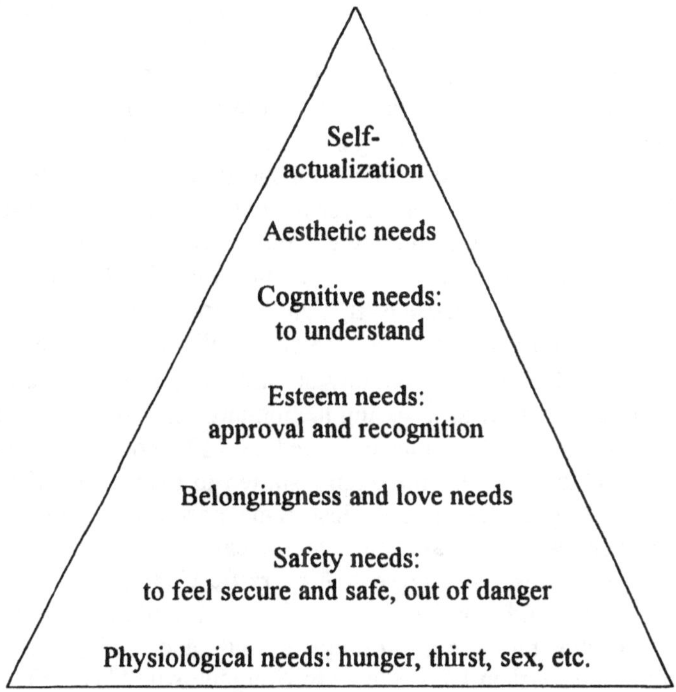

Figure 4.1 Maslow's hierarchy of needs (Maslow, 1954)

Let me illustrate with an example from flying experience relayed to me by a close friend. This gut-wrenching incident occurred well into his flying career after he had done a considerable amount of research work on pilot judgment. He was a careful observer of his own thoughts regarding aviation decisions and in most cases wrote them down before they were forgotten, perchance these introspections might be useful to someone later. This incident began as a planned flight in a Cessna 182 from Washington National Airport to Columbus, Ohio following a presentation of his research to the Aerospace Medical Association annual meeting. He wrote down the details of the flight in a letter to a mutual friend after being told by his girlfriend that she had decided to marry someone else. The following is a quotation from that letter:

> It's Friday morning following the all-day AMA meeting in DC and I'm not in Columbus. That's the bad news. The good news is that I am alive. I left National Airport yesterday evening after the meeting knowing that there was a possibility of icing in the clouds as low as 4,000 feet. I wanted to get out of DC so bad that I would consider landing anyplace and spending the night as long as it wasn't in DC. On that flight, I found out that ice can build up so quickly that options to

> land that you thought you had disappear before you can use them. I came as close to killing myself as I could imagine.
>
> Soon after departing DCA, I discovered ice forming on the wings at 6,000 feet and requested 5,000 in an attempt to get out of it. Because of traffic and because it was the wrong altitude for IFR flight westbound, I did not get it right away and I watched dishearteningly as the ice built up on the wings. For a moment, I considered landing at Martinsburg. Then, I was given 5,000 feet and the ice melted off. That was sufficient altitude to cross the mountains, so I continued west. That was a worse mistake than taking off because then, night fell upon me.
>
> At 70 miles before Morgantown (the closest IFR airport), I began to pick up ice at 5,000 feet, the lowest altitude possible (Minimum Enroute Altitude, MEA) for crossing the mountains. Soon the ice build up slowed me from 130 knots indicated to 110 knots as I tried to maintain the altitude. Then, the engine started to give up RPM as ice built up on the propeller. It also lost some power from my use of carburetor heat which I gave only periodically to conserve power. Then, I began to lose contact with Air Traffic Control, I could hear them but they could not hear me as ice was building up on the antennas. I thought about turning around and going back to Martinsburg but by then I was closer to Morgantown. Then, ATC told me that they had lost me on radar. I thought for a moment about all the accident reports I had read in which the last word of communication was 'lost on radar'. At this point I could not talk to them on the radio and could not communicate through my transponder either.
>
> By now, I was over the highest mountain about 30 miles from Morgantown. I began to lose altitude. I chanced pulling the stick back to 90 knots, the best rate of climb speed for the 182. I had been using full throttle for the last 30 miles. I didn't dare slow any further for fear of stalling. With an ice build up like I had there is no telling what the stall speed is. I knew the trees were very close beneath me though I was in the clouds at night. A stall would put me in the trees and rocks in seconds completely out of control.

At this stage, you might think that he would have been engaged in lots of thoughts about diagnosis, alternative generation, risk assessment, or maybe even his children or his life passing before him. No, these were not his thoughts. Instead, he thought about the ramifications of having an accident or a violation.

> I was reluctant to report my predicament to ATC even if I could. In addition to losing my girlfriend, I was depressed because I had been turned down for promotion to full professor. An accident would make that situation worse. I thought about the fact that I had made plans to pursue a flying job at TWA -- I had just been told that they had begun

hiring people my age. I thought about what would happen if ATC found out I was well below my assigned altitude (not to mention, below MEA) in icing with no deicing equipment onboard. I might be violated! That would knock out any chance for a flight slot at TWA. At this point I could see defeat in romance, academia, and aviation at the same time! Fear of death did not enter my mind.

By then, I had been forced, by the continuing build up of ice, down to 4,300 feet and at this altitude some of the ice began to melt, the propeller began to surge as the ice broke off and I could maintain that altitude. I finally made contact with ATC again (Clarksburg Approach Control). At this point, survival took priority and I reported to Clarksburg Approach Control that I could not maintain 5,000 feet due to icing. I also requested an approach and landing at Morgantown. That report immediately caught their attention and I got a lot of help from that point on. They first told me that the minimum radar altitude for that area was 4,800 feet.

They told me to maintain as much altitude as I could and that in 10 more miles, I could descend to 3,500 feet. I tried to climb but could not get higher than 4,300 feet. As soon as I got even 50 feet higher, the ice build up stopped me and I dropped back to 4,300. This is when the stress built up. That was the longest 10 miles of my life. At times, I could almost feel the tree tops hit the gear.

Finally, Clarksburg Approach said I was cleared to 3,500 feet. I descended immediately and as I did, the engine began to run better, and I gained a bit more airspeed. Clarksburg, who now had me in reliable radar contact and good voice communication, cleared me for the approach into Morgantown. When they handed me to Morgantown tower, what a relief it was to hear a mundane request for the *base of the clouds*! They had no idea what I had been through and I didn't tell them. I continued the approach and broke out at 1,200 feet above the runway. By then most of the ice had disappeared and a normal landing was easy. As I taxied in, I waited for the tower to call and ask to speak with me about the flight but they never did. Another relief!

I spent the night with the friendly folks at the Ramada Inn in Morgantown. The next morning the weather was still cloudy but, after it warmed somewhat, raising the freezing level, I flew on to Columbus IFR and actually felt rather cocky after another ADF instrument approach -- so much so that I made a short-field landing making the 1,000 foot turn off Runway 27L at OSU Airport. How easily, we pilots forget the plights we face!!

In this incident we can see the impact of motivation and associated attitudes. The initial motivation to 'get out of DC' seemed to override any reluctance to fly into forecast icing conditions. Like *The Gambler*, quoted at the top of this chapter, he felt that he *must* succeed. The rationalization that 'I could land and spend the night somewhere else was in part due, not just to wanting to be in Washington for another night, not just because I had a departure reservation from National Airport for a certain time that had an uncertain renewable time, not just to my state of some depression, but also due to the cost of staying in a hotel at personal expense'. Each of these motivations became more important than any safety risk considerations. There was minimal risk in flying as far as Martinsburg, so that part was easily justified, but how could he justify the decision to press on from there? His thinking was that he had established the level of the icing (higher than MEA across the mountains), so why not give it a try? At this stage, his motivation was his commitment to get home that night which overrode or blocked out any rational consideration for the very real possibility of encountering a lower icing level further west. One could go into each of the other decisions along the way in this scenario and discover a host of motivations that overrode rational decision processes and I recommend it for the serious aviator. Figure 4.2 shows what ice can do, even for a jet.

Figure 4.2 Ice on tail of Sabreliner that caused it to crash (photo by John Lowrey)

One incident such as this does not mean that all pilots have similar thought patterns as they face tough situations. However, as I have talked with pilots and read the 'rationalizations' used by pilots for their decisions in incident and accident reports, I have become convinced that this type of thinking is very likely to be the norm. Pilots are reluctant to be this candid because it could reflect badly on the individual and corporate pilot image portrayed to the general public. Certainly, the rational decision processes described in the previous chapter are used in aviation decisions but these irrational thought processes are also more common than many would like to think. Furthermore, these are more likely to be the thought patterns that lead to accidents because they cause lapses in attention that should be focused on rational decision making. Therefore, we must examine them and determine what can be done about them.

Personality and attitudes

Each of us develops strategies over the years to best accomplish our goals of dealing with life and the people around us. As we have seen, the more automatic of these strategies become ingrained in our perceptual systems. Other strategies that become deeply ingrained are known as personality traits. These traits are well established by the age of six and are difficult to change thereafter. We can be quite sure that personality factors can influence a person's flying judgment. Since the beginning of aviation in World War I, there has been an interest in the personality of pilots (Koonce, 1984; Dolgin and Gibb, 1989; Chidester, Helmreich, Gregorich, and Geis, 1991). However, because of the nature of personality traits, it would be futile to attempt to change them in pilot training. Furthermore, most aviation psychologists believe that personality traits are important only in the most extreme cases and can be handled through selection of pilots, at least for airline crew positions.

Attitudes are personal strategies for dealing with the world that are less deeply ingrained than personality traits and can be changed, especially under pressure from several sources at the same time. We are constantly bombarded with attempts to change our attitudes by teachers, theologians, advertising people, parents, peers and superiors who recognize that these attempts can be affective. In the cockpit, attitudes toward risk taking and performance of all other aspects of the flying task can be modified through training.

Attitudes are primarily affected by non-safety background factors such as job demands, convenience, monetary gain, self-esteem, and commitment. If properly developed, this part of aviation judgment would minimize information unrelated to the safety of the flight and direct the pilot's decision toward the use of more rational information and processes. Proper motivational decision making means recognizing that hazardous attitudes are present in every human decision and that these hazardous attitudes should give way to rational thought processes.

Attitudes toward risk taking in aviation are developed within the pilot as a result of flying training and experience. Many of these are developed through watching how our flight instructors deal with decisional issues. Attitudes may be either positive, tending to make us take a safe or cautious approach to flying risk, or negative, causing us to take greater risks than we should.

There is no logical, rational reason why pilots should allow background forces to influence their decisions. However, these forces are a part of human nature and therefore cannot be discounted. They are, after all, one of the major reasons why we fly at all. Power, speed, adventure, image, are motivations to learn to fly and they stay with us after we become professionals. Good judgment says that we should make these forces subservient to our more rational thought processes.

The following true story from the NTSB accident report illustrates how far two young pilots were willing go to continue a flight against all rational safety factors because of the force of a commitment to reach their destination. We sometimes refer to this force as the 'get-thereitis' attitude.

> Two young, male college students, both with instrument ratings and aspirations for airline careers flew a Cessna 172 to Massachusetts for a weekend. On a Sunday afternoon, they were returning to Columbus, Ohio when they made a stop for fuel in Williamsport, Pennsylvania. During their descent and approach, they ran into icing conditions but managed to land safely with about three-quarters of an inch of ice on the plane. They asked the FBO to fuel the airplane while they talked to the people in the FAA Flight Service Station (FSS) on the field about their plans to continue the next leg of their flight to Columbus.
>
> The FSS people were astonished to hear they were even considering a flight back into the icing. There were numerous other eye witnesses including the fixed base operator, a pilot of a Piper Aero, a state police helicopter pilot, and a KingAir Pilot, who tried to convince them not to fly and filed the following reports after the accident.
>
> *The fixed base operator:*
>
> The weather was low ceilings, rain and reported icing from the surface to 6000 feet. At approximately 1400 hours, I observed a Cessna 172 land and taxi to the fuel island at our facility. Through personal observation, ice was seen on the wings, windshield, and tail surfaces. The pilot refueled and remained at our facility for approximately 2 hours. Through conversation, the pilot indicated that he and his passenger were going to proceed to Ohio State University. The pilot was then informed by myself and numerous other pilots in our facility that to attempt the trip with existing conditions was definitely not advisable. The pilot indicated that he was confident he could climb to the 6000 foot level to get out of the ice. He then departed at approximately 1620.
>
> *Pilot of a Piper Aero:*
>
> I had just arrived in a Piper Arrow (PARO). I was scheduled to fly a PARO to PGV but canceled the flight because of the freezing rain and icing conditions. I was explaining this cancellation to another person when one the pilots (of the 172) anxiously asked what it was like up there. I told him that I had rescheduled my flight because of the icing. He then went out and started to knock the ice off the leading edges of the wings, struts, wheel pants, antenna and engine inlet of his airplane. He did not remove the streaks of ice from the back of the leading edge.

State police helicopter pilot:

I left IPT at 1230 westbound, encountered ice at 500 feet AGL over the city and returned to the airport. I saw N1498U land with ice on the leading edges and windshield. Aircraft was parked out in the freezing rain for 1.5 hours. I saw the pilot beating on the leading edges of the wings, struts, and stabilator trying to remove the ice (He had been moved into an unheated hanger). I tried to remove ice from the left side stabilizer (to see how hard it was to do). I struck it with my fist and the ice did not break up. Surface was rough ice. Ice looked about .5 inch thick over the entire stabilizer. I saw both pilots enter N1498U, start the engine and taxi out. Ice covered stabilizer as I watched it leave the ramp.

KingAir pilot:

I returned from PHL in a Beech KingAir C-90. When I reached PIX and started my approach, the aircraft was covered completely with clear ice. After landing, I went to Lycoming Air's office and talked to two pilots getting ready to leave in a Cessna. I told them of the ice and told them they were crazy if they were going to leave in a single-engine aircraft with no de-icing systems on the aircraft.

The following radio transmissions were recorded during the last 13 minutes of the flight:

At 2127:47 98U: *We're at thirty-six hundred feet going up to six.*

At 2128:03 98U: *Be advised we are climbing very slowly due to ice.*

At 2131:54 98U: *We're at thirty-four hundred feet and we need to return to radar vectors (possibly to) Williamsport, unable to climb due to icing conditions.*

At 2132:02 Center: *Turn left and maintain 4,000 feet direct Williamsport.*

At 2132:09 98U: *Unable to maintain four thousand, the best we can do is thirty-four hundred feet.*

At 2132:23 98U: *Last transmission was just an acknowledgment.*

At 2140 98U went off the radar screen.

The result was a crash into the trees on top of a hill near Williamsport. Amazingly, neither pilot was seriously injured. They spent a cold night in the plane and were found by hunters the next day.

Can you picture yourself in the position of these pilots? They were committed to getting back to Columbus for classes the next day. There may have been other reasons as well that we don't know. But, whatever they were, these forces were sufficiently strong to override the rational decision process and attempt a known suicide flight. Even after expressing that they were climbing slowly at 2128:03, they still did not turn back. Finally, at 2131:54, they gave their first request for help -- and then they were still not sure they needed to go back to Williamsport though they did follow the ATC instructions to do so.

In the icing accident scenario given above, these pilots had, to some extent, five hazardous attitudes. First, they had a strong sense of *invulnerability*. They did not think the ice would affect them as it affected the other pilots that day. Second, they also had an extreme *anti-authority* attitude, the regulation about not flying into known icing conditions was not for them. They would not to be told by someone else what to do. Third, they appeared to have a major *macho* attitude in that they wanted to show those folks on the ground that they could do it. Fourth, they were *impulsive* in that they could not wait for the weather to improve. Fifth, at the very end they seemed to give up and became *resigned* to crashing into the trees.

These pilots were driven by an extreme case of 'get-thereitis'. This attitude has led to many aviation accidents. It occurs when a pilot has made a commitment either to himself or to others to be somewhere. In our society, because most other forms of transportation are conducted in all types of weather, we have a well established mental expectation to complete any and every trip we take, more or less on time. It is often difficult to avoid the same mentality in private aviation even though we know that safe completion is highly dependent on the weather. When our ego, or self-concept, is tied up in completing our flying commitments, we have the sometimes fatal illness known to all in aviation as 'get-thereitis'.

Hazardous attitudes: The attitude model

What would cause pilots to behave so irrationally? How can we as pilots learn to prevent ourselves from making such decisions? The five hazardous attitudes mentioned above have been identified as *the attitude model* to explain much of the irrational behavior sometimes found in pilots (Berlin *et al.*, 1982). An awareness of these attitudes, that are found to some extent in everyone, can help to develop a more positive and rational approach toward flying decisions. These five attitudes are:

1 *Anti-Authority:* This attitude is found in people who do not like anyone telling them what to do. They would say, 'I don't need to follow the regulations', or 'You can't tell me what to do'.

2 *Impulsivity:* This attitude is found in people who frequently feel the need to do something, anything, immediately. They would say to themselves, 'Do something -- NOW'.

3 *Invulnerability:* This attitude is found in people who believe that accidents always happen to others, not to them. They would say, 'It won't happen to me'.

4 *Macho:* This attitude is found in people who always try to show how good they are. They think, 'I'll show you, I can do it'.

5 *Resignation:* This attitude is found in people who feel they have no control over their fate. They would say, 'What's the use'. They are likely to go along with unreasonable requests just to be a 'nice guy'.

At the end of this chapter, there is a questionnaire called the 'Pilot Decisional Attributes Questionnaire'. This questionnaire contains five situations each followed by five responses to each situation. It is designed to show the relative amount of each of the five hazardous attitudes in your own thinking. It is not a test of your judgment. We all have these attitudes to some extent. This questionnaire simply shows which one may be more important to you than the others. For that reason, it may be useful to you to help you see which may be worth looking at when you are faced with a difficult decision in aviation. As you fill it out try to place yourself into each situation as best you can and decide what thoughts would come to your mind in each.

Countering hazardous attitudes

We have now identified five hazardous attitudes in pilots as well as several psychological traps into which pilots can fall into. Being aware of these is the first and most important factor in countering them to improve your own judgment. In addition, there are some helpful hints one can use as 'antidotes' to hazardous attitudes. The Table 4.1 presents a list of five hazardous attitudes and their respective antidotes to help you to develop better attitudes.

The poor judgment chain

We know that, in most accidents, there are a series of events leading up to the accident. In 'pilot error' accidents nearly all result from a series of poor pilot decisions which may be called, '*the poor judgment (PJ) chain*'. One erroneous decision increases the probability of another and as the PJ chain grows, the chances for a safe conclusion to the flight decrease. Research has shown that errors lead to workload increases (Jensen and Chappell, 1983; Hart and

Bortolussi, 1983) which compounded with the knowledge of the error greatly increases stress - and the probability of more errors. The following are the principles of the poor judgment chain:

- 'pilot error' accidents are caused by a series (chain) of poor decisions
- one poor decision increases the probability that another will follow
- as the poor judgment chain grows, the alternatives for safe flight decrease
- breaking the poor judgment chain is essential to safe flight

Table 4.1
The five antidotes

Hazardous Attitude	Antidote
1 *Anti-Authority:* The regulations are for someone else'.	'Follow the rules. They are usually right'.
2 *Impulsivity:* 'I must act now, there's no time'.	'Not so fast. Think first'.
3 *Invulnerability:* 'It won't happen to me'.	'It could happen to me'.
4 *Macho:* 'I'll show you. I can do it'.	'Taking chances is foolish'.
5 *Resignation:* 'What's the use'?	'Never give up. There must be something I can do'.

A chain of poor decisions can be seen in the following accident involving pilot/doctor flying a Cessna P210 from Boise, ID to Las Vegas, NM. Critical to the doctor's judgment, was the fact that his passenger was his nurse (with whom he was having an intimate relationship), who wanted desperately to get to Las Vegas to visit her dying mother. A mechanical malfunction, the failure of the vacuum pump which powers the attitude and direction indicators, was critical as well, but we will focus on the human aspects of the case which could have prevented the accident.

> The weather was terrible (sky obscured, 1/4 mile visibility, clouds to 20,000 feet) for IFR flight single-engine in the mountains. With such weather there was no chance for an approach back if something were to

> go wrong. Flight in such conditions is made either not thinking or thinking that nothing mechanical affecting power or flight control will go wrong because if it does, no one onboard will have a chance to survive.
>
> The pilot/doctor had checked the weather and filed an IFR flight plan the night before and knew the type of weather to expect. Yet he arrived at the airport the next morning late which required that his decisions regarding the flight had to be made in the presence of his passenger. Witnesses said that the preflight was conducted quickly and that he had made one last call for weather prior to departure.
>
> After takeoff, he climbed successfully to about 13,000 feet at which time he reported that he had 'lost power' but did not clearly indicate what power he had lost. In fact, he had lost vacuum power and his attitude and direction indicators stopped working. But on top of that, his turn coordinator did not work either! The communication pattern between the ATC and the pilot seems to indicate that he may have known that the turn coordinator was not working prior to the vacuum failure which means he probably took off knowing that he had no backup turn gyro. The ATC controller asked him to make a turn which he tried to do but soon lost control and crashed killing both himself and his passenger.

In this case the chain of poor decisions was very long. It began with the doctor's decision to commit himself to a flight with a desperate passenger in IFR conditions, single-engine over the mountains. Second, he got to the airport late forcing him to make his flight decisions in the presence of his passenger. Third, he did not do a proper pre-takeoff check and missed the malfunctioning turn coordinator. He should have checked this instrument while making taxiing turns to check for its proper operation. He may have made the takeoff knowing that the turn coordinator was not working betting that the vacuum pump and artificial horizon would not fail. Fourth, he made the decision to takeoff in weather below IFR approach minimums for that airport so he had no 'out' in the event of a mechanical malfunction.

Fifth, at the instruction of ATC, he agreed to make a turn, which in his situation essentially gave up control of the airplane to the ATC. He was already on a southerly heading, the only heading on which he could maintain control with no gyros. He should have known that if you lose all of gyros, as a last resort, you can maintain directional control on a southerly heading using the compass. On a southerly heading, the compass leads the turn, thus making it more sensitive than the aircraft. By making very small corrections, you can determine your turn rate, and thus make a close approximation of your attitude from the compass. It takes careful control but it can be done.

As part of the litigation of this case, I simulated the accident situation in actual flight. I flew a C210 into actual IMC, established a southerly heading, and had the safety pilot cover all gyros. I then proceeded to fly for more than 30 minutes, maintaining my southerly heading and even made a climb of 2,000 feet and a descent back to my altitude -- entirely with no gyros. I proved, to myself and to the court, that it can be done quite easily if you are already established on a southerly heading, the air is fairly smooth, and you can give it your full attention. It is quite important that ATC must not ask for any turns and the pilot's only option is to climb or descend to VFR conditions. Our doctor could have climbed to 20,000 feet to VFR and flew to VFR weather. The best strategy for dealing with an error, whether decisional or flight control, is to break the PJ chain by acknowledging the error and taking action immediately to correct the error.

Three other hazardous attitudes

In my experience in recent years, I have come to the conclusion that there are three more hazardous attitudes that should be mentioned. They are deference, denial, and external locus of control. These are all somewhat related to each other and to the resignation attitude mentioned above but need further distinction. *Deference* in aviation judgment refers to the tendency to give up decision making control to another person. This attitude may have been present in the doctor case mentioned above. It is common in cases in which a passenger is a close friend or superior. The pilot says, 'OK, if you say so we will go'. It is often the cause of 'get-thereitis'. *Denial* is a term often used in psychological counseling referring to people who deny that the situation is as bad as everyone says it is. This is common in pilots who find it easy to ignore bad news and focus only on the good news. *External Locus of Control* is a psychological term referring to people who tend to attribute their success or failure to external factors rather than their own abilities (or lack thereof). The opposite, one who has an internal locus of control, feels that he or she is in command of every situation and tends to take the credit or blame for whatever happens. It is essential that pilots be internally controlled. This concept will be discussed at greater length in Chapter 9 on the assessment of judgment.

Psychological factors

There are several flying phenomena with psychological implications that are known to effect judgment that are fairly well known to pilots. Most of these are covered in the five hazardous attitudes but in pilot language, I like to call them 'other psychological traps'.

Other psychological traps

There are a number of other psychological traps into which pilots have been known to fall. These can generally be avoided or corrected by adhering to the principles advocated earlier for combating hazardous attitudes and the PJ chain.

For example, to the naive public, pilots often hold the 'Macho' image. One of the defects in many pilots is that they feel it is their duty to uphold that image. However, the fact is that a large segment of the pilot population does not have this attribute. Even those who do, can have it shattered quite easily by a noticeable failure or error in flight. The much talked about 'Right Stuff' is a fragile image revealed in Allen Shepard's prayer on top of the Redstone rocket where he said, 'Please God, don't let me #### up' (from the movie, *The Right Stuff*).

The tentative pilot. There are numerous examples of incidents of poor aeronautical decision making in which the pilot simply failed to take charge in a situation where action was needed. These types of situations are difficult to reveal in scenarios. They are best placed under the hazardous decisional attitude, 'Resignation'. However, the explanation offered for this attitude, 'What's the use', fails to describe the problem. The following story from my flight training at Moody illustrates a feeling most pilots have faced at one time or another in their flying careers.

> I was in 'transition' training to tailwheel airplanes from a tricycle gear airplane in which I received my private certificate. Although I had soloed the tailwheel airplane (a Cessna 140) for several hours, I had not really learned the proper deliberate control technique for making 'wheel' landings in a strong crosswind. I found myself attempting a crosswind landing on the north runway at Elgin, IL with a moderate wind from the west. As the airplane bounced down the runway, I had that helpless feeling as the aircraft became quite out of control for some moments prior to and after the first two or three bounces. Luck was with me that day because I bent no metal and recovered with a very receptive attitude toward more dual instruction in crosswind landings. In my next dual flight with Van Dam, he demonstrated the positive, deliberate control technique required in wheel landings.

In the sport of tennis, coaches often use the word 'tentative' to describe this attitude. As in tennis, tentative decision making in pilots can lead to disaster just as easily as the macho attitude. Although most pilots become less tentative with increased flying experience, such experience alone contains no absolute preventative, particularly when their fragile egos have been shattered. It can also be the result of life stress events, such as divorce or a death in the family. These events can cause one to have 'tunnel vision' which can obscure relevant information that otherwise would clearly point to a problem. You must be aware of this possibility and take special care to be deliberate in your decision making when such feelings arise - whatever the source.

Get-thereitis. As we have seen, one of the most often mentioned factor (or excuse) for flying into adverse weather or taking on challenges beyond one's capability is known as 'get-thereitis'. Pilots on their way home from a trip feel a very compelling pressure to ignore mounting flight adversity to satisfy a need to get home. The pressure may come from within one's self or from commitments made to other people. From the story of the two flight students on their way to Columbus who crashed after taking off from Willimington, Pennsylvania into icing, one can catch a glimpse of the extent of the pressure that get-thereitis can place on the pilot. Get-thereitis is real and we must be prepared for it. The best way to counter get-thereitis is to mentally remove yourself from the pressure and focus your thinking on the safety factors in the flight.

Duck-under. As we saw in the Allegheny Airlines Convair 580 accident in Chapter 3, on instrument approaches pilots sometimes experience the compelling urge to descend, 'just a little bit', below the approved minimum for the approach in an attempt to get into the airport. This urge is especially strong on second and third tries for an airport. The pilot feels falsely secure having an illusory feeling of being in a world of his own in the clouds quite apart from the buildings, trees, and rocks that exist out there as well. This is a case of denial of the existence of the real situation. It is best controlled by fully realizing that the ground does exist out there, probably within the clouds and not necessarily below the clouds and that the regulations are there for your protection.

Scud running. VFR pilots are more likely to fall into the trap of scud running than commercial/instrument pilots but even the most experienced instrument pilots sometimes fall victim, either because their equipment is VFR only or they don't want to go through the trouble of filing IFR. Scud running is the practice of flying as low as necessary over the ground to avoid the clouds. The danger is that the ceiling and/or visibility may change for the worse at any moment. At the altitude of most scud running, these changes may occur undetected by the pilot which greatly increases the chance of running into the terrain. Pilots, who have an invulnerability attitude, may convince themselves that they can make it through visually just by going a little lower or a little slower, especially when present conditions are reported VFR at the weather stations. They go about it ignoring the fact that the weather between reporting stations may be well below what it is at the stations. There may be pressures to get somewhere with less than IFR equipment on board. Scud running is best countered through the realization that you are putting yourself at great risk when challenging unknown weather.

Continued VFR flight into IMC conditions. The most frequent cause of general aviation accidents is 'continued VFR into IMC conditions. It is not as frequent in commercial operations because more of these flights are made IFR. However, it can happen in these operations as well. It is most likely to happen in conditions such as those mentioned for scud running in which there is a strong need to get somewhere in a less than IFR equipped aircraft or the pilot fails to file IFR. It is a symptom of invulnerability because the pilot believes that no one is likely to be in the airspace where he is flying and he'll be visual again momentarily. The drive to continue visually may be countered by realizing that the practice is, literally, a game of Russian Roulette. The risks are very high and the penalty for being wrong -- fatal.

Falling behind the aircraft. Anyone can get into circumstances in which they find themselves 'behind the aircraft', but it is most likely to occur as a result of lack of experience, especially in the type of aircraft being flown. The phenomenon consists of too many tasks to complete in the apparent time allotted, perhaps, coupled with the failure to understand or complete satisfactorily one or more of the tasks. A commercial pilot could fall behind his aircraft if a new piece of avionics, such as RNAV or FMS, is added and he depends on it without fully understanding it. Another common time for commercial pilots to get behind the aircraft is when encountering an emergency, e.g., an engine failure in a multi-engine aircraft. Lack of recent experience may result in the pilot being over stressed in the situation to the point of causing a reduced capability to complete the basic tasks of navigation and flight control. The best way to counter falling behind the aircraft is not to accept flight tasks until you are thoroughly familiar with the equipment involved in all modes of operation including failures.

Loss of position or situational awareness. Loss of situational awareness is most often associated with inexperienced pilots but can also happen to the most competent commercial pilots. One particular time this may occur is while on radar vectors. An instrument pilot may not continue to maintain an awareness of his position with his navigation system because under radar vectoring, the controller is doing that task for him. Pilots then get into trouble when cleared for the approach and are asked to resume normal navigation. Equipment failure can cause one to be drawn so deeply into the solution of the problem that one forgets everything else about the situation. This is exactly what happened in the Eastern Airlines L-1011 that crashed into the Florida Everglades on December 29, 1972, while the three-member crew were all engaged in the repair of a gear light. Situational awareness is maintained through careful flight planning, disciplined headwork, and vigilant monitoring of the flight progress at all times.

Takeoff with less than minimum fuel. The practice of taking off with less fuel than the regulations or SOPs require is a case of an anti-authority attitude. It may be caused by a hurried pilot not wanting to take the time to refuel or one who has overloaded the aircraft and cannot fuel to the level required plus reserve, and still get off the ground. It may be a pilot who plans for a flight that is beyond the range of his aircraft with proper reserves but he wants to make it non-stop. One can also run low on fuel when the winds turn out to be different than forecast. Finally, some pilots may elect not to take on 'holding fuel' because they can save money for the company. This invulnerability attitude then gets them into trouble when they are asked to hold for weather or traffic at their destination. One must realize that the regulations are written for achievable circumstances. Minimum fuel requirements allow but a small margin for error. Taking off with less than minimum fuel indicates an invulnerability and an anti-authority attitude problem and is correctable through proper understanding and respect for the regulations as minimum criteria to maintain safety. One of my Moody flight instructors, Reed Berry, used to say 'there is nothing so useless as the runway behind you, the altitude above you, and the fuel that's not in your tank'.

Flight outside the certified flight envelope. Except for overloading the aircraft, this is a fairly rare occurrence in commercial operations. Some believe that one particular model business jet was falling out of the sky because pilots were flying it outside of its acceptable flight envelope. Again this is a case of an anti-authority attitude. Design standards, operating limitations, and flying regulations are not aimed at curbing use, but rather at enhancing safety. Jim Greenwood (1983) says that some Learjet operators illegally installed 'go buttons' which disconnect overspeed warning systems, enabling pilots to fly faster without the annoyance of noisy horns, flashing lights, or stick shakers. Overspeeding is believed to be the cause of at least one and probably all of the series of mid-flight Learjet breakups several years ago.

Casual neglect of flight planning. Experienced pilots often fall into the trap of complacency concerning flight planning, preflight procedures, and especially the use of checklists. Some may even think it is macho (and therefore 'professional' appearing) to just hop in the airplane and go. However, this practice, which indicates an invulnerability attitude, is conducted at great risk. There are many factors that can change in short periods of time. Assumptions about fuel state, aircraft state, chart availability, and weather conditions may be incorrect. This is best countered by judgment attitude training in the area of invulnerability.

Conclusion

Good judgment can be very elusive. While on the ground it is easy to make good decisions. But when there are pressures on you to be somewhere at a certain time, it can be difficult to put all aspects of the flight in perspective. The situation is often made even harder by the fact that bad judgment frequently does not result in a bad outcome. Often we make a bad decision, but nothing happens. This makes us regard the poor decision as not being so bad. The next time we are more likely to take the same risk -- nothing bad happened before, nothing will happen this time. It is important to remember, therefore, that the outcome does not dictate whether a decision was good or bad. A decision is good or bad irrespective of the outcome.

To this point, we have seen how people (and pilots) deal with information and misinformation. Good aviation judgment is the ability to distinguish between these, to weigh the good with the bad, to set aside emotional pressures, and to make rational decisions that place safety as the primary consideration. Making good decisions consistently takes a great deal of discipline, but the effort is worth it since the alternative can be disastrous.

Good pilot decision making depends upon an understanding of, and an ability to use, an orderly decision making process. Being a good decision maker in aviation means not acting hastily based upon 'gut' feelings. With an accurate assessment of the risks associated with each flight, you can arrive at decisions that ensure successful conclusions to all flights, even if this means not taking off. Success in decision making is measured very simply: maintaining yourself, your passengers and your aircraft in good condition no matter the airport at which the takeoff and landing occurs.

PILOT DECISIONAL ATTRIBUTES QUESTIONNAIRE

Instructions:

The following is a questionnaire designed to assist you in determining your decisional attributes as a pilot. Please answer all of the questions as honestly as you can. There are no right or wrong answers.

1 Read each of the situations and the five alternatives. As you read, place yourself in a familiar aircraft under the given circumstances.

2 Choose your *most likely* thought pattern in response to the situation and place a rank of '5' beside it. Then rank each of the other responses in descending order of preference from 4 through 1.

Remember - This questionnaire has no right or wrong answers. Some (or all) of the alternatives given may not represent the way you would think at all. For the purposes of this questionnaire, you should assign a rank to all of them. However, if none of the responses are so far from your way of thinking that you would never consider them in the situations provided, then fill out the blanks with the response you would use.

Situation 1

You were on a flight to an unfamiliar, rural airport to show to your best friend your favorite spot on earth, a beautiful waterfall. Before takeoff, flight service said that VFR flight is not recommended, because heavy coastal fog is forecast to move into your destination airport area about the time you expect to land. You decided to takeoff hoping to beat the fog. Enroute you could see the fog coming in off the ocean. You considered returning to your home base where visibility was still good but decided instead to continue as planned and you landed safely. What thoughts would you have had as you made these decisions?

___a. You hated to admit that you could not complete your original flight plan.
___b. You resented the suggestion by flight service that you should have changed your mind.
___c. You felt sure that things would turn out safely.
___d. You reasoned that, because your actions would make no real difference, you might as well continue.
___e. You felt the need to decide quickly so you could beat the fog.

None of the above. Instead your decision would have been: ____________

____

because you would have been thinking: ______________________

____

Situation 2

While taxiing for takeoff, you noticed that your right brake pedal was softer than the left. Once airborne, you were sufficiently concerned about the problem to radio for surface wind information at your destination. Finding strong winds reported at your destination, an experienced pilot who was a passenger recommended that you abandon the flight and return to your departure airport where the runway was more closely aligned with the wind. Instead, you chose to continue the flight and landed with no further difficulties. What were you thinking as you made these decisions?

___a. You felt that suggestions made in this type of situation are usually overly cautious.
___b. Your brakes have never failed before, so you doubted that they would this time.
___c. You felt that, if you had to, you could ask the tower at your destination airport for assistance.
___d. You immediately decide that you must continue.
___e. You were sure that if anyone could handle the landing, you could.

None of the above. Instead your decision would have been: ____________

____

because you would have been thinking: ______________________

____

Situation 3

When you arrived at the airport for an important business trip, you were told that your usual airplane had been grounded because of an airframe problem. Instead, you were scheduled in another airplane of the same type but with different avionics. The weather was forecast to be heavy IFR. After your preflight, you decided to takeoff on your trip as planned. What were you thinking as you made these decisions?

___a. You felt that a difficult situation would not arise so there was no reason not to go.
___b. You told yourself that if there were any question, you would not have been offered the plane.
___c. You were in a hurry and did not want to take the time to think of alternate choices.
___d. You did not want to admit that you may have trouble flying an unfamiliar airplane.
___e. You were convinced that your flight instructor was too conservative when he cautioned you to be thoroughly checked out in an unfamiliar aircraft.

None of the above. Instead your decision would have been: ____________

____

because you would have been thinking: ____________________

____

Situation 4

You were briefed about possible icing conditions but did not think there would be any problem since your departure airport temperature was 50°F. As you neared your destination, you encountered freezing precipitation which clung to your aircraft. Your passenger, a less experienced pilot than you, began to get very concerned. You considered turning back to the departure airport but continue instead. What were you thinking about as you made these decisions?

___a. You felt that, having gone into it that far, there was nothing more you could do.
___b. The concern of the passenger made you 'commit yourself' without thinking the situation over.
___c. You did not want the passenger to think you were afraid because that would make him even more concerned.
___d. You were determined not to let the passenger think he could influence what you do.

___e. You did not believe that the icing could cause your plane to crash in these circumstances.

None of the above. Instead your decision would have been: ________________

____

because you would have been thinking: ________________________________

____

Situation 5

It was a beautiful day and you did not bother to check weather conditions enroute and at your destination. Enroute to pick up three important business associates, you encountered surprisingly strong headwinds. Your fuel supply was sufficient to reach your destination, but there was almost no fuel reserve for an emergency. You decided to continue the flight to your destination and landed with almost no fuel in your tank. What were you thinking as you made these decisions?

___a. You didn't want to be late so you make a quick decision to continue.
___b. You didn't want your friends to hear that you had to land because of poor planning.
___c. You felt that flight manuals always understate the safety margin on fuel tank capacity.
___d. You believed that all things usually turn out well, and this will be no exception.
___e. You reasoned that you had no control of the situation because you had to be there on time.

None of the above. Instead your decision would have been: ________________

____

because you would have been thinking: ________________________________

____

Pilot decisional attributes questionnaire: Scoring key

Scoring instructions

From your Pilot Decisional Attributes Questionnaire answers, write in the table below your rank for each alternative shown. Sum the ranking scores for each scale and enter at the bottom. If you provided a rank for all alternative, the sum of your scores across should be 15 for each situation. If it is not, go back and make sure that you transferred the scores correctly and check your addition. The grand total should be 75.

PDAQ interpretation

Now that you have your scores for each of the attitudes, you may wonder what they mean. The five scales represent the five hazardous attitudes in the order shown in Table 4.1

Scale I: Anti-Authority
Scale II: Impulsivity
Scale III: Invulnerability
Scale IV: Macho
Scale V: Resignation

The totals at the bottom of each column indicate relative strength of that attitude compared with the other attitudes. The higher the score, the greater the importance of that attitude in your thinking when you encounter stressful decisions.

Situation	Scale I	Scale II	Scale III	Scale IV	Scale V	Total
1	b_____	e_____	c_____	a_____	d_____	15
2	a_____	d_____	b_____	e_____	c_____	15
3	e_____	c_____	a_____	d_____	b_____	15
4	d_____	b_____	e_____	c_____	a_____	15
5	c_____	a_____	d_____	b_____	e_____	15
TOTAL	_____	_____	_____	_____	_____	

You should know that there is no 'ideal' profile for pilots. You should examine your profile and see if one or two of the scales is much higher than the others. If so, you should examine yourself as you are faced with stressful decisions to see if that form of thinking may cross your mind. If you think that it does, try to correct it by moving your thoughts toward the more rational ideas that lead to a *safe* conclusion. Remember, everyone has all of these attitudes at some level. The pilot with good judgment has learned to make these attitudes subservient to more rational thought processes.

References

Bandura, A.L. (1971). *Social Learning Theory*. New York: General Learning Press, p. 359.

Berlin, J.I., Gruber, E.V., Jensen, P.K., Holmes, C.W., Lau, J.R., Mills, J.W., and O'Kane, J.M. (1982). *Pilot Judgment Training and Evaluation* (Vols. 1, 2, and 3). Washington, DC: DOT/FAA/CT-82/56.

Chidester, T.R., Helmreich, R.L., Gregorich, S.E., and Geis, C.E. (1991). 'Pilot Personality and Crew Coordination: Implications for Training and Selection'. *The International Journal of Aviation Psychology*, **1**(1), 25-44.

Dolgin, D.L. and Gibb, G.D. (1989). 'Personality assessment in aviator selection'. In R.S. Jensen (ed.), *Aviation Psychology*, London: Gower Publishing Company, Ltd.

Greenwood, J.R. (1983). 'Judgment and discipline: Recipe for survival'. In R.S. Jensen (ed.), *Proceedings of the Second Symposium on Aviation Psychology*, Columbus, OH: The Ohio State University, Aviation Psychology Laboratory.

Hart, S.G. and Bortolussi, M.R. (1983). 'Pilot errors as a source of workload'. In R.S. Jensen (Ed), *Proceedings of the Second Symposium on Aviation Psychology*, Columbus, OH: The Ohio State University, Aviation Psychology Laboratory.

Hilgard, E.R., Atkinson, R.C. and Atkinson, R.L. (1975). *Introduction to Psychology*. New York: Harcourt Brace Jovanovich, Inc., p. 609.

Jensen, R.S. and Chappell, S. (1983). *Pilot Performance and Workload Assessment: An Analysis of Pilot Errors*. Columbus, OH: The Ohio State University, Aviation Psychology Laboratory, Contractor Final Report for NASA Contract No. NAG 2-184.

Koonce, J.M. (1984). 'A brief history of aviation psychology'. *Human Factors*, **26**(5), 499-508.

Lee, W. (1971). Quotation from: F. Dostoevsky, *The Gambler*, New York: John Wiley and Sons, Inc., p. 123.

Maslow, A.H. (1954). *Motivation and Personality*. New York: Harper and Row, p. 334.

5 Can judgment be taught?

This aircraft is flown by a superior pilot. A superior pilot is one who stays out of trouble by using superior judgment to avoid situations that might dam well require the use of his superior skill.

Royal Canadian Air Force (quoted in Greenwood, 1983)

We have seen that a major problem exists with judgment in aviation. The question on the minds of pilots is 'What can be done about it'? Can judgment be taught? In this chapter we look at a series of experiments conducted in various places in the USA, as well as several other countries, that provide convincing proof that judgment can be taught to aviators and that such training can have a major impact on safety. Most of the experiments were conducted using beginning flight students, but two used pilots with private licenses. Two other studies have clearly demonstrated that teaching judgment concepts to operational pilots can have a positive impact on safety. Crew resource management training, which is closely related, has been conducted with operational corporate, airline, and military pilots with considerable success as well (Diehl, 1992, p.46).

The first research in judgment specifically applied to aviation was a study for the US Air Force by Thorpe *et al.* (1976). Situational Emergency Training (SET), as they called it, was developed in response to changing the number of crew members from two (in the F-4 fighter) to one (in the equally complex F-15). There was undoubtedly a further economic concern because the F-15 was considerably more expensive than the F-4. The SET program was intended to teach the pilot to judge the relevant dimensions of the emergency, maintain control of the aircraft, make a decision as to the handling of the problem, and get the aircraft on the ground as quickly as possible. This method replaced the traditional use of memorizing boldface checklist items and was reported to be well received by the Air Force.

After Russ Benel and I presented our initial report on teaching and evaluating judgment (Jensen and Benel, 1977), a series of studies were conducted to determine whether or not our suggestions would work. Berlin *et al.* (1982), with FAA funding, provided the first prototype judgment training manuals for students and instructors which became the widely used model for teaching judgment in a number of studies that followed. The manuals were designed to improve judgmental behaviors using a modified behavioral approach applied through instructor pilots. Five 'hazardous thought patterns' were identified and taught as the core of judgment instruction. These five hazardous thoughts (later called hazardous attitudes by other investigators), were those described in the Attitude Model given in the previous chapter. The following is a description of some of the highlights of the research that has been done directly on pilot judgment training. From these descriptions you can see convincingly that judgment can be taught and you can see ways to teach and evaluate judgment.

The attitude model experiments

The Embry-Riddle study

Berlin *et al.* (1982) reported the first experiment using these teaching techniques. In several lecture/discussion periods an experimental group of 25 Embry-Riddle Aeronautical University (E-RAU) students working toward their private pilot license learned how to identify the five hazardous thought patterns and to replace them with more rational thinking. They were given scenarios which included the hazardous thoughts and were asked how they would respond to those situations. A 'control group', of 25 equally experienced students, received the same flight and ground instruction but received no judgment training.

The judgment training manuals. To support their judgment training E-RAU prepared student and instructor manuals for use in the lecture/discussion periods (Berlin *et al.* 1982). The student manual introduces judgment and the subjects about which flight decisions are made. By means of examples, the student is introduced to the mental processes necessary for safe flight. The student is required to practice with a series of flight decision scenario exercises to gain familiarity with the concepts. The chain of poor decision judgments, which is the usual case in pilot error accidents, is introduced as 'the poor judgment chain' and emphasis is placed on the need to break the chain and suggestions for breaking it. Since those first manuals were developed, a series of judgment training manuals (referred to as Aeronautical Decision Making or ADM) following the same format and model were developed for other categories of pilots including: instrument (Jensen, Adrion and Lawton, 1987), commercial (Jensen and Adrion, 1988), helicopter (Adams and Thompson, 1986), and crew resource management (Jensen, 1989).

Each of the five hazardous thoughts or attitudes is introduced and defined during a brief, easily learned phrase. Students are asked to complete a forced-choice inventory of judgment situations. Then they are shown how to score the answers. The results are kept confidential unless the students choose to discuss them with the instructor. Antidotes to the hazardous attitudes are presented. Practice exercises are given for students to complete to learn to recognize hazardous attitudes and their antidotes.

An instructor manual was also developed, highlighting the instructor's role and providing materials for guidance. The emphasis is on instructor autonomy in the choice and scheduling of appropriate activities. Guidance is given in the choice of objectives and ways of teaching good judgment. Some examples of scenarios are provided, with further examples of exercises which can be integrated into the private pilot training syllabus.

Evaluation and results. When the students in both groups were ready for their license check, they were first given a flight check for judgment in which they were observed in preflight (malfunctions were set for students to find) and during a short cross country flight. Results showed that students who had been given judgment training made 16 per cent fewer judgmental errors than those in the control group who had not received the specialized judgment training (Berlin *et al.* 1982).

The Canadian study

Soon after the E-RAU study, Georgette Buch (Buch and Diehl, 1982) conducted a similar transfer of training study using Canadian Air Cadets from two different schools. The judgment manuals were modified slightly from those used in the Berlin *et al.* (1982) study but the basic attitude model was retained. In this experiment, the two groups of pilots were kept separate from each other (they were from different but equally select schools in Canada) to avoid possible information exchanges between students and instructors representing the two groups. Knowledge and practical tests of the two subject groups were done showing that the control and experimental groups were equal in skills and knowledge both prior to and after training for the private pilot license.

Method. In this experiment the 25 students in each group first received their private pilot license. During their training the 25 students in the control group received judgment training in several special lectures in the classroom and they received some special judgment training in the air during flight training. The basic material for judgment training was the modified attitude judgment manuals. Soon after receiving their license, each student was approached by a judgment examiner posing as a photographer who wanted aerial pictures. He offered to pay for the flight. Each student was quite pleased to accommodate. On the ensuing observation flight, students were carefully observed as they made decisions both during the preflight and in flight. Eighteen different items were scored during the flight on the basis of the decisions made.

Results. As shown in Table 5.1, of the 18 items scored, the judgment trained students performed equal or better than the control group in every case. Some of the differences were quite large (e.g., preflight inspection, fly low along beach). The results show that the experimental group (E) who received the judgment training consistently made 'better' decisions than the control group (C) of pilots who received the conventional training. Overall, the judgment trained pilots scored 83.5 per cent correct decisions; the conventionally trained pilots scored 43.5 per cent correct decisions. These results for a carefully controlled experiment clearly show that attitude judgment training has a significant affect on the decisions pilots make for flights made soon after they have received their private pilot license.

Table 5.1
Canadian observation flight results in per cent of correct decisions

Observed Item	E	C	p
Flight planning	32	16	NS
Preflight inspection	61	8	$<.05$
Use of checklists in flight	96	67	$<.05$
Taxi controls for wind	81	68	NS
Taxi across active runway	97	91	NS
Steep turns at low altitude	72	12	$<.05$
Stall at low altitude	100	59	$<.05$
Fly low over congested area	96	69	$<.05$
Fly low approach to unfamiliar airport	89	57	$<.05$
Fly beyond glide distance to shore	100	41	$<.05$
Fly low along beach	100	13	$<.05$
Fly unplanned into unfamiliar airport	76	21	$<.05$
Attempt landing from high approach	100	64	$<.05$
Attempt landing from low approach	92	33	$<.05$
Attempt to land on too short strip	83	10	$<.05$
Fly cross-country at wrong altitude	93	62	$<.05$
Off-airport procedures	66	44	NS
Traffic awareness	69	69	NS

Source: Buch and Diehl, 1982

The Australian experiment

Another experiment using the attitude judgment training manuals was performed by Ross Telfer (1989) using Australian flight students at Cessnock Airport in conjunction with the University of Newcastle's Institute of Aviation. This study was designed similar to the two mentioned above, but in addition to the experimental and control groups, there was a third group called the 'academic' group who were simply given the judgment attitude manuals and received no other instruction. The experimental group was given the judgment attitude manual, provided with special judgment instruction and they received conventional flight training. The control groups received only the conventional flight training offered by the school. Pre- and post-written tests of manual content plus a flight-test were used to gather data.

Method. As in the Canadian study, the observation flight was made after the students received their Unrestricted Private Pilot License. Also, in similar fashion, the observer posed as a realtor friend of the student's instructor who wanted to take pictures of land and airports. The observer actually took photographs of the landing strips while structuring, indirectly, the flight so that a series of judgment situations were presented to the student. Scoring was not completed until after the flight, when the instructor rated the pilot performance on each judgment situation according to a five point scale ranging from 4 (sound judgment) to 0 (dangerous judgment).

Results. The most interesting result of the written tests was a decline in the awareness of the importance of judgment in the six months between pre- and post-testing of the control group. At the beginning of their training, student pilots recognized that their judgment could be a source of danger to themselves and others. After accumulating flying hours over half a year they were less prepared to concede that their judgment could be defective. Telfer concludes that this perceptual shift might be indicative of an optimistic view of their invulnerability and may be the reason for the fairly universally well known high accident rate for pilots with 100-300 total flying hours.

The flight test results corresponded roughly with those of the two previous studies. Table 5.2 gives details of the flight test results. Again there were items in which the experimental group performed much better that the control (e.g., Hazards affecting the aircraft), but there were some items where the reverse was true (e.g., checks interference). Overall, the performance of the experimental group was better than the control by 68.4 per cent to 46.0 per cent. The performance of the academic group was 57.9 per cent. This experiment again confirms by a wide margin the effectiveness of motivational judgment training.

The E-RAU instrument pilot experiment

The next experiment on attitude judgment training explored two new areas, instrument pilot students and the use of flight simulators for training (Connolly, Blackwell, and Lester, 1987). Subjects in this experiment were a homogeneous group of E-RAU aeronautical science students pilots working toward the instrument rating with an average age of 21 years and flight experience of 200 hours randomly assigned to experimental (16 subjects) and control (13 subjects) groups.

Table 5.2
Australian observation flight results in per cent of correct decisions

Judgment Area	Experimental	Academic	Control
Weather check	87	100	67
Passenger briefing	50	83	50
Hazards affecting aircraft	85	33	30
Passenger - propeller hazard	50	83	33
Checks interference	63	42	67
Departure procedures	75	42	33
Terrain and cloud clearance	56	38	42
Infringement of controlled airspace	75	33	50
Arrival procedures	75	67	42

Source: Telfer, 1989

Method. The judgment training was conducted in two ways. First, four hours of classroom instruction was given using the attitude judgment manual *Aeronautical Decision Making for Instrument Pilots*, (Jensen, Adrion and Lawton, 1987) which emphasizes the five hazardous attitudes. The training in this case was geared for private/commercial pilots working toward the instrument rating. Second, four simulator (GAT-1) sessions, lasting in length from 15 to 60 minutes, were conducted in which judgment was taught from the following list of 19 decision problems requiring VFR night time scenarios:

- Destination airport closed due to an accident
- Severe turbulence
- Drop in oil pressure; no change in oil temperature
- Departure field goes IMC
- Suction failure; attitude and horizontal indicators inoperative
- Loss of navigation station signal
- Loss of two-way radio communication
- Pitot-static system failure
- Partial loss of engine power
- Radar contact lost during radar vectoring
- Low fuel
- Flight vectored into restricted area
- 30-minute delay for landing upon arrival at destination airport
- PCL lighting inoperative upon arrival at destination airport

- Pilot given vectors opposite the desired direction of flight
- Pilot cleared for takeoff immediately behind B-727
- Pilot advised FSS does not have a copy of flight plan
- Scattered cloud deck at cruising altitude
- Smoke or burning odor in cockpit

The control group received the same amount of training but instead of judgment training in the classroom, they received basic instrument instruction and, in the simulator they received instrument training. The same instructors were used to teach both the experimental and the control groups.

Judgment evaluation. Both groups of subjects received simulator tests of their judgment before and after their training. The test flight was a simulated night cross-country line oriented flight training (LOFT) exercise. During the test flight, judgment scenarios were given to the subjects to handle and two types of ratings were made on the pilots. The first was called the 'checklist score' and consisted of the number of items checked from the following list of ten items:

1 Requested preflight weather briefing
2 Activated flight plan
3 Checked fuel
4 Requested radar advisories
5 Checked weather enroute
6 Accurately computed ETA for first intersection
7 Activated flight plan to alternate field
8 Initiated DR procedures following equipment failure
9 Elected cruise altitude above MEA
10 Requested assistance and/or confessed problems

The second measure, called the flight rating, consisted of a combined examination of the graphic track record of the flight, performance of the judgment scenarios, and an examination of judgment exhibited in performing the checklist items (from the above list) on a scale of -5 (worst possible judgment) to +5 (best possible judgment). Raters were told specifically to avoid rating the 'skill' of the pilot.

Results. In the before training test no significant differences between the experimental and control groups were found in either the checklist score or the flight rating. However, in the after training test there was a very big change. As shown in Table 5.3, the experimental group performance was significantly better than the control group. The experimenters also indicate that, when the experimental group is used as its own control, the differences are even more striking. This experiment clearly shows that the judgment lectures combined with simulator training are highly effective methods for judgment training and that judgment training is effective well beyond the initial stages of flight training.

Table 5.3
Results of judgment test before and after training

	Experimental Group	Control Group
Before training:		
Checklist score	4.06	4.31
Flight rating	-2.75	-2.92
After training:		
Checklist score	8.63	3.46
Flight rating	1.63	-2.54

The OSU simulator test of the DECIDE model

The next study was a first attempt to develop and test a model that teaches both rational and the motivational aspects of pilot judgment (Jensen, Adrion, Maresh, 1986). The DECIDE model on which it is based, provides a simple acronym upon which a pilot can depend when faced with a decision that requires more than a passing thought. It is a six-element decision making process that, when followed, can help to organize one's thoughts and prevent overlooking facts that may be important. It has been used effectively in training personnel who fight hazardous materials fires -- a complex decisional situation involving numerous items of probabilistic information and many possible alternatives for action (Benner, 1975). When faced with any decision involving uncertainty, pilots are taught systematically to organize their thoughts using the acronym 'DECIDE' as follows:

D - Detect: The decision maker detects the fact that a *change* has occurred that requires attention.

E - Estimate: The decision maker estimates the *significance* of the change to the flight.

C - Choose: The decision maker chooses a *safe outcome* for the flight.

I - Identify: The decision maker identifies *plausible actions* to control the change.

D - Do: The decision maker *acts* on the best options.

E - Evaluate: The decision maker evaluates the *effect of the action* on the change and on progress of the flight.

The DECIDE model is a 'closed-loop' process, meaning that as soon as it is completed, thinking goes back to the beginning again. In practice, the pilot is a vigilant monitor of all factors that could produce change during the flight. When such a change occurs, the DECIDE process is put into action. In using this systematic procedure, pilots begin with decisions that have some element of uncertainty (weather forecasts, fuel remaining, engine or navigation system reliability, etc.). As pilots become practiced with the model in these circumstances, the procedure becomes second nature causing them to use this system in all decisions.

Subjects. The subjects in this preliminary investigation consisted of 10 pilots from the Columbus area, each with between 60 and 233 hours of total flight time, in possession of a private certificate and current medical, and working actively toward an instrument rating. Five of these (three males, two females) were used as experimental subjects (received the DECIDE training), the other five (four males, one female) were the control subjects (received no training). The experience levels of the experimental and control subjects were very similar.

Flight simulator. The equipment package was designed to provide a realistic flight environment and yet maintain a high level of experimental control. The experimental equipment used in this study included a Frasca 141 flight simulator, Silicon Graphics visual display, and audio/visual recording equipment located at the Advanced Flight Simulation Laboratory at The Ohio State University Airport, Columbus, Ohio. The Frasca 141 flight simulator was configured to approximate the flight and handling characteristics of a single-engine, general aviation aircraft, much like a Piper Archer (PA 28-181). The simulator was a stationary fiberglass cockpit enclosed at the rear with a heavy curtain. The cockpit was equipped with all of the usual controls and instruments of a standard general aviation aircraft.

The instructor console of the Frasca 141 acted as the ATC/research station for the experiment. Communications with the subject pilot were made via headsets, equipped with push-to-talk switches. Position data for all of the flights were recorded on the X-Y plotter located at the research station. The plotter also provided a useful real time display of the progress during the flight. A video camera placed at the rear of the Frasca 141 was used for real time viewing and recording the events within the cockpit. A large screen video machine was used for viewing the flights with the subject/pilot during debriefing.

Methods. The experimental subjects were taught to use the DECIDE model in marginal aviation decisions in a five-hour interactive lecture/discussion format by

Russ Lawton of the AOPA Air Safety Foundation, and Rick Clarke of Events Analysis, Inc. The unique feature of this training method is that real scenarios were used to illustrate the DECIDE model and to teach all aspects of the model.

Three accidents were used in this training program. The first was the Pilgrim Airlines Twin Otter that crashed into Long Island Sound in heavy fog after running out of fuel. The second was the Air Illinois Beech 99 mentioned above. The third was the crash of the 'Condor Aero Club' Cessna 172 after encountering headwinds and failing to choose a suitable landing field before running out of fuel at night.

A fourth incident which turned out successfully for the pilot was used to illustrate the proper use of the DECIDE model. In this incident a private pilot got 'caught' above the clouds in the Columbus, Ohio area and after seeking help from the Air Traffic Controller, was brought down through the clouds safely.

In each of these accidents and incidents the instructor led the students through a chart similar to that shown in Table 5.4 for the Air Illinois accident. This accident, which involved an HS-748-2A, occurred in 1983 and killed all 19 people on board. Soon after departure from Springfield, Illinois at night, they experienced a generator failure. In attempting to correct the problem the pilots inadvertently shut the other generator down also. The captain elected to pass up their departure airport and six other airports along the way in a vain attempt to make it to their home base at Carbondale. Each 'change' identified in the incident was followed through the DECIDE model and the 'Do' or action of the pilot/flight crew identified. Correct and incorrect action within the model are shown as Y (yes) and N (no) responses, respectively. Poor judgment chains can be easily seen in a series of no responses. By going over several of these illustrations, the model elements were learned and the correct method for responding to similar decision situations was taught.

After each subject/pilot finished a familiarization flight, any other questions were answered and he/she was given a short break. The subjects were also given time to review their cross-country flight they were to make from Port Columbus to Frederick, Maryland. The motivation for the flight was established via reading the subject/pilot an indoctrination story putting them in the position of needing to arrive in Frederick to compete in a spot landing contest.

The actual experimental flight was designed with three experimenter-induced changes programmed into the scenarios as check points to observe the decision making process: 1) the failure of the attitude indicator, 2) the introduction of carburetor ice, and 3) deteriorating weather. Each of these changes occurred at 10-minute intervals as the flight progressed. The scenario was designed to force the subject to turn around and return to Columbus. The subject was free to decide when to make the turn.

Table 5.4
Illustration of DECIDE Model chart of Air Illinois accident

CHANGE	D E C I D	ACTION	E
Left generator fails after takeoff	Y Y N N Y	Copilot misidentifies failed generator, disconnects good generator	Y
Copilot tells Departure Control, of 'slight' electrical problem	Y Y Y Y Y	Departure Control offers return to Springfield Airport	Y
Crew gets Departure Control offer to return to Springfield	Y Y N N Y	Captain rejects offer and continues to Carbondale	Y
Right generator doesn't take electrical load	Y Y Y Y Y	Copilot tells Captain of loss of right generator	Y
Copilot tells Captain of generator failure	Y Y N N Y	Captain requests lower altitude for VFR conditions	Y
Copilot tells Captain battery voltage dropping fast	Y N N N Y	Captain tells Copilot to put load shedding switch off	Y
Copilot reminds Captain of IFR weather at Carbondale	Y N N N N	No reaction	N
Copilot turns on radar to get position	Y Y N N Y	Copilot tells Captain about dropping voltage	Y
Copilot tells Captain battery is dropping	Y Y Y Y Y	Captain turns off the radar	Y
Copilot warns Captain about low battery	Y Y Y Y Y	Captain starts descent	Y
Cockpit instruments start to fail	Y Y Y N Y	Captain asks Copilot if he's got any instruments	Y

The primary data were obtained in a structured debriefing interview conducted immediately following each flight. During this interview, the experimenter viewed the video tape of the flight with the subject and attempted to get from the subject his or her thought process surrounding each change that occurred using the DECIDE model to provide structure. These interviews were audio-taped and later analyzed to determine whether or not the subject used the model in making his or her decisions. Each change identified, whether experimenter or subject induced, was analyzed. The completion or non-completion of the DECIDE model was evaluated for each change noted.

Results. As an overall result, each of the four experimental subjects who chose to fly, landed safely after turning around. However, each of the three control (C) subjects who choose to fly, crashed; one of these could not handle the attitude failure, and two failed to diagnose correctly the carburetor ice problem. The experimental (E) subjects had no trouble handling either of these problems.

At the first experimenter-induced change (the attitude failure), one of the three E subjects who chose to fly turned back. None of the C subjects made an attempt to turned back. One of the Es turned back before the first experimenter-induced change making this flight a wash. At the second experimenter-induced change (the carburetor ice), no one turned around in either group, although both C subjects failed to diagnose the problem and consequently crashed.

At the third experimenter induced failure there were no C subjects remaining. The one E subject who was still flying chose to attempt to continue to the destination by descending below the clouds. As planned, she did get below the clouds, but when she continued, she again entered the clouds. She then successfully returned to the departure airport.

Discussion. The objective of this study was to provide a preliminary indication of the value of the DECIDE model for pilot judgment training. Although this objective was met, there was not enough time or funds to run enough subjects to show statistically reliable results. Given the variability in the data shown in this study, one should expect to require about 20 subjects in each group to show reliable results.

Nevertheless, this study did show that the DECIDE model has great potential as a judgment training tool. The DECIDE trained subjects demonstrated clearer thought patterns in diagnosing the carburetor ice problem and the attitude failure problem. They showed greater concern for the safe outcome by successfully landing.

It should be noted that this was a preliminary test of this training technique. Although the AOPA personnel did an excellent job presenting the material to the experimental group, we do not believe that enough time was allocated to make a real, lasting difference in judgment. More could have been gained through the use of role playing to permit subjects to experience the scenarios themselves. Also, a two-day course during which the subjects were required to take home some work to develop their thinking followed by written tests would have aided their learning the material and actually changing their behavior.

A secondary objective of this study was the development of judgment evaluation techniques. We believe that this study demonstrated the potential for testing pilot judgment in a ground-based simulator. The use of realistic scenarios with a recording capability on computer and video tape is an excellent evaluation technique. Furthermore, the debriefing that included the viewing of the video tape was the first chance many of the pilots had to view themselves flying. The response was universally favorable among those who chose to fly the experimental flight. They indicated that it was among the best flight training they had ever received.

This study points out that there is a very strong need for continued research on pilot judgment training and evaluation. The majority of pilot-caused accidents continue to be made because of faulty pilot decision making. Training in this area, together with the development of judgment evaluation techniques are the essential factors needed to reduce these accidents. Finally, this study concerned with process judgment, combined with the previous highly successful studies teaching attitude judgment, suggests that a systematic approach to judgment training can make pilots with less flying experience, safe pilots.

Attitude research in the airlines

Robert Helmreich has been the leading scientist in the development of training and measurement tools for CRM in the airline industry in the 1980s (Helmreich, 1984; Helmreich, 1987; Helmreich and Wilhelm, 1991). His Cockpit Management Attitudes Questionnaire (CMAQ) has been completed by more than 15,000 airline and military pilots who have taken CRM courses (Helmreich and Wilhelm, 1991). Using these tools, Helmreich and his colleagues have become very sophisticated in their ability to isolate the effects of certain aspects of management attitudes including the determination of personality effects on pilot attitudes (Chidester, Helmreich, Gregorich and Geis, 1991), the effects of fatigue on pilot attitudes (Foushee, 1985), the effects of the organizational culture on pilot attitudes (Helmreich and Wilhelm, 1991), and the effects of pilot attitudes on performance in the cockpit (Helmreich, Foushee, Benson and Russini, 1986). CRM research has been very extensive and most goes beyond the scope of this book (See Wiener, Kanki and Helmreich, 1993 for a thorough discussion of CRM). We will examine some aspects of CRM training and evaluation in a later chapter to show how it affects aviation judgment. At this stage, we will see the effectiveness of CRM training as further proof that pilot attitudes can be changed through training.

To place it in the context of pilot judgment, CRM is the extension of judgment to multi-person crews. The primary medium for management in the cockpit (as well as in the earth-based corporation) is through interpersonal communication. Thus much of CRM training is focused on this aspect of crew performance. Crew attitudes are closely linked to communication because they represent one of the primary forces that cause communication to occur or not to occur.

The nature of the attitudes with which we are concerned in CRM varies somewhat from those we spoke of above in judgment training. In CRM our concern is with the attitudes pilots have with regard to management of the flight deck where there usually is more than one crew member. Nevertheless, just as hazardous attitudes affect decisional behaviors, as we saw in the judgment discussion above, they also affect management behaviors in CRM. Thus, just as it is important to affect attitudinal change to improve decisional behavior in judgment, it is equally important to affect attitudinal change in order to affect management behavior in CRM. Weaknesses in both behaviors have been equally detrimental to safety.

In one study, Helmreich and Wilhelm (1991), using the CMAQ, identified attitude shifts in airline crews in two ways. First, the CMAQ was administered prior to any interventions to provide a baseline set of data which revealed that many crews resented the idea of CRM training. The pilots feared that CRM training might involve some type of brainwashing. Through a set of briefings stressing the need for CRM training, the content of the course, and indications about how crew performance would be evaluated, many of these fears were relieved and a second administration of the CMAQ just prior to the CRM course revealed a significant shift toward more favorable attitudes on all three scales as a result of the perceived shift in the organizational culture. After the formal CRM course, attitudes shifted again even more favorably indicating that the CRM training had the desired affect of increasing awareness and acceptance of crew coordination concepts.

In another study (Helmreich, Foushee, Benson and Russini, 1986), the linkage between attitudes and performance was established. In this study, 114 airline pilots who had taken the CMAQ were observed performing their line duties by two different experienced raters. Their overall flightdeck management performance was rated on a 5-point Likert scale from 1 'extremely poor' to 5 'outstanding'. A second rating was also made by examiners who knew each of the pilots and rated them as 'outstanding' or 'extremely poor' cockpit managers. The results of this study showed that the score on the CMAQ was successful in classifying 95.7 per cent of the pilots in these management categories.

Operational experiences

There is a growing number of operational experiences that we can cite following the introduction of judgment and CRM training into the curricula in which pilots are now being trained. Fortunately, accident rates are low and as a result they typically do not make good measurement criteria until a great deal of time has elapsed. Diehl (1992) cites two anecdotes from flying experiences that speak to the effectiveness of CRM training. The first involved a US Navy A-6 crew which experienced a total hydraulics failure, but was able to make a safe landing.

Aircrew coordination training was cited as a factor in this 'save'. In a similar remarkable save, the NTSB credited United's CRM training with assisting the crew of the United Airlines DC-10 accident at Sioux City, Iowa (NTSB, 1990). Finally, Table 5.5 shows how ADM and CRM training has had an effect on accident rates in both civil and military applications.

Table 5.5
ADM/CRM operational experiences

Organization/Subjects	Materials	Accident Rates
Bell Helicopters Inc. US Only, Jetranger Pilots	ADM	36% Decrease
Bell Helicopters, Inc. US Only, Jetranger Pilots	ADM	48% Decrease
Petroleum Helicopters, Inc.	ADM and CRM	54% Decease
US Navy	CRM	28% Decrease
US Navy A-6 Intruder, Crewmembers	CRM	81% Decrease
US Air Force MAC Transports, Crewmembers	CRM	51% Decrease

Source: Diehl, 1992

Conclusion

In this chapter we have seen that since the initial judgment study, many different experiments have been conducted in various locations around the world to show the effectiveness of judgment training. All of these studies have reported positive results in reducing judgment errors in those who were trained. In addition, the organizations that have taken judgment and CRM training to the operational setting report very positive safety results. Although we do not yet know the long-term effect of the judgment and CRM training, it is clear that this training can make a significant difference in the short term. These short-term results clearly indicate, at the very least, that judgment training works and that pilots can learn judgment through training. It seems clear that such training should be incorporated in pilot training as a safety measure.

References

Adams, R.J. and Thompson, J.L. (1986). *Aeronautical Decision Making for Helicopter Pilots*. Springfield, VA: National Technical Information Service, FAA Report No. DOT/FAA-86.

Benner, L. (1975) 'D.E.C.I.D.E. in hazardous materials emergencies'. *Fire Journal*, **69**(4), pp 13-18.

Berlin, J.I., Gruber, J.M., Holmes, C.W., Jensen, P.K., Lau, J.R., Mills, J.W. and O'Kane, J.M. (1982). *Pilot Judgment Training and Evaluation*. DOT/FAA Report CT-82-56, Vols I, II, III, Washington, DC.

Buch, G.D. and Diehl, A.E. (1982). 'An investigation of effectiveness of pilot judgment training'. *Human Factors*, **26**(5), 557-564.

Chidester, T.R., Helmreich, R.L., Gregorich, S.E. and Geis, C.E. (1991). 'Pilot Personality and Crew Coordination: Implications for Training and Selection'. *International Journal of Aviation Psychology*, **1**(1), 25-44.

Connolly, T.J., Blackwell, B.B. and Lester, L.F. (1987). 'A simulator-Based Approach to training in Aeronautical Decision making'. In R.S. Jensen (ed.) *Proceedings of the Fourth International Symposium on Aviation Psychology*, Columbus, OH: The Ohio State University.

Diehl, A.E. (1992, January). 'Does cockpit management training reduce aircrew error'? *ISASI Forum*, **24**(4).

Diehl, A.E. and Lester, L.F. (1987). *Private Pilot Judgment Training in Flight School Settings*. DOT/FAA Report AM-87/6, Washington, DC.

Foushee, H.C. (1985). 'Pre- and post-trip crew performance in short-haul flight operations: A full-mission simulator study'. In *Proceedings of the 38th International Air Safety Seminar*, Arlington, VA: Flight Safety Foundation, Inc.

Greenwood, J.R. (1983). 'Judgment and discipline: Recipe for survival'. In R. S. Jensen (ed.), *Proceedings of the Second Symposium on Aviation Psychology*, Columbus, OH: The Ohio State University, Aviation Psychology Laboratory.

Helmreich, R.L. (1984). 'Cockpit management attitudes'. *Human Factors*, **26**, 583-590.

Helmreich, R.L. (1987). 'Theory underlying CRM training: Psychological issues in flight crew performance and crew coordination'. In H.W. Orlady and H.C. Foushee (eds.), *Proceedings of the NASA/MAC workshop on cockpit resource management* (NASA Conference Publication No.2455) (pp. 14-21). Moffett Field, CA: NASA-Ames Research Center.

Helmreich, R.L., Foushee, H.C., Benson, R., and Russini, W. (1986). 'Cockpit resource management: Exploring the attitude-performance linkage'. *Aviation, Space, and Environmental Medicine*, **57**, 1198-1200.

Helmreich, R.L. and Wilhelm, J.A. (1991). 'Outcomes of Crew Resource Management Training'. *International Journal of Aviation Psychology*, **1**(4), 287-300.

Jensen, R.S., Adrion, J., and Lawton, R. (1987). *Aeronautical Decision Making for Instrument Pilots*. Springfield, VA: National Technical Information Service, DOT/FAA/PM-86/43.

Jensen, R.S. and Adrion, J. (1988). *Aeronautical Decision Making for Commercial Pilots*. Springfield, VA: National Technical Information Service, Report No. DOT/FAA/PM-86/42.

Jensen, R. S. (1989). *Cockpit resource management training*. Springfield, VA: National Technical Information Service, Report No. DOT/FAA/PM-86/46.

Jensen, R.S. and Benel, R.A. (1977). *Judgment Evaluation and Instruction in Civil Pilot Training*. Springfield, VA: National Technical Information Service, Final Report FAA-RD-78-24.

Jensen, R.S., Adrion, J. and Maresh, J. (1986). *A Preliminary Investigation of the Application of the DECIDE Model to Aeronautical Decision Making*. Columbus, OH: OSU Aviation Psychology Laboratory, Final Report AOPA Air Safety Foundation Grant.

NTSB (1990). *Aircraft Accident Report, United Airlines, Inc., DC-10, Report No. AAR-90-6*, Washington, DC.

Telfer, R.A. (1989). 'Pilot decision making and judgment'. In R. S. Jensen (ed.), *Aviation Psychology*, Aldershot, England: Gower Technical.

Thorpe, J.A., Martin, E.L., Edwards, B.J. and Eddows, E.E. (1976). *Situational Emergency Training: F-15 Emergency Procedures Training Program*. Williams AFB, AZ: Air Force Human Resources Laboratory, Technical Report No. AFHRL-TR-76-47[1]).

Wiener, E.L., Kanki, B.G. and Helmreich, R.L. (Eds) (1993). *Crew Resource Management*. New York: Academic Press.

6 Crew resource management

In the past year and a half, an alarming number of safety investigation boards have concluded crew error was causal. Equally alarming is the realization that often the information missing from the critical, often final, decision was available to the crew. In some cases at least one crew member had the answer (that would have prevented the mishap). While mishaps are always tragic, those in which the resources to prevent catastrophe were available and either unrecognized, unused, or simply not offered represent an especially intolerable category.

Aufderheide, 1987

To this point, the focus of our attention has been on aircraft requiring only one person in the crew. Although everything we have said so far about pilot judgment applies equally well to multi-person crews (called 'multi-crews' in the industry), the psychological and social dynamics of the situation changes when there is more than one person in the flight crew. This chapter is an examination of judgment training and evaluation in multi-crew aircraft.

Multi-crew aircraft in the airline and general aviation industries have undergone considerably more technological change than have single-person aircraft. In the military, most single-person aircraft also have seen these changes. As a result of these changes, the role and primary task of the flight crew has changed. Formerly, the primary task of the flight crew was navigation and manual manipulation of the controls. In modern airplanes, much of the actual navigation and control manipulation can and is being 'down-loaded' to the on-board computer. The task of managing and supervising the flight deck and all of the resources available (especially other crew members), that the crew always had, is becoming their primary task. This management task is called 'crew resource management' (CRM).

CRM defined

CRM is defined as the effective use of all resources (hardware, software and liveware) to achieve safe and efficient flight operations. Rational judgment in this environment has the advantage of more resources for both the acquisition and handling of information than we have seen in single-person crews. In addition, in most cases, the pressure of fellow crew members is toward rational decision processes. There are some cases in which this is not the case and the availability of other crew makes the situation less rational. In this chapter, we will examine the dynamics of multi-crews and show how CRM training and evaluation can be used to make crews work together more effectively.

The safety of the airline industry is evidence of the efficacy of multi-crew aircraft. Most flight crews operate safely and effectively. However, when accidents occur, in most cases, they are the result of a breakdown in crew coordination and management. In 1979, John Lauber, one of the early leaders in this area, identified more than 60 airline accidents that were attributed to this breakdown in crew activity (Lauber, 1979). In Chapter 1 (see Figure 1.1), we saw that around 70 per cent of airline accidents are caused by crew error. Most of these accidents are due to crew management, not to breakdowns in technical skill.

Technical competence, the cornerstone of effective crew performance, is assumed in CRM training. However, technical competence may have indirect links to resource management. For example, a less technically competent crewmember may be highly defensive in order to preserve a competent self-image. This may result in the crewmember maintaining unrealistic and self-deceptive attitudes of personal competence, resistance to stress and lack of need for support from other crewmembers. This person may project an air of all-knowing confidence and independence when, in fact, the opposite is true. Such behavior may have a highly adverse effect on CRM.

What are the resources?

The basic resources that the flight crew manages can be placed into three categories people, equipment and other. The *people* resources include most especially other crew members, but also include ATC, cabin crew members, passengers, other pilots, dispatchers and other company people. The *equipment* resource includes the instruments and controls but also includes the often forgotten information coming through the windscreen about other traffic. The equipment resource is becoming very complex with the heavy use of computerized flight management systems (FMS) in modern aircraft. These systems are a major workload reliever for the crew when they have been properly programmed prior to flight. However, they can actually increase substantially the workload on the crew if information needs to be inserted during a flight due to changes in the flight plan. The *other* category includes charts, checklists and operations manuals.

Recent CRM history

From a historical perspective, the key event in the development of CRM concepts was the kickoff workshop held at the NASA Ames Research Center (Cooper, White and Lauber, 1979). CRM development owes a great debt to the classic study by H.P. Ruffell-Smith (1979) in which a full mission simulation of a civil air transport scenario was conducted. This study was the first to show how crew coordination problems can result in poor decision making and performance. It was a strong motivator for CRM training and is often cited in the CRM literature. The First International Symposium on Aviation Psychology in April, 1981, was primarily motivated by the need for new ideas concerning the development of CRM for the airline industry (Jensen, 1981). Each of the biennial symposia in Columbus, Ohio since that day have focused the attention of the world aviation community on CRM development (Jensen, 1983; 1987; 1989; 1991; Jensen and Adrion, 1985; Jensen and Neumeister, 1993). Finally, CRM has been brought to the forefront in the airline industry in recent years as a result of the large percentage of accidents resulting from the failure of the crew properly to manage the flight deck.

CRM related accidents

In the recent past, there have been a few highly visible CRM related accidents in the airline industry that have been motivators behind several of the best CRM training programs that have been developed. Some of these accidents have also been the source of case study material in CRM training around the world. These 11 accidents were:

Eastern, L-1011, Miami, 1972
TWA, B-727, Washington Dulles, 1974
Pan American and KLM, both B-747s, at Tenerife, 1977
United, B-727, Salt Lake, 1977
United, DC-8, Portland, 1978
Saudi Arabian Airlines, B-747, Jedda, 1978
Western Airlines, DC-10, Mexico City, 1979
Danair, B-727, Tenerife, 1980
Air Florida, B-737, Washington National, 1982
National, B-727, Escambia Bay, 1978
Delta Airlines, L-1011, Dallas, 1985

For example, the Pan American/KLM accident was the prime motivator behind KLM's development of the first (and still, the industry's most extensive) CRM course in 1979. The two United accidents were the prime motivators behind the extensive United CRM course which began in 1982. The Delta accident was the initial motivator behind their CRM training program developed somewhat later. In each of these accidents the captain failed to make effective decisions because he, or his crew, did not use effective CRM. Because we can learn so much from them, each of these accidents will be covered elsewhere in this chapter.

Objectives of CRM training

Experience, crew coordination training and LOFT were the major sources of CRM training before the introduction of the formal courses mentioned above. The objective of new CRM courses is to improve the performance of crews in the following areas: decision making, communication, leadership, stress and fatigue management and teamwork.

It is important that we understand at the outset the difference between personality and attitude because both are important to CRM performance. The term 'personality' refers to relatively enduring characteristics of the individual acquired during development. These characteristics are quite resistant to change and are modifiable only through the considerable efforts of psychotherapy. We know that personality factors can influence a person's management capabilities. However, because of the nature of personality traits, it would be futile to attempt to change them in pilot training. Furthermore, most aviation psychologists believe that personality traits are important only in the most extreme cases and can be handled through selection of pilots.

On the other hand, attitudes are less deeply internalized components of the individual and are subject to change fairly easily. Attitudes are constantly bombarded by forces in our society such as advertisers, salesmen, teachers, politicians and preachers attempting to change them. In the cockpit, management attitudes as well as attitudes toward risk taking and performance of all other aspects of the flying task can be modified through training. Therefore, a major objective of CRM training is to improve the attitudes of the pilot so as to bring about better decision making.

CRM training is *not* designed to change personality. Instead it is designed to address crew behavior which is a product of knowledge and thought process, personality, attitude and background. In CRM courses we can teach ways to think clearly in decision making in concert with other crewmembers, each with a different personality, and we can have an impact on attitudes which affect each of the areas of concern. Such training may result in more flexible behavioral strategies and more coordinated crew behavior in critical situations when maximum effectiveness is a *life* or *death* issue.

One of the most important keys to good cockpit management, as in any management position, is communication among crew members. Information must be requested, offered and/or given freely in a timely way to permit the captain to make accurate, effective decisions. It also requires an understanding of communication styles used by other members of the crew for interpretation and to determine the proper emphasis for a response. Finally, it requires an understanding and acceptance of the unique role and leadership responsibility of each member of the crew. Therefore, the primary emphasis in CRM training is in interpersonal communications.

Important background research

'Social Loafing'

One of the reasons that special attention must be placed upon the performance of people when they work together has been identified in social research by the term, 'social loafing'. According to this concept, unless strong group identification and cohesion develops, people do not put as much effort into a task when they are working with others as they do working alone. Latene, Williams and Harkins (1979) cite an unpublished study by Ringelmann in 1929 in which different numbers of people were assigned to a rope pulling task. Table 6.1 shows the amount of effort applied per person for different numbers of persons on the team.

Table 6.1
Group size versus effort applied per person

Group Size	Per Person Force Exerted
1	63 Kg
2	58 Kg
3	53 Kg
8	31 Kg

Through motivational communication, coaching and teaching, this phenomenon of social loafing can be overcome and perhaps even reversed. Leadership and group pressure can be applied to encourage people to put greater effort into a task in an effort to support the 'team'. Coaches of sporting teams use such techniques to get greater effort out of their players. Military leaders use similar techniques to achieve maximum efforts from their troops. If the concepts of teamwork and crew cohesion were understood and practiced, it seems likely that they could be equally effective in overcoming social loafing in aviation which sometimes leads to an abdication of individual responsibility because one crew member thinks the other is taking care of a task when that may not be the case.

Management theories

There are many theories concerning the leadership, management, or behavioral styles used in management courses. Three of these, together with their dimensions, that are useful in CRM courses are shown in Table 6.2.

Table 6.2
Management theories useful in CRM training

Theory	Dimensions
Tannenbaum and Schmidt (1975) 'Management Decision Style'	Authoritarian and democratic
Blake and Mouton (1964) 'The Management Grid'	Concern for relationship and concern for people
Hersey and Blanchard (1977) 'Situational Leadership'	Relationship and task

Management decision style

A model representing a continuum of management decision processes is offered by Tannenbaum and Schmidt (1975). As shown in Figure 6.1, this concept deals with the decision making process used by various types of managers on a continuum from authoritarian (on the left) to democratic (on the right). This model shows how the relationship between manager and subordinate changes as the manager's style moves from authoritarian to democratic. In most cases in the cockpit, best crew performance results from the mid-range of this scale. However, one must recognize that there may be times when either extreme may be needed.

The intra-cockpit authority gradient. The concept of the 'intra-cockpit authority gradient' has been introduced in CRM courses to describe the type of management relationship that exists between the airline captain and his or her crew. According to this concept, a strongly authoritarian captain and a weak co-pilot represent a steep intra-cockpit authority gradient as shown in Figure 6.2. This situation is considered dangerous and the reason for several CRM failures that have resulted in airline industry accidents. A moderately strong captain and a moderately strong co-pilot, both of whom know and assume their proper roles, result in a shallow authority gradient. This situation is considered ideal for safety and crew effectiveness.

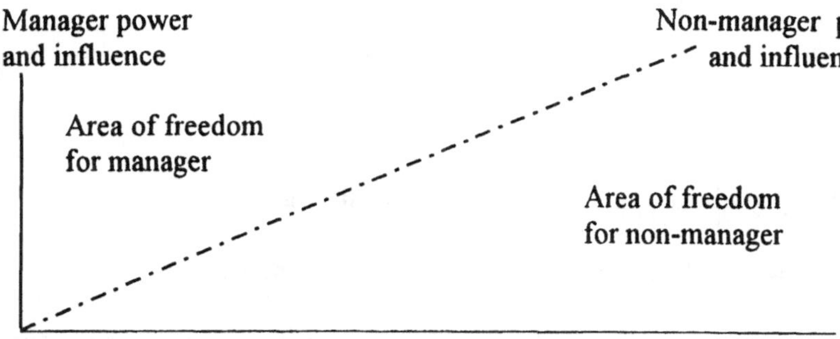

Manager:

makes decision	sells decision	presents decision and responds to questions	presents tentative decision subject to change	presents problems gets input decides	defines limits	makes joint decision with non-manager

Figure 6.1 Range of options for managers and non-managers (Tannenbaum and Schmidt, 1975)

The bottom line of the CRM training effort is better crew decision making. One of the major problems identified in airline accidents involving CRM has been that the captain 'goes solo' in making decisions during an unusual situation. For this reason decision making is the highest objective of CRM courses. Thus, all aspects of CRM courses are developed for the purpose of enhancing decision making. Management attitudes, leadership skills, knowledge, stress and outside forces are identified, modified and/or used where appropriate to improve flight crew decision making in CRM training.

Relationship versus task

Two aspects of attitudes that are very important to CRM are concern for relationships with others and concern for completing the task at hand. Three forms of this model used in CRM courses are the Management Grid (Blake and Mouton, 1964), the Leadership Effectiveness and Adaptability Description (LEAD) model (Hersey and Blanchard, 1977) and the behavioral dimensions.

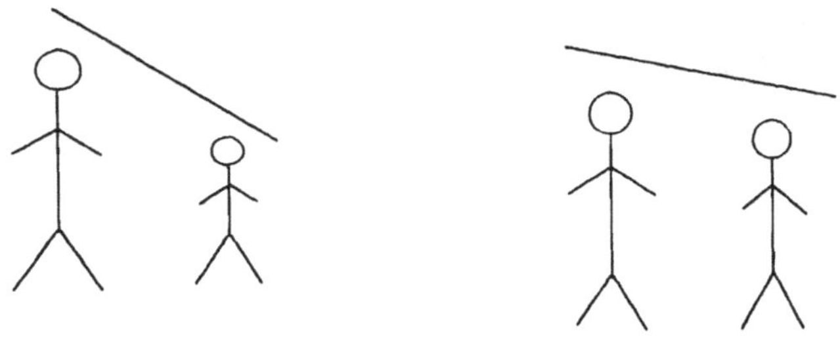

Figure 6.2 Two levels of the intra-cockpit authority gradient

The relationship versus task model is usually presented as a matrix with one side representing the 'relationship' orientation and the other side representing the 'task' orientation. According to this model, one's behavioral orientation, including communication style, can be described by these two independent dimensions. People who are high on relationship orientation tend to consider the feelings of others first believing that the task can be best accomplished when everyone gets along well with each other. On the other hand, those who are high on task orientation tend to consider the accomplishment of the task more important than relationships and would act to get the job done whatever the cost to relationships. The ideal orientation for flight crews in most flight situations is a strong combination of both.

The following relationship vs task model, offered by the Continental Airlines CRM Course (1985), provides a representative form of the model and it provides a theoretical construct that is easily learned in a CRM short course. As shown in Figure 6.3, there are four quadrants in this model:

Autonomous:	Low Relationship, Low Task
Nurturing:	High Relationship, Low Task
Aggressive:	Low Relationship, High Task
Assertive:	High Relationship, High Task

In the application of this model to the flight crew, the assertive style of behavior is advocated because it produces best overall performance in decision making. Although each crew member comes to the cockpit with certain tendencies to be either nurturing, autonomous and/or aggressive, you as a crew member, regardless of your position within the crew, should adapt these emphases to meet the demands of each situation. Extremes of each type are to be avoided because they discourage effective communication.

Some situations may require an aggressive/assertive style (e.g., an emergency in which there is little time to consider relationships or more information). Others may demand a nurturing/assertive style (e.g., helping a crew member through a mistake without destroying his ego). The objective should be to develop each of these styles so as to bring about an adaptable assertive overall style for cockpit communication. Finally, you as a member of the crew must be sure that the other crew member(s) know how you are acting (behaving) at all times, and it is equally important that you know how the other(s) are acting as well.

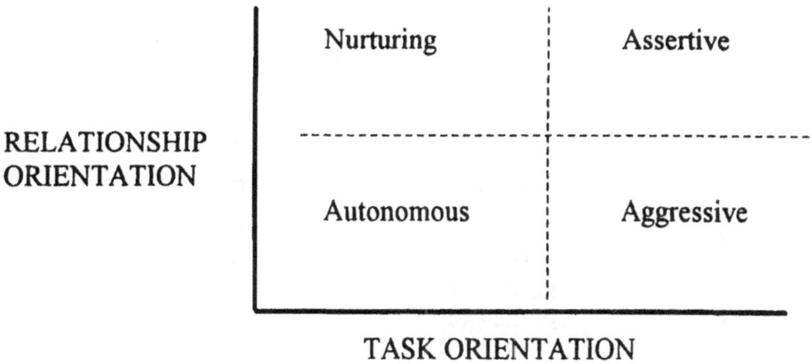

Figure 6.3 Basic behavioral dimensions (Continental Airlines, 1986)

As one can see from an expanded version of the model shown in Figure 6.4, flexibility to move toward one or another of the three alternate quadrants is suggested depending on the situation, yet the basic assertive style is maintained. One can also see that the extreme in each of the non-assertive quadrants can lead to failures in communication and relationships. Those who go overboard in nurturing tend to violate their own rights. Those who are too aggressive violate the rights of others. Those who place too much emphasis on their own autonomy may fail to contribute anything useful.

Theory of the situation: situational awareness

Central to many CRM accidents one finds a discrepancy between what the crew perceives to be true about a given situation (and themselves and their actions) and what in fact is reality. Furthermore, as indicated above, the situation often determines the type of communication style one should use. To help you recognize the situation in which you may find yourself, the following concept called 'Theory of the Situation' is offered. This concept was developed by Dr. Lee Bolman of the Harvard University School of Education (1979).

One of the reasons for discrepancies between perceptions and reality is that a major function of the human perceptual system is to reduce and order the vast amount of information coming in through the senses so that we can understand and respond to it appropriately. Unfortunately, this information reduction and ordering process is not perfect and sometimes leads to mistakes or discrepancies from reality. The process is developed through many years of experience and is therefore different for every person. Perceptions of visual information is fairly consistent with all individuals. However, perceptions of situations based on cognitive material obtained through all the senses is not very consistent.

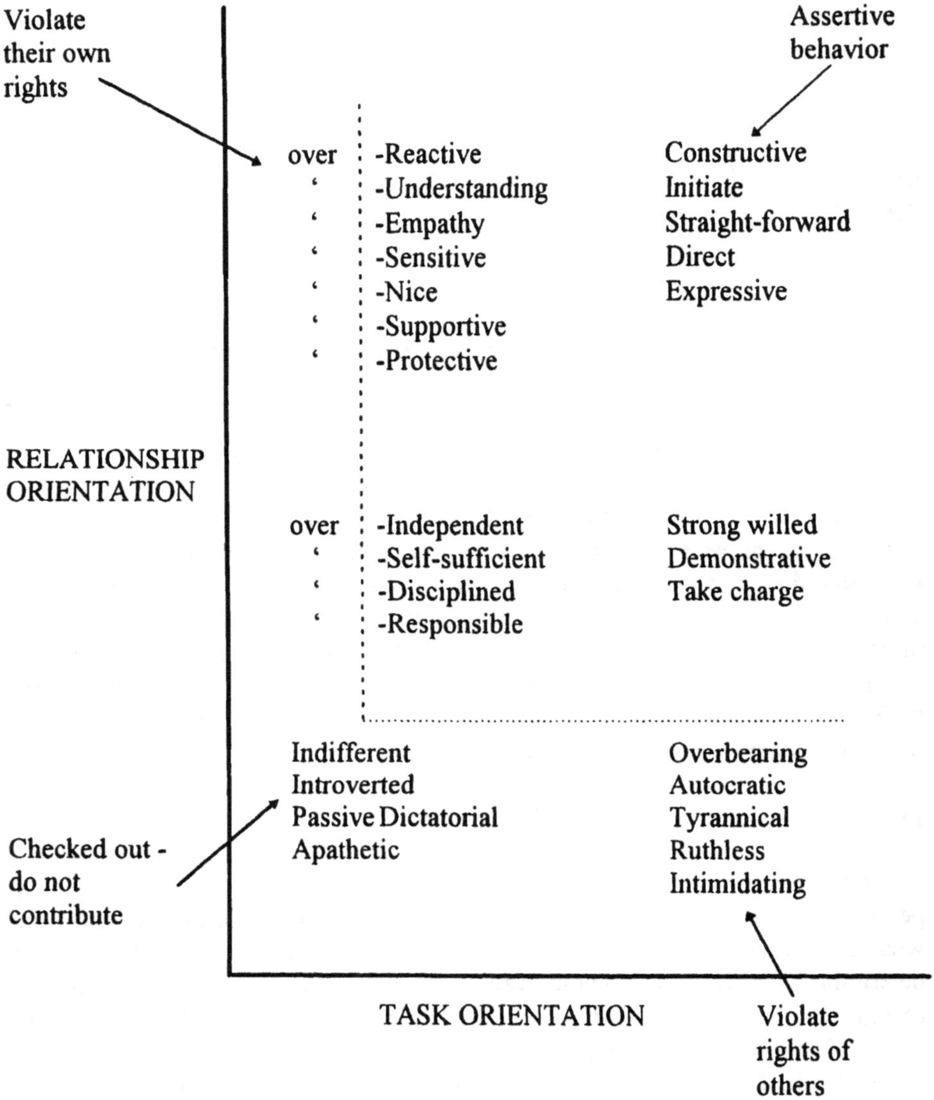

Figure 6.4 Behavior dimensions with descriptors showing how each dimension contributes to assertive behavior

The Theory of the Situation is an attempt to show how the various forces interact as we attempt to gain an awareness of the situation. The following definitions are offered for the elements of this model:

Theory of the situation:	What one assumes to be true based on his/her perception of the facts one has at any point in time.
Reality:	The situation as it is in reality - often fully known only by the 'Monday Morning Quarterback'.
Theory in use:	One's predictable behavior in a given situation that has been developed since birth.
Espoused theory:	An individual's account or explanation of his/her behavior. 'How I say and believe I will behave'.
Theory in practice:	The set of skills, knowledge and experience one calls upon according to his/her theory of the situation.

Figure 6.5 shows the relationships among the elements of the model. If the theory of the situation is in line with reality, the flight crew's assumptions are correct and safe flight decisions are likely. However, if the crew's assumptions are not correct a discrepancy exists between their theory of the situation and reality as indicated at the top of Figure 6.5. The greater this discrepancy, the greater the danger. Discrepancies, such as these, that go unrecognized and are magnified under stress, that underlie most CRM accidents.

The remedy for the discrepancy between the theory of the situation and reality is through testing assumptions. This means actively checking your understanding of the situation with your other crew members, ATC and the instruments and/or computers on board the aircraft.

A second discrepancy can exist, between our espoused theory and our theory in use, which can also lead to decisional errors. If we say we will do something and have no intention of doing it, we create great misunderstandings within the crew and great conflicts within ourselves. The key to reducing this discrepancy is found in courses such as this which help to identify, through communication practice, the extent of this discrepancy and show you how to deal with it. A common response to LOFT exercises as well as the role playing exercise in this course, for example, is 'I had no idea I sounded like that!'

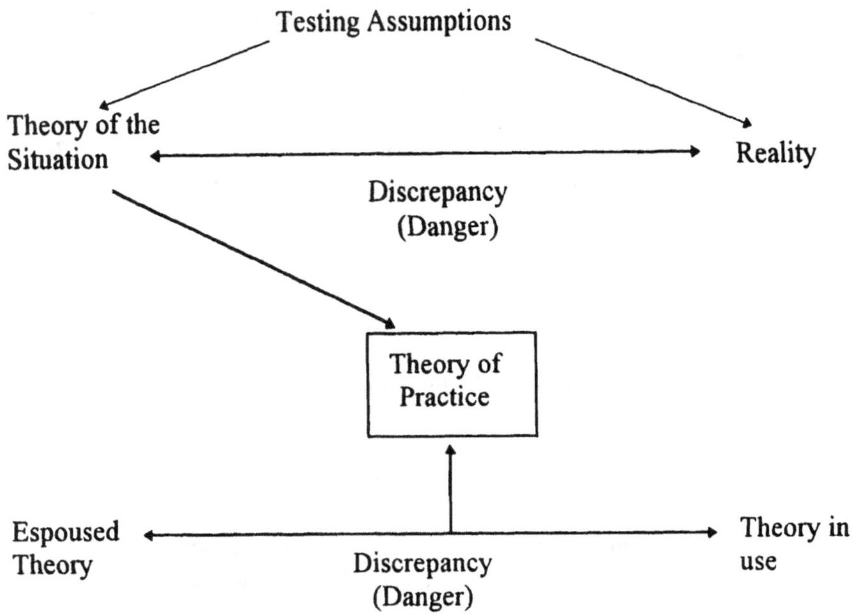

Figure 6.5 The Theory of the Situation (Continental Crew Coordination Concepts course, 1983)

Basic CRM concepts

Communication

The most important aspect of CRM is interpersonal communication. It is through communication that management is conducted. It is the responsibility of all crew members to communicate effectively. Communication means more than speaking clearly with proper phraseology. It also means making the other person understand what you are saying as well as understanding what the other person is saying because it can mean the safety of the flight.

All CRM courses provide some means of determining individual communication style. Most also provide training concerning the identification of the communication style of others. Some advocate a communication style, others leave it up to the individual to determine the communication style that works best for him or her. Some courses also advocate and teach listening techniques. There are five important aspects of communication to which pilots should be proficient to be good resource managers and decision makers: Inquiry, Advocacy, Listening, Conflict Resolution and Critique. Each of these elements will be covered in detail later.

CRM courses list a wide variety of elements essential to good communication. We have selected five of these that are independent of each other and cover the important aspects of cockpit communication very well. None of these five elements deal with semantic clarity or proper grammar. All deal with the transfer of information important to understanding. These five aspects of effective communication are:

- Inquiry
- Advocacy
- Listening
- Conflict resolution
- Critique

Because of the importance of each of these aspects of communication, each will be considered in some detail with cockpit examples. Later, case studies will be presented to further your understanding of these ideas. It is important that you not only understand the concepts presented here, but also that you adapt them to your own communication practices.

Inquiry. Inquiry or information seeking is the first aspect of communication covered because it represents the beginning point to making effective decisions. Good decisions are based on good information. In the cockpit it consists of both a visual scan of the cockpit for the information and questioning other crew members or controllers for information. It also means asking for clarification when the information is not clearly understood.

In the cockpit, crewmembers with fragile egos are often reluctant to ask for clarification because it may reflect badly on either their intellect or their hearing, both of which are important, not only to maintaining flying status, but respect of peers as well. This feeling of an insecure ego, which is often fed by equally insecure peers and/or controllers who fire back a statement about how foolish you look to have to ask such a question, must be overcome if complete understanding and safe decisions are to be made consistently.

An example can be seen in the Air Florida accident when the co-pilot, who was flying, asked the captain, 'Slushy runway, do you want me to (do) anything special for this or just go for it'? This was a clear inquiry about ideas for the takeoff as well as an expression of concern for the takeoff. He then repeated his concern in different, less direct, ways without a response from the captain three more times. The captain's response was first to ridicule the question, 'Unless you got anything special you'd like to do'. He then ignored further expressions of inquiry and concern. Following the captain's response, the co-pilot became less clear in his further inquiries.

Advocacy. Advocacy refers to the need to state what you know or believe in a forthright manner. It means not only stating your position, but maintaining your position until convinced by the facts, not the authority of the other, that it is wrong. There are many examples of airline accidents in which one crewmember had the correct answer (as indicated by the fact that they questioned the actions of the captain) but either did not advocate the position strongly enough, or capitulated far too soon to the authority of the captain. The United accident in Portland in which the flight engineer expressed concern over the fuel state and the Air Florida accident in Washington in which the co-pilot indicated concern for the difference in the engine temperatures are two examples. In both cases the captain's decision would have benefited from a subordinate crewmember who was a greater advocate of his position.

Listening. As demonstrated in the Continental course, it is also important to include a strong element of active listening training as well. One of the greatest reasons for cockpit communication failures is the fact that no one was listening. Listening requires more than passive attention. It requires the listener to open up to the other person, actively inquire through questions and other forms of feedback, and respond appropriately (agreement, acknowledgment, disagreement) - but always accepting that what is said may be true to the other person. Listening is not passive - it is a part of communication for which all members of the crew are responsible. No one must be a sponge.

Non-listeners have the following traits:

- *Pre-Plan*: Intent on what I want to say so I don't listen to what you are saying.

- *Debate*: No matter what is said some people want to take the other side 'I'm going to play the devil's advocate'.

- *Detour*: Like pre-plan, but they wait for a key word to take the discussion to another area of interest to them.

- *Tune out*: Spouses and kids - whatever they have to say, it is not important enough for your attention.

On the other hand, active listeners:

- Ask questions
- Paraphrase - did I hear you right?
- Provide eye contact.
- Use body language.

Active listening results in better communication, safety, efficiency, relationships, decision making and harmony. To summarize active listening the following points are offered:

- It is a basic human need to be heard and understood - active listening serves that need.
- Active listening is a skill that must be learned.
- In an emergency, active listening is a *critical* skill.
- In normal situations, active listening enhances communication, eliminates barriers and lays the groundwork for good communication during emergencies.

Conflict resolution. If you and your other crewmembers are each advocating your position properly, conflict is inevitable. Therefore, an effective process is needed to resolve those conflicts. Conflicts are not necessarily bad as long as they arise over issues within the cockpit. They can become destructive when issues from outside the cockpit, such as taking sides on management policies, personality factors, personal weaknesses, social status, etc., are brought into the argument. It can also be destructive when the argument is over who is right rather than what is right. Such arguments can have a serious effect on the quality of the decisions made because thinking is focused on the wrong issues and when disagreement is not expressed.

Conflict can be constructive if it is handled right. The proper way to resolve conflicts is to:

- Have a policy of crew coordination that is known and accepted by everyone.
- When disagreement arises, keep the discussion on the issues needing resolution within the cockpit.
- Bring out all issues of disagreement.
- Acknowledge and express all feelings that are deep enough to cloud your thinking.

Properly handled conflict resolution is the key to the highest level of problem solving known. It leads to deeper thinking, creative new ideas, mutual respect and higher self-esteem which strengthens team effectiveness. For these reasons, conflict should not be avoided when differences of opinion arise. Rather, it should be recognized as an opportunity to seek better solutions to problems that may not have been thought of previously.

Critique. Even more difficult than conflict resolution is the ability to provide an effective critique of fellow crewmembers. A critique is necessary because it teaches us how we can improve. We make mistakes on all flights. Someone said that straight-and-level flight can be defined as a series of *error* corrections. If we are only told of course and altitude errors (by our gauges), we only improve our ability to track. To improve other cockpit skills such as problem solving, monitoring traffic, communication, etc., we need feedback in the form of a critique from our fellow crewmembers.

Who is responsible for giving the critique? Because of his position, the captain is first responsible for providing feedback to the other crewmembers. However, the captain also makes mistakes that need to be pointed out by his crew. In some CRM classes, captains have lamented the fact that they have *never* received any feedback concerning their performance except on proficiency checks.

How should a critique be done? First, all members of the crew should know to expect a critique. CRM training such as this is useful in bringing about this awareness. Second, to reduce the pain and embarrassment, the critique should be invited by the crewmember getting the critique, especially if it is the captain. Third, the critique should consist of frank discussions among the crewmembers, beginning with flight planning, continuing through the flight, and ending with a debriefing at the end. If properly done, it can become a way of life in the cockpit that resolves conflict and misunderstandings before they arise by preventing issues and important feelings from being covered up.

The attitude and reactions of the person receiving the critique may be just as important as the initiative of the person giving the critique. The receiver's attitude should encourage the other to give feedback. On the other hand, the receiver should realize that feedback from any one person cannot be fully trusted because it comes from one perspective. Everyone who sees you perform, sees it from a different experiential background which influences his feedback to you. Therefore, it is necessary to get feedback from others to get an accurate picture of your performance.

Leadership

An essential part of CRM in aviation is leadership. Each member of the crew must recognize that he or she has a leadership responsibility that is important to effective decision making. An often overlooked aspect of leadership training is the recognition of importance of the 'followers' in management and decision making. Followers not only have an important 'leadership' role in making decisions (e.g., using an assertive style in providing information) but their communication behavior is an important factor that must be considered by the captain in adapting his or her communication style.

The following leadership rules, developed for the US Air Force Aircrew Coordination Training course (Aufderheide, 1987), are most appropriate in the management of civilian cockpits as well.

Thirty rules for getting things done through your people

1. Make the people on your crew want to do things
2. Get to know your crew
3. Be a good listener
4. Criticize constructively
5. Criticize in private
6. Praise in public
7. Be considerate
8. Delegate responsibility for details to subordinates
9. Give credit where it is due
10. Avoid domination or forcefulness
11. Show interest in and appreciation of the other fellow
12. Make your wishes known by suggestions or requests, not orders
13. When you make a suggestion, be sure to give the reasons for it
14. Let your crew in on your plans in an early stage
15. Never forget that the leaders set the style for their people
16. Play up the positive
17. Be consistent
18. Show you have confidence in your people and expect them to do their best
19. Ask subordinates for their counsel and help
20. When you make a mistake, admit it.
21. Give courteous hearing to ideas from subordinates
22. If an idea is adopted, tell the originator why
23. Remember, people carry out their own ideas best
24. Be careful what you say and how you say it; it may be misunderstood
25. Don't be upset by moderate grousing
26. Build subordinates sense of the importance of their work
27. Give your people goals
28. Keep your people informed on matters affecting them
29. Let subordinates take part in decisions affecting them
30. Let your people know where they stand

Traits of a good leader

Technical and professional competence: A cockpit leader is first and foremost a competent airman. Piloting skills must be exemplary and should inspire the confidence of subordinate crew members. Subordinates will give a leader a reasonable period to get their 'feet-on-the-ground', but will not respect the individual who continually relies on others to make decisions or provide guidance.

Communicative skill: Leaders are highly presentable and have a wide-ranging vocabulary; they inspire individual and group confidence. They sense followers' moods and are respected by peers for verbalizing and presenting their views. As one statesman said, 'Men lead with words'.

Listening: A leader listens, but listening involves a great deal more than merely hearing. Leaders interpret and evaluate what they hear, and do not permit personal ideas, emotions, or prejudices to distort what a person says. Disciplined listening prevents them from tuning out subjects they consider too complex or uninteresting. Effective listening is difficult, but it is a key communication skill.

Decision making: An effective leader is skilful at problem analysis and decision-making. All available information should be sought out and used in arriving at a decision. It is easy to make decisions based upon narrow information but the results are generally less than optimum. Limited information which is readily available sometimes presents an incomplete or misleading picture of the situation. Making an extra effort to seek out additional information may place a new perspective upon the situation requiring a decision.

Courage: Leaders view courage as an essential binding influence for unity of action. Followers will usually excuse almost any stupidity, indiscretion, or ill-conceived action, but they will not accept excessive timidity. In holding strong to fundamental principles of leadership, effective leaders see themselves under a continuous challenge to prove by one means or another the quality and character of their person. Courage is indispensable if leaders expect to influence and give direction to the lives of other people.

Risk: A leader must be a risk taker. If they could perform without risk, their jobs would be much easier, but risk taking is inherent in leadership. However, risks involving safety of flight must be balanced against the requirements of the mission. When time permits, decisions involving major risk should only be made after a full consideration of all the factors and facts available.

Perseverance: People who aspire to or have achieved leadership persevere in their work. They stick to tasks and see them through to completion, regardless of difficulties, and they are always optimistic and confident that they can find solutions to problems. They may even be a little bit stubborn when they are convinced of the correctness of a decision.

Sense of responsibility: Personal desires must be subordinated to the needs of the mission. A leader recognizes responsibility and relishes it as an opportunity to display leadership skills.

Emotional stability: Leaders must exercise self-control if they expect to control others, and they must maintain control in the most trying situations. Furthermore, they should strive to keep their personal lives under control and should never allow personal problems to color decisions or reactions to adverse situations.

Enthusiasm: A leader must be genuinely enthusiastic in all the tasks which comprise the mission at hand. Followers will automatically give more of themselves and take more pride in their work when they know their leader is involved and committed. Some leaders are reluctant to delve into functional areas where they have little or no prior experience or qualifications, but at times, it is important to seek new directions and to delve into these areas.

Image: All leaders must have positive self-images formed through objective perception and interpretation of their environment. Self-images are controlling factors in their behavior because all people act as they perceive themselves. Leaders must develop their self-esteem and personal value.

Ethics: Ethics play a key role in the leadership function because they are the basis of all group interaction and decision-making. Professional ethics require leaders to maintain high standards of personal conduct and to adhere to those standards in all situations so that followers can rely on their actions. Leaders should not use their position for personal and special privileges.

Recognition: Leaders recognize the accomplishments of their people. William James, the philosopher, said, 'The deepest principle in human nature is to be appreciated'. The inability to satisfy needs for informal recognition is a common shortcoming of many commanders who are more problem and task oriented than people oriented. Good leaders are aware of the people surrounding them. They know the names of subordinates, their hometowns, family situations, etc. They keep their pulse on the feelings of their subordinates.

Sensitivity: A leader must be sensitive to their own psychological and physiological state and their impact on others. They should be particularly sensitive to departures from their norms occasioned by stress or fatigue. We must also be sensitive to the psychological and physiological states of others, and be prepared to adjust their style accordingly.

Flexibility: A leader must understand that no two people or two situations are ever exactly alike. Yesterday's approach may or may not be the correct approach for today or tomorrow. Effective leaders adapt their approaches to the particular persons or problems at hand. In dealing with problems and situations, leaders should always be ready to redefine and modify their approach in response to new developments.

Humor: Leaders should have a sense of humor because they set the tone for their aircraft. When the leader smiles it is easier for others to smile. Most people prefer to be part of a relaxed and pleasing crew rather than a crew laden with tension. Leaders should not take themselves too seriously. Humor can be a positive and welcome contribution to an efficient and effective cockpit.

Stamina: Leaders must have a high level of physical and mental stamina. Good leaders always seem at the ready and require only normal periods of rest. They know how to pace themselves well and maintain themselves in good physical condition.

The 'team' concept

The cockpit crew is a team and must work together helping each other to do the best job possible for the mission rather than working separately as individuals. An effective method for demonstrating and teaching team performance is through small group attempts to achieve 'synergy' in problem solving. Synergy in CRM training means that the team performance is better than the best individual performance within the group.

Team training, as opposed to individual training, is the key to any good CRM course. The content of team training material must include the essentials of working together to accomplish a team effort. This is somewhat difficult to accomplish in aviation because most evaluations (flight checks) are made on an individual basis. This strong emphasis on individual performance may, at times be detrimental to contributions to the team effort. When accidents occur the crew is evaluated as a team - they go down together, most likely with the same fate. Sadly, this is the only consistent team performance evaluation being used today.

Developing a 'team' concept. One of the most important functions of an effective leader is to develop a 'team' concept within his/her group. By 'team' concept we mean a feeling or motivation to accomplish team goals over and above individual goals. The primary reason for using a team as opposed to an individual is to share the work so as to get the task done in a more timely and effective way. A second reason for using a team is to accomplish synergy or a higher level of performance than the best individual could perform alone.

In aviation we are trained to act as individuals rather than as members of a team. In virtually every flight test, we are judged on our individual performance. It is only very recently that some airline crews have been tested and passed (or failed) as a crew. For this reason, the team concept is somewhat difficult to teach to flight crews outside of the military.

An excellent exercise for learning the team concept is to attempt synergism in the performance of a task. Most CRM courses have exercises that can be used for this purpose. They consist of a problem, either from inside or outside of the aviation context, that each member is asked to perform individually. Then, the small group or 'team' is brought together to attempt the solution of the problem as a team. If the team scores higher than the best individual within that team, it has achieved synergism. Synergism can best be achieved when all five of the communication concepts mentioned above (Inquiry, Advocacy, Listening, Conflict Resolution and Critique) have been used by all individuals within the team.

Delegation. Another essential aspect of a good cockpit leader is an ability to delegate tasks to others, both for division of labor (task orientation) and to provide variety and experience to the other members of the crew (relationship orientation). In the cockpit, the captain is necessarily the leader primarily responsible for delegation. Although he or she gives up the specific task to other members of the crew, he or she still retains responsibility for its completion. Just as the delegator will obtain most of the benefits from the success of the mission, he or she must be willing to accept the risks involved.

Delegation begins with planning but continues as change takes place until the mission is completed. In planning with other crewmembers, the delegator should determine:

- What has to be done?
- Why?
- How well?
- When?
- In what priority?
- With what resources?
- By whom?

Delegation should be made to each according to his interests, capabilities and qualifications - not simply according to his position of authority in the cockpit. The delegator should obtain agreement among the crew for performance standards in terms of the quality and quantity expected, and the time of completion. As the mission progresses and is completed, the following principles apply to the delegator:

- Show interest in what is being done without overseeing so closely that the subordinate feels uncomfortable. Show satisfaction in a subordinate's performance front of others when it is justified.

- Assess results in a non-threatening way.

- Correct mistakes for the safe completion of the mission and to improve the performance of the subordinate - not to show why you have the position you do. Do not criticize the subordinate's failure.

- Live with acceptable differences between the job that is done on your behalf and what you estimate you would have done yourself. People are different and quality, quantity and method of work are bound to vary. Accept this principle as part of the price needed to complete the mission.

Dealing with stress

One of the most significant and universal results of stress is a reduction in verbal communication. Thus, learning to deal effectively with the stress is another important aspect of CRM training. Stress is the term used to define the body's response to the demands placed upon it by physical, physiological and/or psychological forces known as stressors. For example, stress could be imposed by an unexpected windshear encountered during an approach, recognizing low oil pressure during engine run up, losing your wallet, or cutting your finger. Stressors such as these that related to the flight itself are known as acute, or flight related, stressors. Other stressors resulting from life events are known as chronic, or non-flight related stressors. Examples of chronic stressors are financial commitments, job pressures, or family troubles. Learning to recognize and cope with both types of stress in ourselves and our fellow crew members is important to making good decisions.

CRM training method

This section is offered for instructors of CRM courses and is not necessarily needed by the student. It is offered because, although CRM can be taught largely in a classroom, such training requires a highly interactive approach which is different than many conventional courses. It cannot be taught effectively either with a manual such as this chapter or through the conventional lecture technique. The objective of this section is to give some ideas concerning the techniques that have proven effective in teaching CRM.

The recommendations for CRM training offered here were discovered through a matching of the objectives of various CRM courses with the approaches that seemed to best meet those objectives in the most economical way. Suggestions are made concerning teaching methodology, the follow-up to the course and evaluations of CRM training effectiveness.

Teaching methods

There are many different teaching methods used in CRM courses. I would suggest that certain of these methods are uniquely essential to the communication and team training aspects of CRM. Among these essential methods are interactive discussion or role playing with feedback and small group discussion to demonstrate synergism within the group, and case studies. A certain amount of student preparation prior to attending the CRM course is essential and materials should be provided for that exercise.

Small group discussions are necessary to bring about a significant amount of contribution from all members of the class. Communication styles are practiced and feedback from group members or observers can be very instructive and motivating. This method is particularly useful if reports of conclusions are made to the class as a whole. Accident reports are a good source of case study material which can be assigned for small group discussion. For example, a group could be asked to search for the use of certain communication styles by the crews involved in an accident and to determine the effectiveness of those styles.

Preliminary studying. To accomplish the CRM course objectives a considerable amount of preliminary study material should be provided to participants prior to attending the course. A minimum of 10 hours of preliminary individual study is recommended. Furthermore, participants must be required to prepare themselves with this material as measured by written tests prior to attendance in the course. The content of the study material should include accident information as motivators, the basic concepts of CRM, self-assessment tests and material on what is expected of the individual in his or her participation in the course. Preliminary study material should be written so as to motivate experienced flight crews to study them. Ideally, the results of that study should be used in the classroom later.

Role playing. A very effective method of teaching the identification of communication style is through role playing. This method is most effective when the role playing exercises are video taped for play back to the small group as well as the class as a whole. For communication training the response is often the same as reported during video taped LOFT exercises, namely, 'Do I sound like that'? Students who put forth the effort in such exercises report a great and lasting benefit.

Case studies. Cases studies can be conducted effectively by individuals and small groups, provided both report their findings to the class as a whole. They can also be used effectively, by a good teacher as illustrations during a lecture.

Video tape. The presentation of video tapes of CRM accident simulations is a very effective instructional strategy. Tapes are available of the Eastern L-1011 accident in the Everglades, the Saudi Arabian accident in Jedda and other accident simulations. Students should be asked to respond either individually to the tapes with questionnaires concerning CRM concepts or in small groups to discuss failures of the crew and ways to improve the CRM performance.

Slide-tape. A less expensive, though effective audio visual technique is to use audio tapes of cockpit voice recorded final conversations of the flight crew of the

CRM accidents augmented with slides of the cockpit and other aspects of the flight crew's environment. One such presentation has been made of the United DC-8 accident near Salt Lake City. Students should be asked to respond either interactively, through questionnaires, or in small group discussions based on the CRM principles being advanced.

Lecture. Because many CRM concepts are new to pilots, an effective, consistent introduction of the concepts is best achieved through the use of good lecturing with appropriate visual aid materials. Although some fairly effective current CRM courses avoid lectures, their students may lack consistent knowledge and behavioral change because they do not really know the basic concepts very well. Lecture/discussion delivered in an interesting way for the purpose of introducing the concepts is very important to CRM training.

Follow-up

Because of the nature of the material being taught, i.e., attitudes, mental processing and communication, there is a tendency to forget and/or revert back to one's original style very quickly. Therefore, CRM must be refreshed frequently throughout the pilot's career. Furthermore, to assure a behavioral change it is recommended that the refreshment of the behavior be conducted in LOFT training during recurrency annually. Such training must include the consistent use of CRM concepts taught in the course during the discussion prior to the LOFT as well as during debriefing following the LOFT. Finally, because of its known impact on behavior, the LOFT session should be video-taped and played back to the crew during the debriefing. The debriefing should be structured so as to bring out all important aspects of the CRM and the use of CRM in the LOFT session.

CRM course outline

The following outline is offered as an example for a two and one-half day course in CRM. The course runs from 0800 through 2200 on the first two days and 0800 through 1200 on the third. The course should be taught by two instructors working together. It is further suggested that one of the instructors should be a line captain from the organization and the other an experienced educator who is able to communicate basic behavioral concepts.

Much of the background information for the lectures can be found in this book. However, it is expected that other information specific to the aviation organization will be provided. Other material needed include accident simulation videos, team synergy exercises (e.g., the Caribbean Survival Exercise) and personality instruments to help students learn their own communication styles (e.g., Hersey and Blanchard's Strength Deployment Inventory). It is expected that this course will be adapted to fit the particular needs of the organization using it.

Day 1

0800 Arrangements

Put participants at tables in groups of five or six

0800 Introductions

0815 Lecture 1

CRM related accident data
CRM concepts and definitions
Behavioral Dimensions

0900 Introduce case study video (e.g., Portland Accident or Dryden simulation)

Show video tape of Portland accident simulation

0945 Small group discussion with handouts of accident report

Portland accident
Identify communication styles of each member of crew
How did communication styles contribute to accident?

1000 Report of small group discussion

1030 Break

1045 Lecture 2

Communication styles
Verbal behavioral categories
Listening

1115 Small group case study of accident report (e.g., Air Florida)

1200 Lunch break

1300 Lecture 3

Relationship versus Task
Personal Characteristics Inventory

1500 Coffee Break

1515 Lecture 4

Theory of the Situation
Use video simulation of Everglades accident

1545 Small group discussion of Everglades accident

Theories of the situation of Everglades accident crewmembers

1630 Reports of small group discussion

1800 Dinner

1930 Lecture 5

Developing the team concept
Synergism

2000 Small group exercise (e.g., Caribbean Survival Exercise)

Day 2

0800 Lecture 6

DECIDE model
Small group case study using accident report (e.g., Air Illinois or UAL's Salt Lake City accident)

1030 Coffee break

1100 Lecture 7

Hazardous attitudes
Poor judgment chain
Personal Characteristics Inventory

1200 Lunch

1300 Lecture 8

Leadership/followership

1330 Small group discussion

Delegation exercise

1500 Coffee break

1800 Dinner

1930 Lecture 9

Introduce role playing exercise
Have participants develop a scenario from the flight deck

2000 Small group role play on video

2030 Small group report - viewing of video and critique

Day 3

0800 Lecture 10

Strength Deployment Inventory
Stress

1000 Coffee Break

Lecture 11

Evaluation
Summary

1200 CRM Seminar concluded

Communication evaluation

An important part of communication training is the evaluation of the style of communication being used both by self as well as others. There are four ways to evaluate one's own communication behavior: written tests, peer feedback, video tape and expert observer feedback. In the case of flight crew communication evaluation all four methods are useful because each tends to confirm the accuracy of the others. The following techniques for evaluation of one's own communication style are recommended:

- *Written test:* The Person Characteristics Inventory (PCI) given at the end of this chapter shows your communication style in terms of 'expressivity'.

- *Peer feedback:* As suggested above, the critique is a useful method consisting of fellow crewmembers providing feedback concerning communication styles that you are using. These should be stated in terms of the 'behavioral dimensions' mentioned above.

- *Observer feedback:* This evaluation method consists of an expert observing and recording the communications being offered in 13 different categories. This technique has been used successfully by the KLM CRM course both during small group discussion during the CRM course as well as in the cockpit. The recorded information is later shared with the student.

- *Video tape:* Another very effective communication evaluation technique is to have small groups of students make video tapes of their discussion in the resolution of a conflict. They, and other students, can observe the communication styles being offered. This technique has been used very successfully as a LOFT exercise in the simulator as well as in the CRM classroom.

There is also a need to be able to recognize communication styles being used by others and situations that call for modifications to communication style. For example, some situations in which timing is critical require the use of an assertive style that moves toward aggressive. Recognition of other's style enables one to appropriately adjust one's own style as well.

Personal Characteristics Inventory

The following questionnaire called the 'Personal Characteristics Inventory' (PCI) is given at this point to focus the CRM training on the relationship versus task orientation which is important in multi-pilot crew decision making. Please answer the questions according to the instructions provided as honestly as you can. The scoring method and interpretation of the PCI will be provided later.

The items below inquire about what kind of a person you think you are. Each item consists of a *pair* of characteristics, with the letters A-E in between. For example:

Not at all artistic	A..B..C..D..C	Very artistic

Each pair describes contradictory characteristics -- that is, you cannot be both at the same time, such as very artistic and not at all artistic. The letters form a scale between the two extremes. You are to choose a letter which describes where you think *you* fall on the scale. For example, if you think you have no artistic ability, you would choose A. If you think you are pretty good, you should choose D. If you are only medium, you might choose C, and so forth. Circle the letter that best describes you. Be sure to answer every question.

1.	Not at all aggressive	A..B..C..D..E	Very aggressive
2.	Very whiny	A..B..C..D..E	Not at all whiny
3.	Not at all independent	A..B..C..D..E	Very independent
4.	Not at all arrogant	A..B..C..D..E	Very arrogant
5.	Not at all emotional	A..B..C..D..E	Very emotional
6.	Very submissive	A..B..C..D..E	Very dominant
7.	Very boastful	A..B..C..D..E	Not at all boastful
8.	Not at all excitable in a *major* crisis	A..B..C..D..E	Very excitable in *major* crises
9.	Very passive	A..B..C..D..E	Very active
10.	Not at all egotistical	A..B..C..D..E	Very egotistical
11.	Not at all able to devote self completely to others	A..B..C..D..E	Able to devote self completely to others
12.	Not at all spineless	A..B..C..D..E	Very spineless
13.	Very rough	A..B..C..D..E	Very gentle
14.	Not at all complaining	A..B..C..D..E	Very complaining
15.	Not at all helpful to others	A..B..C..D..E	Very helpful to others
16.	Not at all competitive	A..B..C..D..E	Very competitive
17.	Subordinates oneself to others	A..B..C..D..E	Never subordinates oneself to others
18.	Very home oriented	A..B..C..D..E	Very worldly
19.	Very greedy	A..B..C..D..E	Not at all greedy
20.	Not at all kind	A..B..C..D..E	Very kind

21. Indifferent to other's approval	A..B..C..D..E	Highly needful of other's approval
22. Very dictatorial	A..B..C..D..E	Not at all dictatorial
23. Feelings not easily hurt	A..B..C..D..E	Feelings easily hurt
24. Doesn't nag	A..B..C..D..E	Nags a lot
25. Not at all aware of feelings of others	A..B..C..D..E	Very aware of feelings of others
26. Has difficulty making decisions	A..B..C..D..E	Can make decisions easily
27. Very fussy	A..B..C..D..E	Not at all fussy
28. Gives up very easily	A..B..C..D..E	Never gives up easily
29. Very cynical	A..B..C..D..E	Not at all cynical
30. Never cries	A..B..C..D..E	Cries very easily
31. Not at all self-confident	A..B..C..D..E	Very self-confident
32. Does not look out for self, principled	A..B..C..D..E	Looks out only for self, unprincipled
33. Feels very inferior	A..B..C..D..E	Feels very superior
34. Not at all hostile	A..B..C..D..E	Very hostile
35. Not at all under-standing of others	A..B..C..D..E	Very understanding of others
36. Very cold in relations with others	A..B..C..D..E	Very warm in relations with others
37. Very servile	A..B..C..D..E	Not at all servile
38. Very little need for security	A..B..C..D..E	Very strong need for security

39. Not at all gullible	A..B..C..D..E	Very gullible
40. Goes to pieces under pressure	A..B..C..D..E	Stands up well under pressure

Scoring the PCI

To score your PCI, 14 questions are relevant. Seven are used to indicate 'instrumentality' or task orientation and seven are used to indicate 'expressivity' or relationship orientation. On the score sheet given below, write down your letter response to the question numbers indicated. The letter response should be converted to a number using the following formula: A=0, B=1, C=2, D=3, E=4. Next to the letter response, write your number response. Total your scores for Task and Relationship at the bottom.

Task Questions			Relationship Questions		
Quest #	Letter	Number	Quest #	Letter	Number
3	____	____	5	____	____
9	____	____	11	____	____
16	____	____	13	____	____
26	____	____	15	____	____
28	____	____	20	____	____
31	____	____	25	____	____
33	____	____	35	____	____
40	____	____	36	____	____
Total		____			____

Interpreting the PCI

To determine where you stand on the Task versus Relationship matrix, place an X on the position of your scores on the PCI results graph shown in Figure 6.6. If your Task score is 21 or above, you would be considered high on task orientation. If your Relationship score is 23 or higher, you would be considered high on relationship orientation. Lower scores than these would be considered low on either scale.

Your score on this test should not be considered positive or negative as far as your capability as a pilot is concerned. However, if you are very low on either scale, you might consider looking carefully at the remainder of this course to improve your thought patterns in the cockpit. In tests of professional pilots, about 90 per cent score high on task orientation and about 50 per cent score high on relationship orientation.

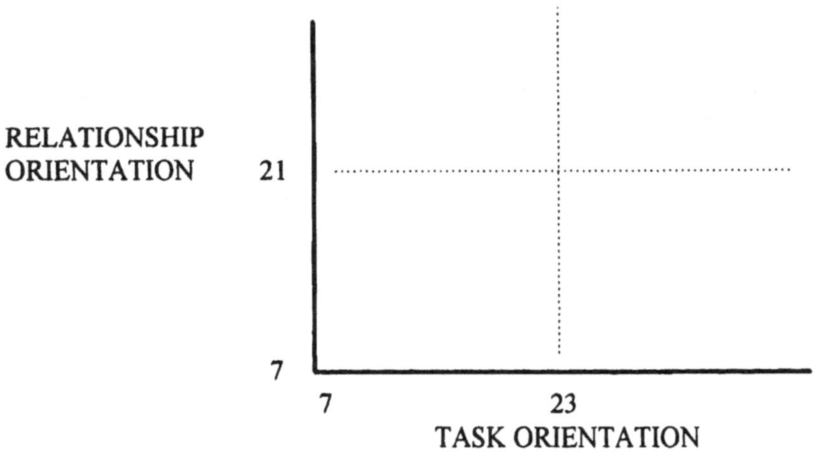

Figure 6.6 PCI result matrix for relationship versus task orientation

Conclusion

A comprehensive book on CRM has recently been published offering the whole range of perspectives from theory to assessment, to advanced cockpit CRM, to cultural problems (Wiener, Kanki and Helmreich, 1993). This book offers a great resource for those interested in the development of a CRM course. I would also offer that the proceedings of the Biennial International Symposium on Aviation Psychology mentioned earlier provide a wealth of information useful to the development of CRM courses.

Once a course has been developed the task of CRM training development is not over. The assessment of the course must provide feedback for continual modification and improvement. The target audience must be continually 'refreshed' with new CRM illustrations to make each repeat of the course meaningful. Therefore, CRM training is a never-ending process but requires sharing information across the world as new ideas are developed. The Symposium on Aviation Psychology is one place where such exchanges take place. The Ohio State University offers a repository of CRM information for course developers to gain insight.

As a result of the observation of 114 airline pilots who had taken the management attitude questionnaire, Dr. Helmreich has concluded that the effective pilot/manager is one who (Helmreich, Foushee, Benson and Russini, 1986):

- recognizes his or her own personal limitations
- recognizes his or her diminished decision capability in emergencies
- encourages other crew to question decisions
- is sensitive to personal problems of other crew
- discusses personal limitations
- verbalizes plans while acting as pilot flying
- provides training for other crew
- has a relaxed and harmonious flight deck
- adapts management style to crew and situation
- coordinates cabin crew activities

References

Aufderheide, E.H. (1987). *Aircrew Coordination Training: A Military Airlift Command workshop on human resource management in the aircraft*. Scott AFB, IL: Headquarters Military Airlift Command.

Benner, L. (1975). 'D.E.C.I.D.E. in the hazardous materials emergencies'. *Fire Journal*, 69(4), 13-18.

Berlin, J.I., Gruber, E.V., Holmes, C.W., Jensen, P.K., Lau, J.R., Mills, J. W. and O'Kane, J.M. (1982). *Pilot Judgment Training and Evaluation* (DOT/FAA/CT-82/56). Daytona Beach, FL: Embry-Riddle Aeronautical University, (NTIS No. AD-A117 508).

Blake, R.R. and Mouton, J.S. (1964). *The Managerial Grid*. Houston, TX: Gulf Publishing.

Blanchard, K. and Hersey, P. (1982). *Management of Organization Behaviors: Utilizing Human Resources*. Englewood Cliffs, NJ: Prentice Hall.

Bolman, L. (1979). 'Aviation accidents and the "Theory of the Situation"'. In: G.E. Cooper, M.D. White and J.K. Lauber (eds.) *Resource Management in the*

Cockpit. Moffett Field, CA: NASA Ames Research Center (NASA Conference Publication 2120).

Buch, G. and Diehl, A.E. (1984). An investigation of the effectiveness of pilot judgment training. *Human Factors*, **26**, 557-564.

Continental Airlines (1986). Notes from the Crew Coordination Concepts course.

Cooper, G.E., White, M.D. and Lauber, J.K. (eds.) (1979). *Resource Management in the Cockpit*. Moffett Field, CA: NASA Ames Research Center (NASA Conference Publication 2120).

Diehl, A.E. and Buch, G. (1986). 'Developing an international program to improve pilot decision making'. In: G. Buch (ed.), *Proceedings of the Pilot Decision Making Conference*, Ottawa, Ont, Canada: Transport Canada.

Diehl, A.E. and Lester, L. (1987). *Private Pilot Judgment Training in Flight School Settings*. Springfield, VA: NTIS, DOT/FAA/AM-87-6.

Diehl, A.E. (1986). 'Past and future initiatives'. In: G. Buch (ed.) *Proceedings of the Third Pilot Decision Making Conference*, Ottawa, Canada: Transport Canada.

Hart, S.G. and Bortolussi, M.R. (1983). 'Pilot errors as a source of workload'. In: R.S. Jensen (Ed) *Proceedings of the Second Symposium on Aviation Psychology*, Columbus, OH: OSU Aviation Psychology Laboratory.

Helmreich, R. (1983). 'Cockpit management attitudes'. In: R.S. Jensen (ed.) *Proceedings of the Second Symposium on Aviation Psychology*, Columbus, OH: OSU Aviation Psychology Laboratory.

Helmreich, R., Foushee, H.C., Benson, R. and Russini, W. (1985). 'Cockpit resource management: Exploring the attitude-performance linkage'. In: R.S. Jensen and J. Adrion (eds.) *Proceedings of the Third Symposium on Aviation Psychology*, Columbus, OH: OSU Aviation Psychology Laboratory.

Helmreich, R.L., Foushee, H.C., Benson, R. and Russini, W. (1986). 'Cockpit management attitudes: Exploring the attitude-performance linkage'. *Aviation, Space and Environmental Medicine*, **57**, 1198-1200.

Hersey, P. and Blanchard, K.H. (1977). *Management of Organizational Behavior: Utilizing Human Resources*. Englewood Cliffs, NJ: Prentice-Hall, Inc.

Jensen, R.S. (ed.) (1981). *Proceedings of the First Symposium on Aviation Psychology*. Columbus, OH: The Ohio State University, Aviation Psychology Laboratory, Technical Report APL-1-81.

Jensen, R.S. (ed.) (1983). *Proceedings of the Second Symposium on Aviation Psychology*. Columbus, OH: The Ohio State University, Aviation Psychology Laboratory.

Jensen, R.S. (ed.) (1987). *Proceedings of the Fourth Symposium on Aviation Psychology*. Columbus, OH: The Ohio State University, Aviation Psychology Laboratory.

Jensen, R.S. (ed.) (1989). *Proceedings of the Fifth Symposium on Aviation Psychology*. Columbus, OH: The Ohio State University, Aviation Psychology Laboratory.

Jensen, R. S. (ed.) (1991). *Proceedings of the Sixth Symposium on Aviation Psychology*. Columbus, OH: The Ohio State University, Aviation Psychology Laboratory.

Jensen, R.S. and Adrion, J. (eds.) (1985). *Proceedings of the Third Symposium on Aviation Psychology*. Columbus, OH: The Ohio State University, Aviation Psychology Laboratory.

Jensen, R.S. and Adrion, J. (1986). *Aeronautical Decision Making for Commercial Pilots*. Springfield, VA: NTIS, DOT/FAA PM-86/42.

Jensen, R.S., Adrion, J. and Maresh, J. (1986). *A Preliminary Investigation of the Application of the DECIDE Model to Aeronautical Decision Making*. Columbus, OH: OSU Advanced Flight Simulation Laboratory, Final Report AOPA 86-3.

Jensen, R.S. and Benel, R. (1977). *Judgment Evaluation and Instruction in Civil Pilot Training*. (FAA Report No FAA-RD-78-24), Savoy, IL: U. of I. Aviation Research Laboratory.

Jensen, R.S. and Neumeister, D. (eds.) (1993). *Proceedings of the Seventh Symposium on Aviation Psychology*. Columbus, OH: The Ohio State University, Aviation Psychology Laboratory.

Latene, B., Williams, K. and Harkins, S. (1979). 'Many hands make light the work: The causes and consequences of social loafing'. *Journal of Personality and Social Behavior,* **37**(6), 822-832.

Lauber, J.K. (1979). 'Resource management on the flight deck: Background and statement of the problem'. *In*: G.E. Cooper, M.D. White and J.K. Lauber (eds.) *Resource Management in the Cockpit*. Moffett Field, CA: NASA Ames Research Center (NASA Conference Publication 2120).

Lauber, J.K. (1984). 'Resource management in the cockpit'. *Air Line Pilot*, September, 20-38.

Lawton, R. (1986). 'The DECIDE: An approach to training pilots in PDM'. In: G. Buch (ed.), *Proceedings of the Third Pilot Decision Making Conference*, Ottawa, Canada: Transport Canada.

Ruffell-Smith, H.P. (1979). *A Simulator Study of the Interaction of Pilot Workload with Errors, Vigilance and Decisions* (NASA Report No. TM-78482). Moffett Field, CA: NASA Ames Research Center (NTIS No. N79-14769).

Sweeney, A.B. (1981). *Leadership: Management of Power and Obligation*. Wichita, KS: Test Systems International, Ltd.

Tannenbaum, R. and Schmidt, W.H. (1975). 'How to choose a leadership pattern'. *Harvard Business Review*, May-June.

Wiener, E.L., Kanki, B.G. and Helmreich, R.L. (1993). *Cockpit Resource Management*. New York: Academic Press.

7 Stress, fatigue and nutrition

Therefore, I tell you, do not worry about your life, what you will eat or drink; or about your body, what you will wear. Is not life more important than food, and the body more important than clothes? Look at the birds of the air; they do not sow or reap or store away in barns, and yet your heavenly Father feeds them. Are you not much more valuable than they? Who of you by worrying can add a single hour to his life?

Matthew 6:25-27

Stress, fatigue, and nutrition are interrelated factors that can have significant effects on pilot judgment. High levels of stress for an extended period of time can cause fatigue and poor eating habits. Fatigue can lead to situations causing stress and poor eating habits. Poor nutrition can lead to stress and fatigue. The problem with each of these factors is that they affect each person differently and generally only have a detrimental effect on judgment at extreme levels. Not being a medical doctor, I do not have an extended background to deal with these topics. For those who wish to look at the topics in greater detail I will direct your attention to others who have more experience than I. My credentials consist of my personal experience with each of these factors as they apply to aviation and my study of the work of others.

Stress

Dirk Van Dam, the Director of flight training at the Moody Institute of Aviation used to say that safe flying requires that while flying, the pilot should be in a mental state of 'relaxed vigilance'. Safe and efficient flying requires a heightened state of alertness at all times, but the pilot's mind should be confident and at ease. The above quote from Matthew not withstanding, the pilot is in control and is responsible for the situation while in flight. The pilot is the most important element of the flying system and the safety of any flight rests on his or her shoulders. This state of vigilance must not be given up until the engine is shut down and the chocks are in place.

Why then, you may ask, should we be concerned with stress and relaxation in flying? First, maintaining a high level of vigilance, of necessity results in a certain level of stress in the pilot. Stress is one of the most important forces affecting our ability to make logical decisions. High levels of stress over extended periods of time can cause fatigue which must be relieved through periodic rest and relaxation. While technical flying skill is somewhat immune, judgment is often adversely affected by high levels of stress. It causes us to have 'tunnel vision', or a narrow point of view, as we gather and process information. We do not see all of the information that may be in front of us and we have difficulty making choices from among alternatives. At high levels, stress can adversely affect the pilot's ability to make decisions. At extremely high levels, stress can lead to panic and even loss of motor control.

Second, scientists have found that decision making (and often the lack thereof), is one of the leading causes of stress (Janis and Mann, 1977). This is particularly true of decisions to put off hard decisions such as the decision to turn around (the 180 degree turn), to go to an alternate, or to wait out weather. The simple commitment to make a flight, whether self-imposed or placed on us by others, can cause a great deal of mental stress that can lead to all of the problems mentioned above when the weather is marginal for our skill level. Such pressure is one of the leading causes of workload in the cockpit and can cause us to fail to allocate the necessary attention to the task of flying the airplane.

Third, one of the most significant and universal results of stress is a reduction in verbal communication. As we have seen, failure of the crew to communicate properly is the leading cause of accidents in airline operations. Thus learning to deal effectively with the stress is another important aspect of CRM training.

Fourth, pilots, being human, may bring the stress built up in their lives into the cockpit. Even though the source of this stress may not be related to aviation, its effects on pilot decision making are just as strong as those stressors that are developed in the cockpit. Stress has a cumulative effect. If it persists over a long period it can severely affect our performance and health. Some medical people say that over half of our illnesses are due to stress in our lives.

The growing interest in stress in our society reflects the widespread awareness that stress is related to many physical and mental disorders, and to a large number of accidents in homes, automobiles and aircraft. In this chapter, we examine stress as it affects our lives, in general, and our flying performance, in particular. You will find tests to determine your level of stress both momentarily and in life generally. Simply recognizing the involvement of stress does not necessarily solve the problem. It is necessary to understand how to cope with it as well. Suggestions are made to help you deal with stress more effectively.

What is stress?

Stress is the term used to describe the body's nonspecific response to demands placed upon it, whether these demands are pleasant or unpleasant (Sellye, 1974). Strictly speaking, from the field of engineering where the term originates, *stress* refers to the force being applied (e.g., the weight on a bridge). In engineering terms, the word *strain* is used to refer to the response to the force being applied. In the bridge example, strain is the give in the bridge due to the weight being applied. Because of the widespread use of the word by psychologists and physiologists 'stress' to refer to human body's response, we will use that term here rather than the more technically correct term.

There are two broad categories of stress: chronic and acute. Chronic stress is the result of long-term demands placed on the body by life events both positive (e.g., marriage) and negative (e.g., divorce). Acute stress results from demands placed on the body by the task at hand (e.g., maintaining a state of high vigilance or heavy use of muscle strength). Short-term demands could be an unexpected windshear encountered on final approach, a higher than expected headwind forcing you to consider a different destination for your flight, losing your wallet, or cutting your finger.

Relationship between stress and performance

Everyone operates most effectively at some moderate level of stress (Yerkes and Dodson, 1908). The relationship between stress and performance that has been verified in numerous experiments is illustrated in Figure 7.1. At very low levels of stress, motivation and attention are minimal and, as a consequence, performance is poor. As the level of attention and motivation increase (and stress), so does performance. However, at very high levels of stress, panic ensues and performance deteriorates dramatically. Complex or unfamiliar tasks require a higher level of attention than simple or over-learned tasks and thus produce greater stress. Such tasks also are more adversely affected by increasing levels of stress than those which are over-learned or simple.

Stages in the body's reaction to stress

The human body responds to these and all other demands in three stages: first, there is an alarm reaction; then resistance; and finally, exhaustion (if the demand continues). This three-stage response is part of our primitive biological coping mechanism which would have prepared our ancestors for 'fight or flight'. In the *alarm stage*, the body recognizes the stressor and prepares for fight or flight by activating a part of the brain which stimulates the pituitary gland to release hormones. These hormones trigger the adrenal glands to pour out adrenaline. Adrenaline increases heartbeat and rate of breathing, raises blood sugar level, increases perspiration, dilates the pupils, and slows digestion. If the alarm results in *fear*, the body reacts with low blood pressure resulting in a pale face. The process results in a huge burst of energy, greater muscular strength, and better hearing and vision. In the short term solution to the immediate problem causing the fear, the production of adrenaline causes a greater level of alertness (to a certain point) which permits a greater capability to find a solution. The long term results are not harmful unless they are very severe and lasting.

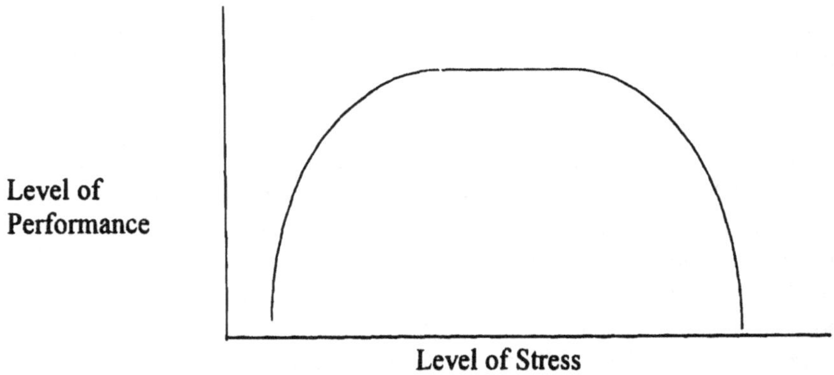

Figure 7.1 Relationship between stress and performance

The body's reaction to anger in this stage, however, is quite different. Contrary to the alarm reaction to fear, the *anger* reaction results in the secretion of nor-adrenaline which results in a physiological reaction of high blood pressure as can be seen in the red face. In some ways the stress on the body produced by anger is much more dangerous than that produced by fear. On the other hand, the effects of anger, secretion of nor-adrenaline, cause high blood pressure which in the short term does not assist in the development of a solution to an immediate problem and in the long term can be very dangerous to one's health. In aviation, anger should be avoided or minimized.

In the *resistance stage*, the body repairs any damage caused by the stress and may adapt to some stresses such as extreme cold, hard physical labor, or worries. Fortunately, most physical and emotional stressors are of brief duration and our bodies cope with the physiological demands of the stress. During our lifetime, we go through the first two stages many times. We need these response mechanisms to react to the many demands and threats of daily living.

However, if the stress continues (for example, if you were caught above clouds flying VFR or realize that you may not reach your destination because of a fuel shortage), the body will remain in a constant state of readiness for fight or retreat. It may be unable to keep up with the demands, leading to the final stage which is *exhaustion*. With exhaustion almost all control is lost as the mind is no longer able to keep a proper perspective. Sometimes pilots will *resign* themselves to their fate at this point. This reaction can be seen in the Air Florida pilots as one said to the other, 'We're going down, Larry!' The response was, 'I know!'

In flying, accidents often occur when the task requirements exceed pilot capabilities, especially when stressors such as fatigue, illness, and emotions are involved. The difference between pilot capabilities and task requirements are shown in the 'Margin of Safety' diagram shown in Figure 7.2. The margin of safety is minimal during the approach when task requirements are highest and pilot capabilities due to fatigue are reduced. If an abnormality or distraction occurs to make the task unexpectedly higher, or if the pilot's capabilities are further reduced due to strong emotions such as anger, lack of sleep, illness, alcohol, etc., an accident is risked.

The effects of personality

There is no question that personality influences the way we react to stress. Some people have personality styles that may contribute to stress related disorders. They may feel so fearful of making mistakes, of being criticized, of doing less than a perfect job that they withdraw from challenging situations or avoid confrontations, which result in feeling unfulfilled, frustrated, incompetent. As children, they may have learned that expressing feelings, such as anger, can get them into trouble. Thus they express their anger indirectly or deny it altogether.

Psychiatrists have described two personality types which have been linked with certain diseases. Type A behavior has been seen as a major cause of coronary heart disease and is characterized by a competitive, aggressive, achievement-oriented, time-dominated orientation to life. Type A people are usually unaware that their behavior creates problems for others or is detrimental to their health and well-being, since this behavior is condoned and applauded by our achievement-oriented society.

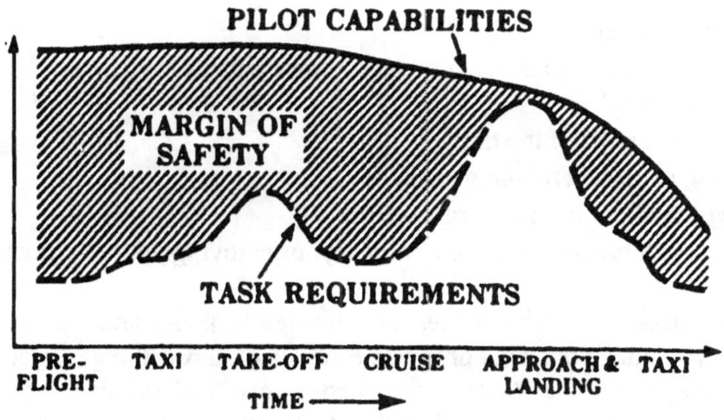

Figure 7.2 Conceptual diagram of margin of safety over the duration of a typical flight

The behavior of a Type B person, in contrast, is everything Type A people reject. Type B individuals have found a comfortable, more relaxed pace. They look at scenery with enjoyment, allow time for frequent refreshment and rest stops, really enjoy being alone or with friends and family. Type Bs work more slowly and thoughtfully, which can permit greater creativity. They allow themselves the leisure to develop more fully as people, and have a number of outside interests, activities, and friendships. Many Type Bs have plenty of drive and achievements, but time is scheduled with a calendar, not a stopwatch. If you recognize the Type A pattern in yourself, you should consider modifying your life style. Not only will it make you a safer pilot but you will live longer as well.

Chronic stress

Just as it is important that you know the symptoms of hypoxia, it is important that you know your body's specific response to high levels of stress. Unfortunately, the symptoms of chronic stress tend to be more subtle and less visible. Under high levels of stress, people often feel desperate, trapped, miserable, and often helpless but, unless they are well in tuned with their own *feelings* they are likely to miss these signs. It is important for everyone, but especially pilots, who must preflight themselves before every flight, to recognize that these may be warning signs that could be used showing that they are overly stressed to the extent that their performance could be adversely affected. Some symptoms of abnormally high levels of chronic stress include:

- rushing your speech
- completing other people's speech
- hurrying when you eat
- disliking waiting in line
- seeming never to catch up
- scheduling more activities than you have time available
- detesting 'wasting' time
- driving too fast most of the time
- often trying to do several things at once
- becoming impatient if others are too slow
- little time available for relaxation, intimacy, or enjoying your environment.

Most of us go back and forth between such hurried behavior and a more relaxed schedule. If you find yourself doing some or all of those things listed above, greater distress is not inevitable. Some people can and do live faster lives, because their bodies and minds can handle a faster pace. Others learn to adjust to a faster pace. You can learn ways to remain healthy while living faster. But the chances of distress are greater, especially if you are not aware of the dangers or do little or nothing about them.

The life stress scale. If you hope to succeed at reducing stress associated with crisis management in the air, or with your job, it is essential to begin by making a personal assessment of stress in all areas of your life. You may face major stressors such as loss of income, serious illness, death of a family member, change in residence, or birth of a baby, plus a multitude of comparatively minor positive and negative stressors. These major and minor stressors have a cumulative effect which constitutes your total stress-adaptation score which can vary from year to year. To enhance your awareness about the level and sources of stress in your life, complete the questionnaire below (Table 7.1).

Interpreting the life stress scale. The more change you have, the more likely you will suffer a decline in health. In a pilot study (Holmes and Rahe, 1967; Alkov, Borowsky and Gaynor, 1983), it was found that of those persons who reported LCUs (life change units) that totaled between 150 and 199, 37 per cent had associated health changes within a two-year period of such life crises. Of those who had between 200 and 299 LCUs, 51 per cent reported health changes, and for those with over 300 LCUs, 79 per cent had associated injuries or illnesses to report. On the average, health changes followed life crises by one year.

If you have a high LCU score, it does not necessarily mean you will get sick or have an accident. Each of us has a personal stress adaptation limitation. When we exceed this level, stress overload may lead to poor health or illness. To avoid exceeding your personal limit, learn to recognize the warning signals from your body and mind that tell you when your stress level is getting too high. When you observe warning signs, it is time to take preventative action.

Acute stress

The urgency of time drives most of us. Nowhere is this more evident than in piloting an aircraft. Time stress has its own special mental aspects. Multiple tasks must be performed simultaneously to get them done at all. Fuel remaining is directly related to time. Passenger requirements and economy of operation are often directly related to time. Demands often exceed the time available and overloading means that the stress response is dangerously aroused. Irritability, impaired judgment, hypertension, headaches, and indigestion are frequent early signs of distress and potential illness. This is of crucial concern in the field of general aviation where one person must often make all of the decisions and perform all of the tasks alone.

Each person has a fairly well-defined sense of time urgency within which he or she works effectively and gains a sense of accomplishment. Beyond this comfort zone of reasonable time pressure, deadlines threaten, time seems to run out, there is not enough recovery time for a change of pace, and the person begins to feel over-stressed. Some typical symptoms of high levels of acute stress are:

Table 7.1
Life Events Stress Profile

Write in the number of Life Change Units to the left of each event listed that you have experienced in the last year. Total your score (the sum of the scores you have written in the blanks) at the end of the questionnaire.

Life Change Units		Life Event
100	____	Death of spouse
73	____	Divorce
65	____	Marital separation
63	____	Jail term
63	____	Death of close family member
53	____	Personal injury
50	____	Marriage
47	____	Lost your job
45	____	Marital reconciliation
45	____	Retirement
44	____	Change in health of family member
40	____	Pregnancy
39	____	Sex difficulties
39	____	Gain of new family member
39	____	Business setback
38	____	Change in financial state
37	____	Death of close friend
36	____	Change to different line of work
35	____	Change in number of arguments with spouse or partner
31	____	Mortgage or loan over $10,000
30	____	Foreclosure of mortgage or loan
29	____	Change in responsibilities at work
29	____	Son or daughter leaving home
29	____	Trouble with in-laws or partner's family
28	____	Outstanding personal achievement
26	____	Spouse or partner begins or stops work
26	____	You begin or end work
25	____	Change in living conditions
24	____	Revision of personal habits
23	____	Trouble with boss or instructor
20	____	Change in work hours or conditions
20	____	Change in residence
20	____	Change in school or teaching institution
19	____	Change in recreational activities
19	____	Change in church activities
18	____	Change in social activities
17	____	Mortgage or loan less than $10,000
16	____	Change in sleeping habits
15	____	Change in number of family social events
15	____	Change in eating habits
13	____	Vacation
12	____	Christmas
11	____	Minor violations of the law

Total Life Change Units _____.

(Holmes and Rahe, 1967)

- tense voice: high pitch, high speed, air in speech
- sitting at edge of seat
- rapid breathing or holding one's breath
- pupil dilation
- sweat
- higher than normal heart beat
- dry mouth
- skin color may show red spots (flush)
- clenching teeth
- tense muscles
- white knuckles
- body movements: jerky or motionless
- eyes move rapidly or not at all
- giving the 'cold shoulder' - showing back of shoulder
- attention is difficult to focus

A frequent reaction to time pressure is juggling, attempting to cram several activities into insufficient time. The human brain seems to lack the capacity to perform many simultaneous conscious operations efficiently because one task may interfere with another. Too many pressures can lead to distress and into the poor judgment chain.

Lead-time and after-burn. Associated with any intense human activity are two necessary phases or time periods -- lead-time and after-burn. Consider, for example, a pilot facing an upgrading examination. Anticipatory stress or lead time is often useful in moderate amounts, because it prepares both your body and mind for what is about to happen. It increases sharpness and motivation, but it can also be an interference. A person pays more attention to what *may* happen rather than to what *is* happening. This reaction can cause pilots to 'get behind' their aircraft causing more time pressure and distress.

After-burn is the time needed after the test to think about results and to set the experience to rest. If there is not enough time to 'wind down' -- to relieve tensions built up during the anticipatory stage and the pressures of the review -- then the energy that surged during the experience will not be released, and the body and mind will remain stressed. A fast pace, especially if it is led by someone who needs quite a bit of lead- and after-burn time, can be a significant source of tension, stress, and disease. Pilots need some time for after-burn following a flight. The acute stress built up during a flight due to the noise, turbulence, uncertainties about the flight requires a recovery time after the pilot gets inside the base structure. Such time should be allowed by the pilot's organization and taken by the pilot for his or her health and for the safety of the operations to follow.

Acute stress and flying

In flying, there are three classes of acute stressors: physical, physiological, and psychological. *Physical stressors* include conditions associated with the environment such as temperature and humidity extremes, noise, vibration, and lack of oxygen: often encountered in flight. *Physiological stressors* include fatigue, lack of physical fitness, sleep loss, missed meals which have led to a low blood sugar level, discomfort associated with a full bladder or bowel, and disease. *Psychological stressors* are the social or emotional factors related to commitments of the flight, or they may be precipitated by mental workload such as analyzing an aircraft or navigational problem in flight. When you need to consider only one thing at a time to reach a decision, you generally will have no difficulty making a decision. In flight, however, you will frequently have to deal with many sources of information simultaneously. Frequently, decisions are based on incomplete information received within a short time period.

For example, in a cross-country flight, you may realize that you are much lower on fuel then you expected. The clouds ahead appear to be building, and there is static on the radio making it almost unreadable. You are off course and you can't seem to find a familiar ground reference point. On top of this, you failed to take a comfort stop before the flight and you now have a full bladder. The cabin heater is not functioning properly, and you are now starting to encounter turbulence. You now have many things on your mind. You begin to worry about arriving at your destination on time and missing an important appointment. You begin to worry about a forced landing and damaging the aircraft which a friend was not keen on lending you in the first place. Your palms are now beginning to become sweaty and your heart is starting to pound. You feel a growing tension, and your thinking is becoming confused and unfocused. You may give too much attention to the 'what if' questions which should be ignored. You are reaching, or have already reached the overload state. It is probable that you will begin to make poor decisions. These might include pressing on into bad weather, or over-flying good landing areas until you are almost out of fuel.

There can be plenty of stress with which to cope in the flying environment itself without adding to them the burden of your life stressors. On the other hand, your life stressors may be sufficiently great already. That one poor (initial) in-flight decision can lead to a dangerous compounding of stress-creating conditions. Stress effects are cumulative. They will eventually build to a point where the burden is intolerable unless you know how to cope.

Coping with stress

Performance can be enhanced and health can be protected by using constructive coping responses to balance stress. For example, you can simply take a five-minute break and relax. You could just take a deep breath or change positions, or say something to someone in the cockpit. Deep muscle exercises and other relaxation techniques that are well practised outside the cockpit can be done effectively inside the cockpit as well. Because the instrument approach is such an intense maneuver, the appropriate time for a relaxing minute might be after reviewing the approach procedure and before contacting the approach controller.

Some less than ideal coping strategies. Flexibility and a creative range of coping behaviors enable people to handle considerable stress. A limited coping repertoire may be harmful. For instance, if eating or drinking is the primary coping response to stress, obesity or alcoholism are likely results which present their own problems. It is a bad idea to use 'addictive' solutions to handle life stressors. Other potentially destructive responses include violence, procrastination, drug abuse, overwork, poor sleeping habits (sleep disorder), compulsive spending sprees, total withdrawal, and caustic remarks. They make a problem worse or initiate a new one rather than solve anything.

Living with distress is perhaps the least acceptable approach to dealing with stress. It may be necessary for short periods of time, but it does not promote long-term health. For instance, the intensity with which a concert pianist prepares for a performance is both invigorating and grueling. The temporary sacrifice is made to reach a goal, but does little lasting harm. Unfortunately, some people are almost addicted to stress and go to great lengths to create distressful situations. These people not only live with distress, they wallow in it, playing out 'loser' and 'poor me' life scripts or showing how tough they are.

Withdrawing from the stressor is another coping response that enables one to get away from a distressful situation when other approaches have not reduced the distress. Leaving the scene for a while by recessing a meeting or taking a walk, a day off, a nap, or a vacation can be healthy responses to restore vitality and relieve overload. If you are in a commercial flying job in which you are constantly facing a superior who places all the responsibility on you for 'no-go' decisions, it may be time to withdraw.

However, like other coping responses, this approach can be constructive or destructive, depending upon when and how it is used. It can provide a change of pace and renewal or it may merely be a means of escape and, in fact, create more stress. This avoidance technique can create further problems unless it is used as a means of reducing the present level of stress, and not a way of withdrawing from the problem or its solution.

A more permanent move, such as changing your job, place of residence, or relationship, should be considered as a last resort in most cases. When the stressor cannot be changed (for instance, the crowding and noise of city life) and to remain in the same environment would mean living with perpetual distress, withdrawal may be the most viable coping response. But usually there are alternative means by which you can moderate the negative stressors in your life without leaving the stressful situation.

Changing how you relate to the specific stressors in your life means altering the relationship between yourself and the demands. The source of your tension may be an over-demanding chief pilot who assigns you a flight late in the day when you are already fatigued and then berates you for falling behind schedule. Being more assertive and candid with your boss about your work schedule and time demands, suggesting alternatives, and understanding his scheduling problems can defuse the daily tension. Other ways of changing the stress link are:

- developing new skills
- establishing a support network of close friends
- being more diplomatic
- being tolerant of others' imperfections and your own
- broadening your perspective

Although some situations are beyond our control (for example, the erratic movement of a low pressure system into your flight path and incomplete maintenance release), this is not true for all stressors. You can change many features in your environment that adversely affect your safety, health, and self-development. Distractions or discomfort associated with heat, light, or noise can often be simply controlled. Frequently changing jobs, cities, homes, or partners could overload your coping mechanism. The ideal is to change those things you can and accept those you cannot change by tolerating them and recognizing your own limitations.

Ways to lower stress. There are several ways to lower the stress experienced when a situation cannot be changed. Some of these can be used even during a flight. Others would not be recommended in flight, but would require a more long-term solution. For short-term relief, Dr. Michael Thomas (1989) recommends a six-step pilot stress relaxation exercise to be done at least one-hour before a flight. The six steps must be done in the order given to benefit from the exercise.

1. Sit in a comfortable position. I do not recommend doing this exercise while lying down unless you wish to fall asleep.
2. Take three deep breaths. As you do this, slowly inhale while thinking of the word 'calmness', and imagine yourself exhaling 'tension'. The simple act of taking deep breaths will initiate the relaxation process.
3. Close your eyes and relax every muscle in your body while saying the words 'calmness and comfort'. As you do this you will be associating calmness and comfort with relaxation. In time this association will become very strong and you will be able to create calmness and comfort in your body by merely saying the words 'calmness and comfort' silently to yourself.

4 Scan your body for any indication of tension. When tension is found, focus your concentration on the tense muscle and imagine calmness and comfort entering it. This should be done while allowing yourself to release the tension. If you find it difficult to release a tense muscle, you can exaggerate the tension on that muscle, hold that tension for a few seconds, and release the tension while thinking the words 'calmness and comfort'.
5 Once the muscles are all calm and relaxed, keep scanning your body while saying the words 'calmness and comfort' to yourself. Do this for a few minutes so that you will be sure to remove all tension from your body.
6 When you are ready to end the exercise, place your feet flat on the floor and feel them making contact with the ground. This is very important because when pilots are stressed they tend to lack a firm, grounded, stable feeling. By physically feeling your feet making contact with the ground, you will feel more firmly planted. Once this feeling is experienced, open your eyes.

A similar form that I have found effective is called 'progressive relaxation'. In this exercise you begin your thought process at your furthest extremity, your feet. Focus all of your attention on your big toe. Think to yourself, 'my feet are completely relaxed, my feet are completely relaxed', while consciously relaxing foot muscles. As you feel relaxation in your feet work your thoughts up to your ankles and legs until all of your body is relaxed.

The third method, called 'deep breathing exercises', can reduce tension by producing a deep state of calmness and relaxation. When you breathe softly and slowly, it is difficult for your emotions to become aroused out of a tranquil state. Several disciplines include breathing exercises as part of their relaxation strategies. In yoga, 'pranayama', or control of the life force, is an important study. Since we all have to breathe anyway, controlling your breathing is a quick and easy way to help relieve tension.

For long-term relief from chronic stress changes to your approach to life may by required. First, you should approach life from the standpoint of moderation. Because much of our stress is the result of doing things and taking risks to excess, a change to a moderate life style will reduce the stress. Second, physical exercise on a regular basis is very important. In the short term vigorous exercise takes your mind off your tension and as you gain in stamina from the exercise you are better able to handle the stress. Third, communication with confidant or professional may be needed. Vocalizing the sources of stress has the effect of at least partially relieving it. It can be just as effective to talk to a friend as to a professional provided the friend is one with whom you can place your complete trust. Fourth, many have found relief from stress through religious faith and practice. This form of relaxation permits the unburdening of your tensions on

God which can be very relaxing. Which ever method you choose, it is important to your safe flight to be involved in some form of relaxation.

Nutrition

There are many books on nutrition and there are many differences of opinion among the experts in the field. On the other hand, there is general agreement on several points concerning the effect of nutrition on human performance and judgment which we will examine briefly. Some of these facts are mentioned in aviation texts but, in actual practice, they are often ignored by pilots. In commuter airline operations pilots often do not have time for a sit-down meal and they get what they can from a vending machine. Unfortunately, what they get is usually filled with fat, sugar and preservatives.

The effect of blood sugar on mood and alertness is well known. Within minutes after being hit by a sugar intake (such as a candy bar), insulin mechanisms react first decreasing blood sugar levels. Then as the sugar is digested (within 20 minutes), the blood sugar level is elevated abnormally high. About one hour after eating the candy bar, the insulin again overreacts sending sugar levels abnormally low. After about two hours the system equalizes and the body is back to normal. Each of the down swings in sugar is accompanied by downward shifts in the alertness and it may be very difficult for the person to stay awake. Performance in flight is highly questionable during these downward shifts in blood sugar. The following true story is an extreme example but illustrates the point:

> The pilot had obtained his commercial license two months before the flight. He had a practice of abstaining from solid foods on one day each week. Occasionally he would extend his fast for an extra day from the evening meal on Sunday until dinner on Tuesday evening. While fasting he would drink water or tea with lemon juice and no sugar. On the day of the flight, he had been fasting for six days. His wife had mentioned to him that he appeared to be a bit weak and should not fly, but the pilot said that he felt fine. The pilot loaded his aircraft and departed on his chartered flight. The plane was found on a bush-clad slope, where it had first contacted the trees thirty feet above the ground with wings level.
>
> A small quantity of partially digested food in the pilot's stomach was disclosed in the autopsy report. The pilot was not aware of the possible insidious effects of fasting on flying ability and furthermore that he felt in normal health. When he finally ate the candy bar, it only aggravated his already existing hypoglycemia. One of the mechanisms that regulate release of insulin into the blood is activated through the stomach. This insulin release, with its blood sugar lowering effect, precedes by half

an hour or so the increase in blood sugar that results from digestion of the food.

This may be an extreme case, but even missing two meals can cause you to be susceptible to hypoglycemia. The symptoms include light-headedness, disorientation, sweating, salivation and shivering. Poor nutrition in general increases your susceptibility to the stresses of flight. Hypoxia, hyperventilation, and fatigue are some of the stresses which missed meals and inadequate nutrition can aggravate.

Personal experience

I have had an experience with nutrition that has helped me to understand how food affects, not only long-term health, but short term performance and mood. After my younger sister died of colon cancer at the age of 37, my brother and older sister convinced me to try a very low fat diet to see its short term affect on me. It is well known that the long-term effect of a low fat diet is beneficial. I was surprised by the extent of the benefit in the short term.

The diet I tried is called 'The McDougall Program' (McDougall, 1991). It is a starch-centered diet supplemented with fruits and vegetables. Expressed negatively, it does not include *any* meat, poultry, fish, dairy products, or caffeine. The 12-day program diet claimed such major changes in such a short time that I decided to give it a try. If followed strictly, this program claims to result in reduced cholesterol levels of more than 27 points in 12 days. For people with high blood pressure, near normal can be expected and you can expect to feel much more vigorous. Recipes are offered for all foods to be eaten during the 12 days.

I decided to follow the program strictly so that I would know it's true effect. I began from a position of good health at age 51 (weight 160, cholesterol, 137; blood pressure, normal). I had stopped running the year before because of pains in my knees but still rode a bicycle and exercised on machines three times a week. I had chronic back pains whenever I became stressed that were severe enough to be treated with chiropractor manipulations on three different occasions over the previous 15 years. I noticed the following changes as I started into the diet:

- After 3 days, I no longer felt sleepy in mid-afternoon, and generally felt in an even mood all of the time.

- After 4 days, my voice felt better. I found I could sing more clearly (less mucus due to no dairy products).

- After 5 days, I was feeling so good that I tried running. I ran two miles with no pain at all.

- After 6 days, my back seemed to feel better.

- After 7 days, I noticed for the first time in several years, no arthritic pain in my knuckles. Also, an arthritic pain in my neck that I noticed every time I turned to look back while driving disappeared.

- After 8 days, I ran three miles with no pain. After this I ran regularly, and have not had any recurrence of the pain in my knees.

- After 10 days, I began to run faster than I had before because I believed that I was now operating at a lower level of risk for a heart attack. My feeling of stress was noticeably reduced.

- After 12 days, I lost about four pounds and felt much better. There were still noticeable highs and lows in alertness during the day but I was very surprised to feel so vigorous in the afternoon and the cycles were not nearly as extreme.

I found that the diet is very hard to keep when eating with family or when eating out. I was single when I started the diet and have since become married. My wife is supportive and has tried the diet in a limited way. It is also hard to find the right foods while eating out. Nevertheless, I still follow it as much as I can.

After seeing its affect on my own body, I cannot help but think it is vitally important for pilots to be careful of their nutrition habits, especially within four hours of flight. I certainly would recommend not eating meat or other fatty foods within four hours of a long flight and very little sugar and caffeine. In flight, water, fruits and vegetables should be taken to maintain a state of relaxed alertness.

In the McDougall Program, there are only three members of the 'basic food groups': starches, vegetables and fruits. A strong emphasis is placed upon the foods that contain large quantities of carbohydrates including: whole grains (rice and corn), whole grain flours (wheat, rye), whole grain pastas (wheat, soba), tubers (potatoes, yams), legumes (beans, peas), and winter squashes (acorn, hubbard). You may know that marathon runners 'load up' on carbohydrates before an event because these foods slowly release large amounts of energy that is stored. You can have the same benefits for your busy life if you load up daily. I had no weight problem but for those that do, the McDougall Program allows you to eat all the food you want of the type recommended and you can expect to lose weight by reducing your intake of certain foods high in sugar and fat such as fruits and nuts.

The McDougall Program recommends against the consumption of any alcohol. We know that alcohol is responsible for a significant percentage of our accidents. Most pilots know that alcohol can be directly linked to poor flying performance. What they do not always accept is that it takes 24 hours for the body to eliminate its effects -- not the 8 hours required by US regulation for general aviation. Also, many pilots do not realize how little alcohol it takes to have such an effect on judgment. Consider the following true story from a friend:

> The pilot and his fiancee had stopped in at a popular club for forty-five minutes. At the time of takeoff from the unattended airport, a heavy snowstorm was moving through the area. An instrument-rated pilot, he saw no problem in continuing with his plan for the flight. As they broke into the clear above the clouds (he had not filed a flight plan), the pilot proceeded to show his wife-to-be a few tricks. Shortly after a few lazy eights and a couple of stalls, the woman said she didn't feel well and wanted to land. The pilot finally contacted ATC, and asked for vectors to the nearest airport. There was a VFR airport ten miles north, reported to be in the clear. They made their way to the airport with the help of ATC and made three attempts to land in VFR conditions. This entire time the pilot felt he was in control and the wind and other factors were hampering his landing. In fact, the wind was calm. They crashed killing both people on board.
>
> *Analysis.* It was discovered through a toxicology examination after the crash that the pilot had a blood alcohol level of over 0.20 per cent. Alcohol affects judgment even before performance becomes impaired. In this case, the pilot assumed the 'bottle to throttle' rules did not apply to him. He also seems to have had an air of 'invulnerability' about him, mixed in with a 'macho' attitude. The social influence of wanting to impress his fiancee also plays on his decision making. Most importantly, though, is the fact that the alcohol consumption before a flight can result in mental confusion, disorientation, dizziness, exaggerated emotional state, disturbance of sensation, impaired perception of color, form motion, or dimensions, decreased pain sense, impaired balance, lack of muscular coordination, staggering gait and slurred speech. These conditions can appear with as little as one ounce of eighty proof liquor, eight ounces of beer, or eight ounces of wine. The pilot who believes that 'alcohol doesn't affect me' should know that one average drink, producing 0.015 per cent blood alcohol, will result in enough change in mental function to impair pilot performance -- and that does not take into account the effects of altitude.

Fatigue

Fatigue is the state of being tired which can have origins in physical activity, mental activity, or a myriad of mental and physical factors including illness, nutrition, sound, vibration, heat or cold, age, boredom, fear, anger, etc. It is a highly personal feeling (i.e., one person's reaction to a given activity could be stimulation and another's, fatigue). Also, there are wide differences between individual tolerance to various conditions before fatigue has an effect. The study of fatigue is very difficult because, with proper stimulation, people can respond to given situations and perform remarkably well despite being very tired.

Fatigue is known to be a factor in many aviation accidents and incidents. Pilots have fallen asleep on autopilot and flown until running out of fuel. Flying long enroute legs can be remarkably stimulation-free (boring) making it very difficult to stay awake. Flying numerous short flights each with a mentally draining minimums approach (such as we frequently see in commuter aviation) can be very tiring. In these conditions pilots become concerned about a reduction in mental capacity, perhaps resulting in forgetting an essential routine task such as lowering the landing gear.

The Society of Industrial Medicine has a questionnaire and chart that evaluates health from the standpoint of fatigue (Salvendy, 1987). Fatigue is evaluated from the standpoints of physical complaints, mental symptoms, and neurosensory symptoms. Table 7.2 shows the items that are evaluated in the questionnaire. On Figure 7.3 you can see how these symptoms and complaints apply to your level of fatigue and obtain recommendations for taking action. Consider the following:

> To ensure full crop coverage, the pilot often made a special pass across the ends of the fields where buildings and power lines are found. Time is saved if that hard-to-get cotton can be sprayed on normal passes by diving sharply over the power lines or buildings or, whenever possible, by flying under the lines. All crop dusters catch an occasional line, but there is a cutting leading edge on the landing gear and a cable running between the cockpit and vertical stabilizer to ward off wires. Power lines are not always the only things stretched between tall poles and the telephone lines often found hung below them are usually too tough to cut.
>
> The pilot could see the fence under the lines, but not having checked out this particular set of wires, and having a windshield covered with spray, he had no intention of going under. However, at the last second he decided to go under, and saw the pesky telephone line hanging below the main lines just seconds before he hit it.
>
> *Analysis.* This pilot found out the hard way that when you are tired, you make mistakes. It is natural to become fatigued after long hours in the air, or after a long day. This is 'acute fatigue' and can be alleviated

by a good night's sleep. Fatigue is hazardous for a number of reasons. It produces carelessness, forgetfulness, sloppiness, slow reactions, inappropriate reactions, irritability, disinterest, and the loss of timing involved in performing tasks. It erodes judgment and causes disorders of attention, distractions, channeled attention, and inattention.

Table 7.2
Subjective complaints related to fatigue

physical complaints	mental symptoms	neurosensory symptoms
• heavy feeling in head • headaches • tired feeling in entire body • tired feeling in any part of body • pain and/or tension ache in some body part • stiffness • choking • lack of saliva • yawning • cold sweat	• lightheaded and disoriented • inability to collect thoughts • inability to concentrate • irritated • excessive sleepiness • easily distracted • disloyal • failure to remember • lack of confidence and makes mistakes • anxiety attacks	• tired eyes and inability to focus • dry eyes • lethargic inaccurate movements • tottering gait • turning sore • dizziness • twitching of eyelids • ringing in ears and loss of hearing • arm and leg muscle twitches • inability to sit up straight

Source: Salvendy, 1987

Conclusion

The key to stress management is not claiming it for yourself. You can improve your internal environment by understanding, training, and shaping your mind. Thinking deeply, close to your subconscious such as your state of consciousness just as you are falling asleep or just beginning to wake. If you can teach yourself to move into this state, you will know how to put yourself into a beautiful state of relaxation. Loosening up inhibitions, overcoming our limitations, and working on developing a positive attitude toward life are all important aspects of a well-rounded, stress-controlled existence. Consider how you can use these strategies to help yourself deal with stressful concerns. These strategies will be helpful both in dealing with daily stress levels but also reduce the chance of high stress levels and panic in flight situations.

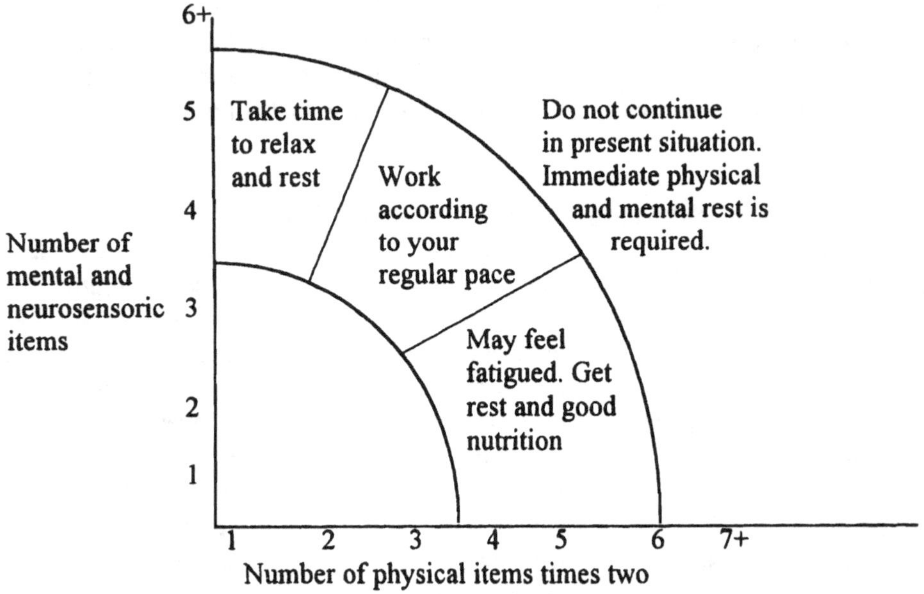

Figure 7.3 Diagnosis of fatigue by subjective complaints

Develop a positive attitude towards life. Put stressors in a favorable context. As we have said, some stress is essential to safe and effective performance. You need to remember and recognize the beneficial aspects of stress, and use the power of positive thinking. Your attitude determines whether you perceive any experience as pleasant or unpleasant.

Many of us take things too seriously and need to learn to take one thing at a time. When we worry too much we need diversion, something to put in the place of worrying (e.g., a pleasant thought). A thorough stoppage (stop negative thought patterns by shouting words like 'stop' or 'no' in the middle of an anxious series of thoughts), or a change of scene (e.g. going to a movie, reading, visiting a friend, doing something to escape from your routine) are useful ways to get out of prolonged stressful events, at least for a time. Often it takes strong mental effort to motivate oneself to do these things while experiencing distress.

Chronic stress is the product of an entire life-style. It is not just the product of an occasional crisis. Consequently, each person must learn to monitor personal internal arousal levels and find ways to relieve stress. Health can be protected by using constructive coping responses to balance stress. One of the best ways to cope with chronic stress is to use the 'Total Body Approach' or the 'Wellness Concept'. The objective of this concept is to attack the problem before it becomes serious. The total body approach takes account of six aspects of well-being.

Physiological. Minimizing or stopping activities detrimental to your health, such as drinking to excess or smoking. Developing an exercise plan including vigorous exercise three time per week can reduce both chronic stress and make you less prone to acute stress in flight.

Nutritional. Diet is extremely important to lowering stress levels. Certain types of food are harder to digest (e.g., red meat) than others (e.g., fresh fruits and vegetables) thus demanding different amounts of body resources. Some foods, such as sugar, cause fairly immediate physical and mental reactions for which you must compensate. Controlling these food intakes can improve your stress levels at critical times.

Environmental. This could mean a five-minute shutdown of immediate environmental sources of stress such as taking the phone off the hook, turning off the TV, turning on relaxing music, and/or moving away from people for a short time. In the cockpit this type of break is more of a challenge but it could be done by handing over the controls to your co-pilot for a short period of time.

Emotional. There are many relaxation programs and techniques that can have an enormous impact on your emotional well-being. The Germans have a series of water pools called Kneiff Wassertretanlage, in various locations around the countryside to help people relax. The French have two-hour lunch breaks. There are at least two relaxation courses connected with CRM training in Germany designed for both pilots and cabin attendants. There is much one can do to help relax in our fast moving society.

Spiritual. Most people find that stress is reduced through contemplation on spiritual matters. The verse from Matthew at the top of this chapter is one example. Worshipping together with people not only takes your mind away from the stresses of the workplace but it puts you in contact with people who accept you unconditionally on a different set of values which can be very stress relieving. The very nature of worship can help to re-establish a new perspective on yourself which can be helpful in making stressful decisions. Taking one day per week off for such activity can be very useful for both mental and physical health. This concept comes from Old Testament times when many practices were given to the Israelites for the sake of their physical well-being. Many of these practices are just as important today.

Life-style. Increasing health-producing behaviors such as relaxing at regular intervals. Using self-regulation and self-control information, such as appropriate time management and thought stopping (deliberately stopping yourself from thinking negatively). Accepting responsibility for your own health, by developing a stress reduction program.

Exercise. Inactivity is a serious health hazard. Exercise is one of the key elements to long life because it protects us by preventing or reversing physical illness, reduces physical tension and anxiety, and increases the quality of our lives. Prolonged vigorous aerobic exercise releases adrenaline which has the effect of making one optimistic (the 'runner's high' is caused by this natural drug). Aviation, for the most part, is a sedentary activity. We sit in an aircraft for long periods without any physical exertion. Health researchers strongly recommend that we build into our weekly activities at least three 30-minute periods of vigorous exercise.

What you do in one of these areas supports, enhances, and capitalizes on action in the other areas. For instance, poor eating habits may increase your stress level, leading to weight problems and lack of vitality. This lack of energy may slow down productivity and lead to increased pressure at home and at work to get things done. The pressure can lower your self-esteem or defensive behaviors, thus throwing your entire life-style out of balance increasing your stress to unhealthy levels.

Most pilots give their aircraft a thorough preflight, yet many forget to preflight themselves. The following I'M SAFE checklist is a short summary of the material presented in this chapter and may be useful to you as a part of your flight preparation:

ARE YOU FIT TO FLY?
THE 'I'M SAFE' CHECKLIST

I llness:	Do I have any symptoms?
M edication:	Have I been taking prescription or over-the-counter drugs?
S tress:	Am I under psychological pressure from the job? Do I have money, health or family problems? Have I had time to wind down since my last flight? Do I feel positive about this flight?
A lcohol:	Have I had anything to drink in the last *24* hours? Do I have a hang-over?
F atigue:	How much time since my last flight? Did I sleep well last night and am I adequately rested?
E ating:	Have I eaten enough of the proper foods to keep me adequately nourished during the entire flight? Have I avoided the foods that cause drowsiness?

References

Alkov, R.A., Borowsky, M.S. and Gaynor, J.A. 'Pilot error as a symptom of inadequate stress coping'. In R.S. Jensen (ed.) *Proceedings of the Second Symposium on Aviation Psychology*, Columbus, OH: The Ohio State University, Department of Aviation.

Janis, I. and Mann, L. (1977). *Decision Making: A Psychological Analysis of Conflict, Choice, and Commitment*. New York: Free Press.

McDougall, J.A. (1991). *The McDougall Program: 12 Days to Dynamic Health*. New York: Penguin Group.

Holmes, T.H. and Rahe, R.H. (1967). 'The social readjustment rating scale'. *Journal of Psychosomatic Research*, **2**, 213-218.

Salvendy, G. (ed.) (1987). *Handbook of Human Factors*. New York: John Wiley & Sons, p. 905.

Sellye, H. (1974). *Stress Without distress*. Philadelphia, PA: Lippincott.

Thomas, M. (1989). *Managing Pilot Stress*. New York: Macmillan, pp 149-150.

Yerkes, R.M. and Dodson, J.D. (1908). 'The relation of strength of stimulus to rapidity of habit-formation'. *Journal of Comparative Neurology and Psychology*, p. 459.

8 Teaching pilot judgment

To teach how to live without certainty, and yet without being paralyzed by hesitation, is perhaps the chief thing that philosophy, in our age, can still do for those who study it.

Bertrand Russell (Hogarth, 1980, p.27)

In contrast to the rest of the book, this chapter is written specifically for persons who are responsible for teaching judgment, including classroom teachers, flight instructors, simulator instructors and computer based instruction facilitators. Although it contains some judgment content material including a few flight scenarios that may be useful to pilots, the focus of this chapter is on teaching methods, not content. Teaching methods such as situational training, questionnaires for scenario building, role playing exercises, knowledge exercises, awareness exercises, diagnosis exercises, risk management exercises and attitude awareness exercises are provided.

Because teaching judgment in aviation is in its infancy, I would like to state at the outset that this is the most tentative chapter of the book. The teaching methods offered are those that have been developed and tried by the author and several other brave aviation instructors in the USA and other countries. All of us believe and hope that new and better ways to teach judgment to aviators will be discovered soon, perhaps making parts of this chapter obsolete. Nevertheless, by offering some of the methods used by the early judgment teachers and a few new ideas, this chapter will promote the development of effective judgment teaching methods.

Introduction to judgment training

Judgment is learned, primarily, from experience. All of our experiences in life mold judgment tendencies that we bring to flight. These experiences can be either useful or detrimental to good judgment in flight. On the other hand, most of our flying experiences contribute to the development of *good* judgment -- although there are some exceptions. This chapter will show teachers and facilitators how to teach pilots to use their experiences and the experiences of others to develop good flying judgment.

There is an element of truth to the saying that good judgment is learned by the cautious and the lucky over a lifetime of flying experiences. However, neither caution nor luck are the best sources of judgment training for pilots. If your students are too cautious, they may never fly near the limits of their capabilities and thus never expand their knowledge of the dangers these limits bring. On the other hand, if they rely on luck, they may develop a belief that they are immortal and invulnerable to some of the real hazards of flight.

In reality, the best way to learn judgment is to discover it from the experiences of others. Although experience is a great teacher, in aviation the experience of others is safer. It is important to realize that good decision-making skills, like any other skill required in flying airplanes, can be learned more quickly and more safely through a systematic training program than through the traditional 'trial-and-error' method.

Whenever we expose our students to new experiences we are helping them to make better decisions. Whenever we make a decision concerning a flight in their presence, we are teaching them by example how to make decisions. We cause 'simulated' emergencies (e.g., engine out, instrument and other subsystem failure and stall recovery) to occur for the purpose of teaching and testing their ability to make decisions in those circumstances. In some ways teaching factual knowledge about aviation is a form of judgment training as well. The more our students know about aviation the better they are at making decisions.

One of the best forms of judgment training is to motivate your students to be constantly learning. The Japanese call it *Kinsa* or the state of continuous learning and improvement. This concept, being taught in Total Quality Management (TQM) seminars in the USA, is one that flight instructors have emphasized for a long time in aviation, but it is worth repeating as the world has begun to use it in other fields to improve safety and effectiveness of human operations.

On the other hand, certain aspects of judgment are not being taught very well. These include intellectual processes such as risk analysis, problem solving and diagnosis and perceptual system weaknesses and biases, and they include hazardous attitudes that can override otherwise effective, rational decision making. Lacking effective teaching techniques for these aspects of judgment, the aviation system has relied on experience, usually measured in thousands of flying hours, to provide judgment in pilots. Accident statistics mentioned above suggest that it is time to discover and develop effective teaching techniques for these aspects of aviator judgment as well. We believe that it can be done through a systematic purposeful program of exposing students to simulated situations representing what they can expect in flight.

A systematic pilot judgment training model

Figure 8.1 presents a systematic model for pilot judgment training. It suggests five stages of the instructional systems design including assessment, selection, development, training and evaluation. A part of the development phase is the TECEP model (Figure 8.2) which is based on a model developed by Micheli (1972) to show how various training media and training principles apply to provide the best training for the least cost. There are several aspects of the design that should be pointed out because they are sometimes forgotten or ignored in training systems. First, the training is based on behavioral objectives. Therefore, strong early emphasis needs to be placed on the development of specific behavior objectives for the training. Second, the training is geared for the trainee level of experience (e.g., private, commercial, etc.). Third, the training begins with the instructor. Because of the nature of judgment training, a strong emphasis must be placed on instructor training to be sure it is taught properly. Fourth, there is a strong evaluation emphasis which provides feedback both to the program and to the behavioral objectives that feed into the program. Because judgment training development will continue to evolve for several years feedback is essential to provide continuous improvement.

Learning principles

Two learning principles, *practice* and *feedback*, are offered here because they have proven to be the most reliable in all human attempts at behavior change. The first, practice, was given to us in an ancient Chinese saying:

> Tell me something, and I will forget it,
> Show me something, and I will remember it,
> Let me say or do something, and I will change.

If you say something to someone else, you become committed to what you have said. This is the basis for the psychological principle called, 'cognitive dissonance' (Festinger, 1975). This principle says that a very strong force is created to believe and follow what we have publicly said we believe. Furthermore, actions that do not follow what we have said publicly, are dissonant to us and we try to avoid them. For this reason, encouraging students to voice or do something in the optimum way tends to change their behavior in that direction. Therefore, it is important for instructors to create innovative ways to get their students to voice and/or practise good judgment behavior.

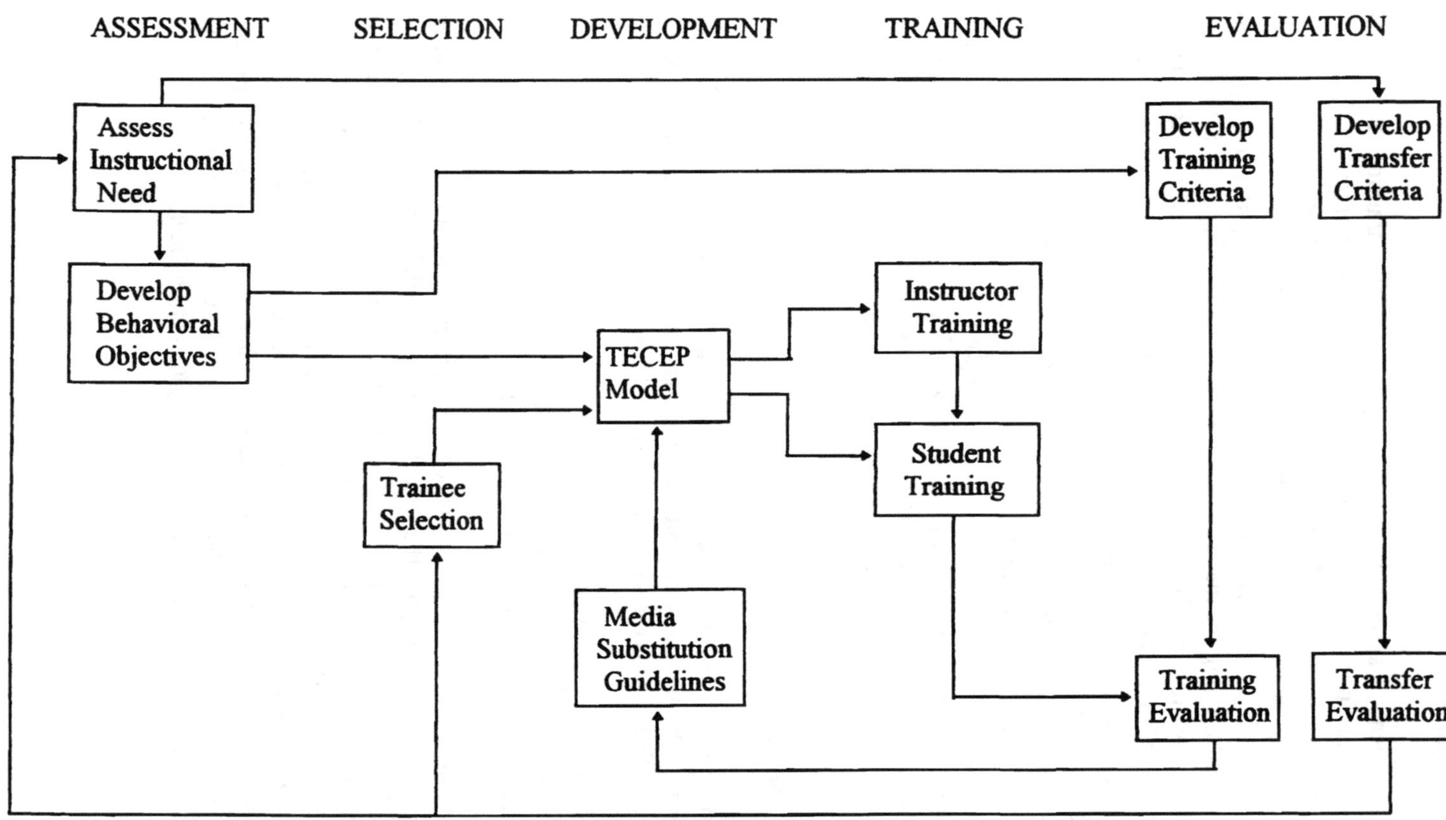

Figure 8.1 An instructional systems design for pilot judgment training.

TECEP MODEL

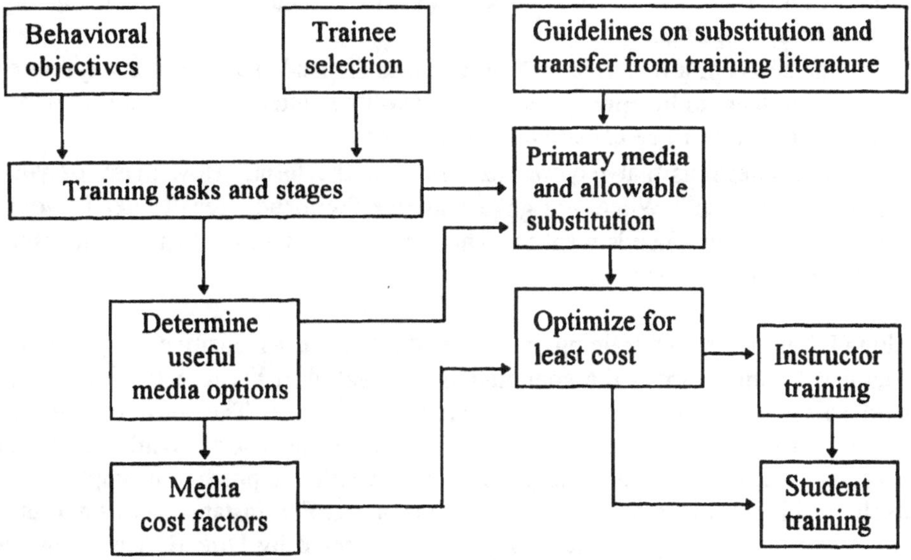

Figure 8.2 Training effectiveness and cost effectiveness model (Micheli, 1972)

The second learning principle, feedback, is offered here, not only because it is effective but because it does not work well in judgment training unless it is given in a particular form. Although it is important elsewhere in flight training, in judgment training it is essential that the student hear *why* a particular action was right or wrong not just that it was right or wrong. When a student hears only that a decision was right or wrong, he or she may consider it a judgmental form of feedback and take it personally. On the other hand, when a student hears why he or she made the right or wrong decision, meaning and ownership are ascribed to the behavior and it becomes a part of new behavior. Therefore, in judgment training, it is highly important that the instructor provide well thought-out feedback in the form of reasons why the student did something right or wrong.

In all instruction, but especially in the classroom, it is vital that the instructor provide some of the motivation to the students by using a delivery form that captures their interest. All of us have had teachers who were 'boring' and most of us have had teachers who were 'good'. The difference is seldom the subject matter being taught. Usually, the difference is in the delivery style of the teacher. This brings to mind a third principle for teaching. In lectures it is very important to *talk directly to the students and hold them accountable for the ideas and questions you raise*. This can be done in a variety of ways. Here are a few:

- Speak directly to the students near the front including eye contact, hand gestures and other body language. Body language includes walking to the back of the room and to bring in those students throughout the room using the same body language.
- Ask questions of a rhetorical nature. Because the students are not always sure you mean them to be only rhetorical, they will pay attention to your questions perchance they will be called upon to answer them.
- Ask real questions that require answers -- of the form, 'How many of you think this is true'? When you get a tentative few hands, recall later in your lecture how many hands went up. This tends to hold all of them accountable for what some of them did.

Finally, in judgment training it is important to be as positive as you can. Judgment training carries the connotation of negativism because the idea is to avoid some pleasurable activities and prevent accidents. The most interesting teaching material on judgment is in the form of accident reports. While these are necessary, whenever possible the good instructor will use positive examples and attach positive conclusions to the examples used. For instance, consider the following practical judgment teaching methods offered by Dirk E. Van Dam of Moody Aviation:

> (Judgment training should) provide a high exposure to practical situations that require choices. 'Live' work and dynamic involvement with operational problems have long been characteristic of Moody training. Apparently, this kind of process removes some of the fear of the unknown and the inhibitions which tend to limit decisions or action. It has often been said that Moody trained pilots graduate with a lack of confidence. This is probably still true. On the other hand, it seems also apparent that these same people very quickly become confident as they deal with new situations not too unlike those which were part of their training. Rather than practice the well-known, it may be that Moody has developed a process which specializes in the 'new and different'. This may not generate high confidence initially, but in the long run confidence occurs easily through successful coping. Given the two alternatives of high exposure *with* the high risk factor involved, or the minimum exposure with a relatively low risk factor, I believe that operational judgment is best advanced by the former.
>
> In both evaluation and training, the particular manner in which decisions occur may be somewhat unique to the process developed at Moody. I think we specialize in 'first time' decisions and procedures or techniques which can usually be followed without too much revision. There is evidence that procedures and habits learned in *initial* pilot training are reverted to even many years later when a pilot is faced with an emergency. Assuming that this is true, it seems vitally important that

initial exposure not only be correct in the procedures used; but that it be the kind of exposure which will return as relevant or typical in later years. Anticipated critical situations as well as those which are unexpected, when dealt with in an effective manner, produce a high degree of satisfaction and long-term learning in the trainee. I believe that through a process of instructor qualification and standardization, this type of training is characteristic of Moody Aviation. Specifically, Moody training has the following characteristics:

1 Substantial use of detailed checklist for normal and emergency procedures.
2 Considerable emphasis on critical situations which has been developed through feedback from field operations.
3 A major portion of flight time in tailwheel type aircraft even though this has a higher risk factor, particularly in student training.
4 Over 50% dual instruction in a pilot's total flight time while in the course.
5 Exposure to mountainous terrain as a context for initial training.
6 Exposure to international flying and cross country experience in the western USA during advanced training.
7 Operational experience in typical flight activity such as ATC operations.

Training method: situational training

A fourth teaching principle that should be mentioned for the classroom is *whenever you present an idea, follow the idea with an example.* From this principle follows the basic teaching method for aviator judgment: situational training or the use of examples or case studies to teach judgment content and behavior. The idea is to provide a variety of interactive flight decisional experiences to the student in a minimum of actual flight hours. This is done through case studies, role playing and simulations of decision situations faced by pilots, some of which resulted in accidents and others which resulted in incidents with positive endings. In each case it is up to the instructor to construct the scenario so that realism is felt by the student. The student is then asked to respond with thought, decisions and action. Following the action the instructor provides feedback including the reasons why the decisions were or were not appropriate.

Training media

What are the tools that instructors can use to teach judgment to aviators using the situational training method? For the most part, they are the same as those we have been using in conventional flight training, namely, the classroom, the simulator and the airplane. With modern technology becoming available to most people in aviation, a highly effective new medium for teaching judgment is computer based training (CBT). Each of these media has certain advantages and flaws. The question is how can we use each medium most effectively in judgment training?

Judgment training in the classroom

The classroom or ground school, for example, is the least expensive of the media mentioned and the only one in which many people are taught together. Here, the knowledge or judgment concepts are taught most effectively. However, the effective instructor can also use the classroom for student practicing and feedback to promote behavioral change. The classroom can be effectively used for judgment scenario introduction and review exercises both individually and in groups. Role playing exercises, which are very effective for behavioral change, can be conducted in the classroom with little or no other materials. Written tests and questionnaires for individuals can be done in the classroom.

Group study of scenarios can be very effective because they permit verbal responses by everyone. Consider the following true scenario from *Aviation Consumer*, May 15, 1980 for group study. Students could be asked to identify the poor decisions made by this pilot that led to his accident. They could first work the problem individually and then in a small group. In this example, the pilot was a 31-year-old western Texas oil field worker who had recently obtained his private license. He had 107 hours total time 3.9 hours on simulated instruments, 2.9 hours night dual and 4.9 hours night solo.

> On December 15, 1978 -- a Friday -- the pilot *worked a full shift in the oil fields* and then had a friend fly him to Midland, Texas, where he picked up his rental airplane, *flew to his home field of Colorado City*, Texas, pick up his wife and children and flew on to Memphis, Tennessee. The FBO at Midland told investigators that he rented the pilot a Piper Archer II, but gently suggested that he might want to stay overnight at home and continue to Memphis in the morning. But the pilot said *he wanted to beat a cold front* to Memphis, and added that, anyway, he '*liked to fly at night*'. He *departed a little after 5 p.m.* with topped-up tanks.
>
> His friend was waiting at Colorado City, and later told investigators it took the pilot three tries to land, since *he did so downwind*. Even so, he barely got the airplane stopped before running off the end of the 3,000-foot runway.
>
> *The pilot then loaded up his passengers*: his wife, five-year-old son, and three-year-old daughter. His friend also suggested waiting until the next morning; this was rejected. The friend asked if he could top off the

tanks, but the pilot said *he was in a hurry* and, besides, had plenty of fuel to reach Texarkana, Arkansas, where he planned to refuel. The Archer *left without filing a flight plan* or *getting an FAA weather briefing*.

The next known contact with Archer 6902J was around 8:35 p.m., when the plane is assumed to have been approaching Texarkana. The pilot called Shreveport (LA) FSS and inquired about weather at various fields in the area. Some had gone down to marginal VFR, others were approaching IFR and it was clear that a ceiling was developing from about 3,000 to 6,000 feet, with fog rolling in underneath. The Archer was apparently on top. Part of the conversation went:

FSS: You are IFR, is that correct?
02J: *It is a VOR, ah VFR.*
FSS: A VFR flight plan, sir?
02J: Hey, 10-4 on a VFR

The pilot inquired about Memphis weather and got a brighter picture there -- 3,000-foot ceilings -- expected to stay that way -- and clear of fog. (However, our calculations indicate that he could hardly make it to Memphis with the fuel remaining.) The pilot thanked the briefer and *left the frequency* at about 8:45 p.m.

He was not heard from again until about 10:15 p.m., when Shreveport heard weak *transmissions from an airplane 'in trouble'*. It took some minutes while controllers and FSS staffers coordinated, trying all channels and asking for relays from airliners, before Fort Worth Center established contact with 6902J.

When identified on radar at about 10:29, the plane was still about 25 miles west of Texarkana Municipal Airport; the pilot had apparently *flown in circles for well over an hour* since talking to the FSS. The pilot told the controller, 'I'm in trouble, I'm in the clouds, I can't see and I'm lost'. The controller issued vectors for Texarkana. The pilot mentioned having about 15 minutes of fuel remaining; the controller reminded him to use his most economical fuel setting. The controller queried, 'You are IFR qualified'? The pilot replied, 'Negatory, Negative'.

There was a handoff and at 10:34 the new controller asked 'How much time you got left'? The pilot responded, 'Uh, I'm ... I'm ... It's hard to say. It's showing empty now. I'm saying about 10 minutes of fuel left' (Archer capacity: 50 gal. total, 48 usable).

Weather at the airport was: Indefinite ceiling 200 feet, sky obscured, visibility one-half mile in fog. The lowest IFR minima at the field were 200 feet and a half-mile. Runway 22 had an ILS, but to use it the flight would have to go past where the localizer (back course) or the VOR approach to runway 4. The controllers chose to issue vectors for the straightest possible course to runway 4.

For a time, 6902J was above a cloud deck at 5,300 feet. This conversation ensued:

Center: How much flying experience do you have?
02J: Very little. Uh, I just got my private license and I'm still a student.
Center: Okay, 02J, have you had any instrument at all?
02J: Uh, *just on private training.*

Later, the controller gave a little pep talk: 'Okay, 02J, we're gonna have to go through these clouds here pretty soon now, so get used to the...get your head inside the cockpit if possible and get used to the artificial horizon again. It's just like you did in the training'.

And so began the approach, the controller giving vectors and keeping up spirits, the young pilot complying precisely. In this manner, the plane was guided right down the runway, but descending as low as 650 feet MSL (250 AGL), the pilot could not see the runway lights. The controller was starting instructions for a missed approach, when the pilot reported, 'My wife said she seen the lights'. The pilot attempted to turn around and come back from north of the field. Responding to a heading instruction, he said, 'Roger, let me just, up just a little bit. I'm at 400 feet and I just seen a house'.

The controller (amazingly) kept his cool: 'all right, 02J, at that altitude you should be below...there's nothing, uh, no restrictions on your approach from the north...uh, remember the field elevation is 399 feet'.

The last conversation:

02J: Roger, now give me the heading.
Center: Okay, heading 160, 160, 02J.
02J: Roger, just a second. Uh, give me a heading now.
I'm hurting. I'm out of gas.
Center: Okay, heading 130, heading 130, you are one mile from the airport, heading 130.

The pilot did not answer. The plane had crashed, nose steeply down, wing low, into heavy trees. Only the girl survived, with serious injuries.

In this accident, there are at least 15 separate and pronounced errors (in italics) in judgment in an unbroken chain. The first poor decision was to fly under fatigue -- this tragic accident shows how fatigue can affect your judgment. Perhaps our best defense against a chain of poor decisions is to develop the sensitivity to recognize when a poor judgment chain is developing. This type of sensitivity, which is a characteristic of knowledgeable, experienced, expert pilots, can be learned with little flying experience through systematic judgment training using examples such as this.

Looking back at the example just given, classroom students can be asked to identify the pilot's judgment errors and to indicate the type of judgment errors they each represent (e.g., rational, motivational and which hazardous attitude). Also, they could be asked to chart the development of the poor judgment chain and identify ways that the chain could have been broken to avoid the accident, perhaps using the antidotes.

Judgment training on computer

It is my belief that in the next few years, the computer will become the most effective device in teaching judgment, not only to pilots, but to a host of other professions in which people make frequent risky decisions. The advantages of the computer over other forms for judgment training may not be obvious but they are pervasive. First, the computer can be programmed to be very flexible, accommodating all types of scenarios including pressures from the job and from passengers. Second, the computer can be 'flown' permitting an element of control and forcing decisions in a timely way. Third, the computer can be programmed to provide feedback concerning the value of the decisions made. Fourth, the pilot does not get the same feeling of being 'put down' when it is the computer talking that he gets when the criticism comes from a human flight instructor. Fifth, the results of the student's performance can be stored on computer as a permanent record, if desired, to show progress over time. Sixth, the computer could even be used to assist in preflight decision making by comparing a pilot's experience level with the current flight plan and weather -- thus providing objective guidance concerning the risks and advisability for the proposed flight.

For the computer to be effective, someone must develop the software including judgment scenarios for teaching and evaluating aviator performance. However, as in any software development effort, this could be expensive. At least one effort to program a computer with judgment training scenarios has begun at Bell Helicopter Textron (Fox, 1993). As in any judgment training, computerized training requires a large pool of scenarios because each time a student uses it, whether for training or for testing, he or she should encounter a new scenario.

Judgment training in the flight simulator

Another medium that is highly effective for teaching aviator judgment is the flight simulator. There are a whole host of flight simulators from inexpensive PC computer software packages to very expensive motion-based devices. Their effectiveness depends greatly upon the subject matter being taught and the effectiveness of the instructor. In the case of judgment material, the type of simulator does not mean nearly as much as the effectiveness of the instructor. Very effective judgment training can be accomplished in low-cost simulators. However, little judgment training will be accomplished even in highly sophisticated simulators if the instructor does not prepare the scenario well and provide appropriate feedback.

What can an instructor do to provide meaningful, behavior changing judgment training in a flight simulator? The answer is, almost everything. Judgment training does not happen when a student flies a simulator 'solo'. Instead, the instructor must teach by playing the part of a drama director. The stage must be set, mostly with words in the form of a written scenario which the student reads prior to arrival for training or they could be offered orally after the student arrives for training. However it is done, it must be believable, and the best way to make it believable is to use real events that have happened to pilots.

The flight itself should take the form of a LOFT (Line-Oriented Flight Training) scenario as it is called in airline training operations. That is, the flight should be represented as close to a real airplane flight as possible. The instructor should act as air traffic control and create the realism of actual flight for the student by his or her actions. Judgment assessment is taken through observation of student activities during the flight. Video taping of the student's performance during the exercise would be very useful for this purpose. Feedback to the student should wait until after the scenario is complete.

Judgment feedback should be offered in a non-judgmental way similar to any feedback concerning the performance of the student. It is most important that the instructor knows and carefully explains *why* the student acted rightly or wrongly in the scenario. A highly effective feedback technique can be accomplished by asking the student for his or her thought patterns while viewing the video tape of the scenario. Verbalizing the thought patterns and reasoning behind decisions can be a highly effective method for causing permanent behavior change.

Judgment training in the airplane

In some ways the airplane is the least effective medium for teaching aviator judgment. In most cases, the scenario must be interrupted before its conclusion to avoid a safety risk. It is considerably more expensive to conduct judgment training in an airplane than it is in ground devices. Verbal communication is more difficult because of the noise level, and misunderstandings between instructor and student are more frequent. Nevertheless, there is no more meaningful place to teach than the airplane and the likelihood of permanent behavior change for lessons learned in the airplane are higher than in any device on the ground. The good judgment instructor will show how to make decisions by his or her example carefully explaining why decisions were made. Also, the student is given responsibility for decisions as soon as they show that they can do so safely.

Using video taping

Virtually, all professional athletes have taken advantage of video taping technology to show themselves what they are doing wrong so that they can see how to improve. Students naturally resist the corrections offered verbally by instructors. The instructor may not see the action quite right or, more likely, the student may perceive that he didn't see it right and either an argument follows or the student reduces his trust in the instructor. Therefore, a video tape of a student performing flight maneuvers, especially those that involve complex decision making can be a highly effective teaching tool. Video taping can be done in the classroom as a part of group activities or role playing in which decision making is to take place. Viewing oneself in such exercises provides powerful visual and auditory feedback to show how one can improve communication and problem solving techniques.

The simulator used for judgment training does not need to have high fidelity to be effective. For example, as we have seen in Chapter 5, Jensen, Adrion and Maresh (1986) did a judgment training experiment in which they used the DECIDE model for teaching judgment to aviators. The data were gathered, in part, by video taping the subject during performance of a cross-country flight in a Frasca fixed base flight simulator with a Silicon Graphics visual display for outside the window scenes. Three 'changes' were programmed into the flight to cause the subject eventually to turn around and head home. The object of the experiment was to determine as much as we could about the thought processes of pilots as they made the decisions. These data were obtained in an extensive, systematic debriefing of the subjects immediately following the flight while viewing the video tape of their own performance. Not only did the relatively low quality simulation and video yield excellent results of pilot thought patterns, but every subject reported that this had been the best flight instruction they had ever received. Even though there was not much on the video tape for anyone else to see, viewed immediately following the experience stimulated the subjects to recall in vivid detail the thinking process that they used in making their decisions.

Another use for video tapes in training involves the use of 'trigger tapes' in a classroom setting. The FAA's 'Back to Basics' program has developed several highly effective video tapes each presenting a scenario of a potential accident situation. A classroom instructor using these tapes could ask students to view a tape and note particular thought patterns in the mind of the pilot that may lead to good and poor decisions. These could be done both individually and in small groups followed by presentations to the class as a whole. These discussions could be used to stimulate discussion and learning of judgment principles.

Judgment scenario development

The development of realistic aviator judgment scenarios is one of the most challenging tasks to the instructor using situational training methods. Every judgment instructor should have a cadre of judgment scenarios to serve as illustrations and teaching tools for all occasions and all student types. This book purposely presents many scenarios that I have found helpful in teaching judgment in the classroom setting. Good instructors can use these scenarios both to gain student attention and to teach judgment principles. Therefore, a constant task of the instructor is to read and listen for new judgment scenarios. The following criteria are offered as guidelines for judgment scenario development:

1 The scenarios should be based on true events in aviation.
2 The scenarios should be varied to serve as illustrations for each decision aspect being taught.
3 Scenarios should be offered representing the level and type of flying experience of the students.
4 Although they are based on real events in aviation, scenarios can be embellished to emphasize a particular point or to make them more interesting and motivating to the student.
5 Scenarios must not be repeated for the same students. This is especially true for examinations used to test for judgment capabilities.
6 Although much of judgment teaching is aimed at avoiding negative situations, not all scenarios should be negative. Seek out scenarios with positive results that also illustrate the points to be made. Another way to show a positive result is to change some aspects of the scenarios found to show positive results rather than negative. Finally, students should always be asked to show how the players in the scenarios could have caused a positive result.

Some sources of scenarios are as follows:

- Your own flying experience.
- Stories told by your flight instructors, fellow pilots and students.
- Stories from popular aviation magazines (e.g., 'I Learned About Flying From That', *Flying;* 'Never Again', *AOPA Pilot*)
- NTSB reports of accidents
- Aviation Safety Reporting System (ASRS) and other confidential incident reporting systems
- Have students write up their own scenarios from their experience or that of other pilots (see Pilot Judgment Questionnaire below).

The development of judgment scenarios is encouraged as a useful exercise to provide the continuous learning experience necessary to achieve expertise in aviators. The following questionnaire is a part of a longer version that we have used at the Ohio State University for several years to encourage students to think about judgment and to gain access to more scenarios for instruction. It could be used or adapted to fit your needs.

PILOT JUDGMENT QUESTIONNAIRE

1. Have you ever made a flying decision which, upon reflection, you would have considered crazy or bad judgment and you survived more because of luck than good sense?

   Yes _____ No ______

2. What types of risky (less than 100% certainty of success) choices have you faced in your occupational flying career?

   Weather_____ Mechanical______ Electrical______ Aircraft Loading _____

   Aircraft Separation (ATC)_____ Runway Conditions______ Other ______

   ____

3. Which of the following pressures have you faced in choosing to make a flight in marginally safe conditions?

   Duty______ Economic______ Peers______ Adventure (Challenge)______

   Superiors (Job)______ Social (Friends and Family)______ Other _______

   ____

4. Briefly describe a difficult decision situation from your own flying experience. What were your decision alternatives? What were the safety factors making this a decision worth thinking about? What pressures did you feel to make or continue the flight in spite of the safety factors?

   ____

   ____

   ____

   ____

5. Optional: What was your decision? Briefly describe:

   ____

____

____

____

Teaching rational decision making

With practice and experience, people gradually learn heuristics or short cuts to get through overwhelming amounts of data found in some situations so that they can quickly make decisions with little effort. Rational expert decision makers are not necessarily lazy but they always seek the shortest route to completion of a tedious task.

Ideas developed in other fields for teaching rational short cuts can also be used in aviation. As mentioned in our detailed model (Chapter 2) the areas of rational decision making that must be taught are vigilance, problem recognition, problem diagnosis, alternative identification and risk assessment. For example the following scenario was developed and used in an experiment to test for aviator problem solving skills (Flathers *et al.* 1981):

> You are on a day trip from Augusta, ME to Lebanon, NH. You fly out of Augusta at 9:00 a.m., cleared Victor 39 to Neets Intersection, Victor 496 to Lebanon. You climb to a cruising altitude of 6,000 feet. After fifteen minutes of routine IMC flying, your instruments indicate a level flight attitude, an increase in airspeed and steadily decreasing altitude.

The solution is that the vacuum pump has failed causing the attitude indicator to slowly drift downward. The pilot tries to keep the attitude horizon level causing the airplane to slowing descend and gain airspeed. Students could be asked to work on problems such as this either individually or in small groups and report their finding to the class as a whole. Feedback from the instructor concerning the acceptability of the report concludes the training session.

The DECIDE model

As described earlier, the DECIDE technique can be used to systematically organize pilot thoughts and prevent overlooking facts that may be important. To teach pilots to use the DECIDE technique, first familiarize them with the meaning of each of the terms. Then offer scenarios that permit the students to fill in each element of the model as changes or new information is made known. Use the generic chart shown in Table 8.1 for a series of work assignments for students to fill out and discuss based on accident scenarios. Show the students how to recognize the pattern (chain) of poor decisions being made. Then show them how they can systematically use the technique when they are faced with difficult decisions.

This model could be used to analyze the thought processes taking place in any accident in which sufficient communication and other behavioral data is available to answer the questions in the chart. Through such analyses the reasons behind many 'pilot error' accidents might be found and the students will be enlightened. Furthermore, the use of this technique through several accident scenarios as well as successful outcomes is an effective way to learn to structure one's thought processes and make better decisions when faced with uncertain situations. This case study technique was developed by Russ Lawton of the AOPA Air Safety Foundation (Jensen, Adrion and Maresh, 1986).

Table 8.1
Generic DECIDE model chart

CHANGE	D E C I D	ACTION	E
1 __________	_ _ _ _ _	________________	_
2 __________	_ _ _ _ _	________________	_
3 __________	_ _ _ _ _	________________	_
4 __________	_ _ _ _ _	________________	_
5 __________	_ _ _ _ _	________________	_
6 __________	_ _ _ _ _	________________	_
7 __________	_ _ _ _ _	________________	_

Teaching pilots to avoid hazardous attitudes

As we saw in Chapter 5, the motivation or attitude component of aviator judgment has proven to be highly responsive to training. What are some effective ways to teach proper safe attitudes in aviation? Pilots must first be aware of their hazardous attitudes and then they must be motivated away from those attitudes. Both of these goals can be met through the introduction and review of various scenarios. The remainder of this chapter is a guide to help your students identify and understand the five hazardous attitudes as defined in Chapter 4 and to see how they can influence a pilot's reaction to situations requiring judgment. These true flying experiences are presented as examples of showing how attitude judgment can be taught in the classroom. As you recall, the five hazardous attitudes are:

ANTI-AUTHORITY	'The Regs are for someone else'.
IMPULSIVITY	'I must act now -- there's no time'.
INVULNERABILITY	'It won't happen to me'.
MACHO	'I'll show you. I can do it'.
RESIGNATION	'What's the use'?

The following hazardous attitude exercises are based on true flying situations involving air-taxi and commercial pilots and managers. At the end of each situation, you will be asked to select the alternative which best illustrates the reactions of a pilot who has a particular hazardous attitude or response pattern. After you select what you feel is the best alternative, look immediately at the next page for the proper response. This page will tell you if your answer is correct. If you answered correctly, go on to the next situation. If you answered incorrectly, you will be told why. Then go back to the situation and select another alternative.

THE ANTI-AUTHORITY HAZARDOUS ATTITUDE

From the five choices following the situation presentation, pick the ONE choice that is the best example of an anti-authority hazardous attitude. REMEMBER -- if you did not choose the correct answer, select another until you choose the correct one.

> *Situation 1:* During a training flight in a medium twin, the instructor pilot fails the inboard engine while in a 60 degree steep turn. After a shaky recovery you continue the training. Which of the following alternatives best illustrates the ANTI-AUTHORITY reaction?

a. You think to yourself, 'if the Feds could see us now!'
b. You are sure the instructor pilot knew what he was doing.
c. You can make these airplanes do a lot more than the manual says.
d. Nothing has gone wrong in training before.
e. You need to get as much training as you can, as quickly as possible.

RESPONSE LIST 1: THE ANTI-AUTHORITY HAZARDOUS ATTITUDE

Alternative a: Correct. 'The Regs are for someone else' attitude is portrayed by this statement. You are thinking in an anti-authority manner.
Alternative b: Here you are assuming someone else has the responsibility for you. This is the resignation hazardous attitude.
Alternative c: No. This is the macho attitude, where you get to show what you can do.
Alternative d: This is the idea that nothing would happen to you, which implies you are invulnerable.
Alternative e: Here you are saying that this possibility couldn't happen to you, as if you are invulnerable.

A GOOD JUDGMENT THOUGHT: Recognize and discuss any potentially dangerous situations that occur during training.

THE IMPULSIVITY HAZARDOUS ATTITUDE

From the five choices following the situation presentation, pick the ONE choice that is the best example of an impulsivity hazardous attitude. REMEMBER - if you did not choose the correct answer, select another until you choose the correct one.

Situation 2: The owner of the jet you are to fly tonight, is anxious to get to his destination to make a presentation at a convention. The tops of the level V thunderstorms were reported to be 5,000 feet above your aircraft's service ceiling. Which of the following alternatives best illustrates the IMPULSIVITY reaction?

a. You are sure you can push the aircraft up over the tops. The service ceiling is a conservative guess and you're light anyway.
b. You can't change the weather, so you might as well go.
c. You just have to show this guy you can get him there.
d. A little storm won't stop you.
e. You want to hurry and get going, before things get worse.

RESPONSE LIST 2: THE IMPULSIVITY HAZARDOUS ATTITUDE

Alternative a: No, this exhibits the anti-authority attitude by disregarding the flight manual's recommendations.
Alternative b: Once again you are resigning yourself to the fact that you cannot control the situation (weather).
Alternative c: You are being macho in your thinking when you want to 'show-off' your skills in this manner.

Alternative d: This kind of attitude reflects the hazardous thought of invulnerability. Go back to Situation 1 and select another alternative.

Alternative e: You're correct. This is an example of the impulsivity hazardous attitude by 'acting now, there's no time'.

A GOOD JUDGMENT THOUGHT: Base your decisions on fact, not fantasy.

THE INVULNERABILITY HAZARDOUS ATTITUDE

From the five choices following the situation presentation, pick the ONE choice that is the best example of an invulnerability hazardous attitude. REMEMBER - if you did not choose the correct answer, select another until you choose the correct one.

Situation 3: While pre-flighting the pressurization system, you discover that the rate-control is inoperative. Knowing that you can manually control the cabin pressure, you opt to disregard this discrepancy and depart on your trip. You'll just have to handle the system yourself. Which of the following alternatives best illustrates the INVULNERABILITY reaction?

a. It's too late to fix it now.
b. You can handle a little problem like this.
c. What is the worst that could happen?
d. You certainly didn't break it, it isn't your fault.
e. You barely need the pressurization at the filed altitude anyway, you can hedge on the regs.

RESPONSE LIST 3: THE INVULNERABILITY HAZARDOUS ATTITUDE

Alternative a: This is the attribute of impulsivity, where 'there is no time'.

Alternative b: Wrong. This is the macho attitude showing through.

Alternative c: Right! Nothing bad is going to happen to you because of your invulnerability. Go on to the next hazardous attitude.

Alternative d: No. This is the resignation syndrome, where someone else is responsible.

Alternative e: This is denying the regulations in an anti-authority pattern.

A GOOD JUDGMENT THOUGHT: Opt for operating systems.

THE MACHO HAZARDOUS ATTITUDE

From the five choices following the situation presentation, pick the ONE choice that is the best example of a macho hazardous attitude. REMEMBER - if you did not choose the correct answer, select another until you choose the correct one.

> *Situation 4:* On an IFR flight plan you emerge from a cloud to find yourself within 300 feet of a helicopter. Which of the following alternatives best illustrates the MACHO reaction?

a. You're not too concerned, everything will be all right.
b. You should fly a little closer, just to show him...
c. It's not your responsibility to know he was there.
d. You quickly turn away and dive, to avoid a collision.
e. With events like this, why should you ever follow the rules?

RESPONSE LIST 4: THE MACHO HAZARDOUS ATTITUDE

Alternative a: No. This is the attitude that you have never had problems before and it would never happen to you.
Alternative b: That's it. Macho hazardous thoughts include 'I'll show you'.
Alternative c: Taking for granted that you are not in control is the hazardous attitude of resignation.
Alternative d: This is acting on impulse. 'I must act now, there's no time'.
Alternative e: An attitude that the rules do not have to be followed shows the anti-authority hazardous thought pattern.

A GOOD JUDGMENT THOUGHT: Fly defensively.

THE RESIGNATION HAZARDOUS ATTITUDE

From the five choices following the situation presentation, pick the ONE choice that is the best example of a resignation hazardous attitude. REMEMBER - if you did not choose the correct answer, select another until you choose the correct one.

> *Situation 5:* Your co-pilot shows up for duty and his behavior is somewhat out of the ordinary. You know he has had a cold and when questioned he said maybe it was the antihistamine he took the night before. Although your aircraft requires a crew of two, you decide to ignore your co-worker's drowsiness and inattention. Which of the following alternatives best illustrates the RESIGNATION reaction?

a. You could fly this jet by yourself anyway.

b. You two have flown together many times, and everything has always worked out fine.
c. What else can you do? He was assigned this flight, too.
d. He might not be within the regs, but he says he's okay and that's what counts.
e. There's really no time to call another co-pilot now.

RESPONSE LIST 5: THE RESIGNATION HAZARDOUS ATTITUDE

Alternative a: This 'I can do it' attitude is one of the macho trait.
Alternative b: 'Nothing could happen to you' is the invulnerability attitude.
Alternative c: Absolutely. When you feel as if it is out of your control, you are exhibiting the resignation attitude.
Alternative d: When the rules and regulations do not apply to you, this is the anti-authority hazardous attitude.
Alternative e: No. This is the 'there's no time' component of the impulsivity hazardous attitude.

Antidotes for hazardous attitudes

After working with the five hazardous attitudes, you should be more aware of, and alert to, these attributes in your own thinking. This is an important first step in eliminating them from your decisions. We now introduce ways to counteract the hazardous attitudes with positive responses to each situation.

Since you cannot think about two things at the same time, one way to keep from thinking in a hazardous manner is consciously to alter your pattern of thought. By telling yourself something divergent from the hazardous thought, you are, in effect, 'taking an antidote' to counteract that hazardous thought pattern. You remove a hazardous thought by substituting the antidote or proper thought process. Thus, if you discover yourself thinking, 'It won't happen to me', mentally tell yourself, 'That is a hazardous attitude'. Recognize the hazardous attitude, correctly label it and then say its antidote to yourself. To do this, you must memorize the antidotes for each of the hazardous attitudes. Know them so well that they will automatically come to mind when you need them.

Hazardous attitude and antidote identification

Each of the following scenarios is a description of a flight situation and, more importantly, the thought pattern running through the pilot's head. Each situation contains several examples of hazardous attitudes. After reading the situation, identify as many hazardous attitudes as you can. Underline and number each sentence that contains a hazardous attitude. Then write the name of the hazardous attitude and the antidote in the space provided at the end of the page.

THE FIVE ANTIDOTES

HAZARDOUS ATTITUDE	**ANTIDOTE**
ANTI-AUTHORITY 'The Regs are for someone else'.	'Follow the rules. They could keep you out of danger'.
IMPULSIVITY 'I must act now, there's no time'.	'Not so fast. Think first'.
INVULNERABILITY 'It won't happen to me'.	'It could happen this time'.
MACHO 'I'll show you. I can do it'.	'Showing off is foolish'.
RESIGNATION 'What's the use'?	'I'm not helpless, I can make a difference'.

Check your responses with those identified in the key *after you have completed the scenario*. Your responses should closely match the hazardous attitude responses found in the key. Although some situations can be interpreted in more than one way, your answers should agree with a majority of the hazardous attitudes and you should be able to write the antidotes in your responses.

> *Situation 1:* The young pilot is 'building time' flying freight at night with the hopes of someday getting on with one of the major airlines. (1) He figures that he will be so good after a year of this type of flying that he will be able to bypass the commuters and proceed straight to his choice of airlines. He is a college graduate with a degree in engineering, though his grades were mediocre.
>
> His nightly route takes him to five major airports in the midwest. This particular December night, the whole area has low ceilings and moderate mixed icing forecast and reported. The aircraft he usually flies is down for maintenance, so he is in a model that he has minimum time in. (2) This really is no concern for him as he knows he can fly it well, regardless of the weather.

It is cold and windy on the ramp, with blowing snow. (3) The pilot performs a brief pre-flight and thinks to himself, 'this is a much nicer plane than the other, I'm sure everything is all right'. He did notice some slush build-up in the wheel wells, but didn't clean them out because they would just get filled-up again anyway. (4) There really isn't anything he can do about the snow on the ramp and his cargo was arriving. (5) He needs to get moving.

Though the snow plows have been working round-the-clock, the wind is such that the taxi-ways and runways are still snow covered. The blowing snow makes it difficult to see. As he taxis out, he discovers the defroster is inoperative. (6) He thinks to himself, 'boy, these guys should take better care of this equipment'.

He is cleared for takeoff and as he takes the active, he looks around the cockpit for his flashlight. As he begins his takeoff roll, he remembers he left it in the hanger. (7) 'Oh well', he thinks, 'nothing is going to happen to the electrical system. It's a dumb rule to require a flashlight on board, especially with redundant electrical systems'. (8) He proceeds into the clouds.

Sentence	Hazardous Attitude	Antidote
1	________________	____________________________
2	________________	____________________________
3	________________	____________________________
4	________________	____________________________
5	________________	____________________________
6	________________	____________________________
7	________________	____________________________
8	________________	____________________________

Key to Situation 1

Compare your responses in Situation 1 with those given below. Remember, your responses may not be identical to those in this key. Still, you should have identified most of the hazardous thoughts indicated.

The young pilot is 'building time' flying freight at night with the hopes of someday getting on with one of the major airlines. He figures that (1) *he will be so good after a year of this type of flying that he will be able to by pass the commuters and proceed straight to his choice of airlines.* He is a college graduate with a degree in engineering, though his grades were mediocre.

His nightly route takes him to five major airports in the midwest. This particular December night, the whole area has low ceilings and moderate mixed icing forecast and reported. The aircraft he usually flies is down for maintenance, so he is in a model that he has minimum time in. (2) *This really is no concern for him as he knows he can fly it well, regardless of the weather.*

It is cold and windy on the ramp, with blowing snow. The pilot performs a brief pre-flight and thinks to himself, (3) '*this is a much nicer plane than the other, I'm sure everything is all right*'. He did notice some slush build-up in the wheel wells, but didn't clean them out because they would just get filled-up again anyway. (4) *There really isn't anything he can do about the snow on the ramp* and his cargo was arriving. (5) *He needs to get moving.*

Though the snow plows have been working round-the-clock, the wind is such that the taxi-ways and runways are still snow covered. The blowing snow makes it difficult to see. As he taxis out, he discovers the defroster is inoperative. He thinks to himself, (6) '*boy, these guys should take better care of this equipment*'.

He is cleared for takeoff and as he takes the active, he looks around the cockpit for his flashlight. As he begins his takeoff roll, he remembers he left it in the hanger. 'Oh well', he thinks, (7) '*nothing is going to happen to the electrical system.* (8) *It's a dumb rule to require a flashlight on board, especially with redundant electrical systems*'. He proceeds into the clouds.

(1) Macho - 'Taking chances is foolish'.
(2) Macho - 'Taking chances is foolish'.
(3) Invulnerability - 'It could happen to me'.
(4) Resignation - 'I'm not helpless, I can make a difference'.
(5) Impulsivity - 'Not so fast. Think first'.
(6) Resignation - 'I'm not helpless, I can make a difference'.
(7) Invulnerability - 'It could happen to me'.
(8) Anti-Authority - 'Follow the rules, they are usually right'.

Review exercises

The following scenarios are based upon actual incidents as expressed by the pilots involved. Students should be asked to identify the risks involved, the hazardous attitudes and the appropriate antidotes to those attitudes. They should also be asked to contribute additional comments about the pilot, environment and aircraft conditions as they apply to the outcome of the flight.

The transition from flight instructor to passenger or cargo hauling duties is one of the most difficult transition periods requiring the highest piloting *and* decision making skills. Most flight instructors log many hours in small, single-engine aircraft. Some instructors, of course do obtain multi-engine ratings, but it is quite difficult to maintain proficiency without actually instructing multi-engine students. Even as a multi-engine flight instructor, the demands and decisions of the commercial pilot are quite different and usually much more involved than those of instruction. New equipment, new working hours, competition for routes and upgrading and the demanding nature of servicing the customer are pressures inherent to the new flight-instructor-turned-freight-pilot. All of these factors combined with routine considerations of flying can create a temporarily unstable environment. Let's listen carefully to pilots in common situations frequently faced while moving 'up-the-ladder' in the aviation world. In the following example, have your students carefully read the scenario first and then answer the questions that follow.

From flight instructor to freight pilot

Great! The waiver just came in today from the insurance company and I can finally fly the run tonight by myself. Good thing the company is in a bad way for pilots. They had to ask for a waiver to get me insured with less than the required hours stated on the policy.

About time, I really need to start making some bucks. Life as a starving flight instructor is over. I really don't know where the insurance companies got the right to require so many hundreds and thousands of hours just to be able to fly these little airplanes. I've had plenty of experience flight instructing and that should sure count for something. How else would I be able to get the experience? There's really nothing I can do about the system. I sure am glad they didn't ask to actually look at my logbook entries.

Now I can get out there and show them I am as good, no, that I'm better than the rest. I should move up quickly, after I get the chance to show them my stuff for awhile. I really can't wait to get my hands on the bigger equipment. Then I'll have a real good chance at the airlines.

The weather's pretty bad tonight. Most of the guys have been picking up moderate and sometimes severe ice. I can't believe all the ice still hanging on the plane that just came in. Oh well, lookout airways!

Discussion

1. In reading this pilots' thoughts on his new job, what risks did he fail to consider?

2. Which hazardous attitudes dominate this pilots' thinking?

3. Indicate at least three specific examples of hazardous attitudes.

4. What are the antidotes for the hazardous attitudes above?

The next scenario addresses that subject we all know so well after having been on the road. What can we learn from this one?

Getting home

It was the end of the week, and the pilot had already put in a full day's flying. He was hitching a ride in a company plane to his home town to visit his girlfriend this weekend. It was his college homecoming football weekend and they had fifty yard line tickets with old friends.

When he got to the aircraft he was to ride home in, he discovered that there were no extra seats installed. The plane was configured for hauling cargo only. He inquired about an extra seat and was told by operations that none was available. The person in charge for the evening shift indicated that a seat-belt was behind the co-pilot's seat and could be attached to the floor of the aircraft. He was sarcastically told to make himself comfortable.

He had not been with this company very long and had never run into this situation before. Unfortunately, the pilots he was to ride with had a reputation for flying an aerobatic freighter. He did his best to try to recall the regs regarding whether he needed an actual seat or not, but, being unsure, decided it must be okay or the company wouldn't let him do it. Besides, he really wanted to get home, and he sure couldn't afford an airline ticket.

When it came time to depart, the crew took their places and he rigged up a seat-belt arrangement that might have kept him in the vicinity of the aircraft if anything were to happen. There was no means for tightening the straps.

Though he was not typed in the aircraft in which he was riding, he did note some apparent peculiarities in this crews' procedures. Being a 'lower' pilot, he did not dare say anything or ask any questions. He wouldn't want them to think he didn't know why they were doing (or not doing) certain things. They did mention that maybe they should keep the shiny side up tonight since there was an 'outsider' on board.

They finally arrived at their destination and, of course, the flying pilot had to give him a show with his 250 knot approach and short field landing technique. The uncomfortable no-seat pilot thought, 'Maybe he thought he impressed me, but just wait, I'll show him!'

Discussion

1. In assessing risk, give one example in each of the following areas:

   a. Knowledge:____________________________________

   b. Physical:______________________________________

   c. Financial:_____________________________________

d. Legal: ______________________________

e. Egocentric: ______________________________

f. Sociologic: ______________________________

g. Benefits vs Risks: ______________________________

2. What is the major decision this pilot had to make?

3. What are the predominant hazardous attitudes showing in his decision making patterns?

4. Identify three explicit examples of hazardous attitudes in this example:

5. What are the antidotes for the above hazardous attitudes?

References

Benner, L. (1975). 'D.E.C.I.D.E. in hazardous materials emergencies'. *Fire Journal, 69*(4), pp 13-18.

Festinger, L. (1975). *A Theory of Cognitive Dissonance*. Evanston, IL: Row Peterson.

Flathers, G. ., Giffin, W.C. and Rockwell, T.H. (1981). 'A study of decision-making behavior of aircraft pilots deviating from a planned flight'. In R.S. Jensen (ed.) *Proceedings of the First International Symposium on Aviation Psychology*, Columbus, OH: OSU Department of Aviation.

Fox, R. (1993). Personal Communication.

Hogarth, R. M. (1980). *Judgement and Choice*. New York: John Wiley & Sons.

Jensen, R.S., Adrion, J. and Maresh, J. (1986). *A Preliminary Investigation of the Application of the DECIDE Model to Aeronautical Decision Making*. Columbus, OH: OSU Aviation Psychology Laboratory, Final Report AOPA Air Safety Foundation Grant.

Micheli, G.S. (1972). *Analysis of the Transfer of Training, Substitution and Fidelity of Simulation of Training Equipment*. Orlando, FL: US Naval Equipment Center NAVTRAEQUIPCEN TAEG Report 2.

9 The assessment of pilot judgment

In general, I believe that such qualities as coping with complexity and ambiguity without frustration and a flexible approach to the 'how' of problem-solving, along with a mild confidence in one's own ability, are some initial indicators of a pilot's capacity for good judgment.

Dirk E. Van Dam (1977)

Perhaps the greatest stumbling block to the formal implementation of judgment training in aviation is the assessment problem. As we saw in the previous chapter, effective training cannot take place without feedback, and feedback cannot be provided without some form of assessment of performance. Unfortunately, the psychological nature of the subject matter of judgment training is not conducive to assessment because many of the training objectives are not directly observable. The problem can be seen clearly in the FAA's published guidelines for pilot certification (FAA, 1994), which have for many years included 'The evaluation of pilot judgment' as Item 4 of the topics to be evaluated by pilot examiners. However, neither a definition nor criteria for the evaluation of judgment have been offered in these guidelines. On the positive side, pilot judgment has been assessed fairly successfully in a number of training research programs. In this chapter an examination of basic assessment concepts is presented followed by several techniques used to assess pilot judgment in research to date.

How should one approach practical judgment evaluation? There are four basic forms of evaluation used in education:

1 Attitude -- Did you like it?
2 Knowledge -- Did you learn anything?
3 Behavior -- Did it change the way you think or behave?
4 Global -- Did it make the system better, safer, more effective?

The first three of these are training evaluations of the individual and can be used to develop and provide feedback to the person being trained (i.e., the ISD Model shown in Chapter 8, Figure 8.1). Each of these can also be useful to the teacher providing the training. The fourth form is a transfer evaluation to show the effect of the program has on the system or world and is particularly important to the sponsors and developers of the training program. Each form requires a different approach. In this chapter we will show how each of these forms have been used in research to date on pilot judgment and suggest how they might be better used in future assessments of judgment.

Can pilot judgment be evaluated?

A cynic might state, 'If it can be taught, it can be evaluated'. However, there is very little evidence to suggest that pilot judgment is being systematically evaluated -- just as there is little evidence to suggest that it is being taught. Fortunately, there is evidence to suggest that complex judgmental behavior can and is being evaluated with varying degrees of success in other contexts. For example, McGuire and Babbott (1967) and Levine and McGuire (1968; 1970) simulated clinical problems that a physician might face and provided carefully structured environments in which to evaluate the physician's performance.

McGuire and Babbott's technique required initial information from a patient, involved sequential interdependent decisions, allowed physicians to acquire relevant additional information, forced physicians to abide by decisions and was adaptive to allow for varying decisions and outcomes of the decisions. Scoring was done by a group of experts classifying the judgments into five categories from poor to excellent. They also provided a summary of responses including a measure of efficiency and both errors of commission and omission. They report that evaluation of clinical judgments can be made using this technique.

Levine and McGuire (1968) demonstrated that judgmental evaluations may be made on the basis of role playing by candidates both before a medical board and before medical examiners. They concluded that this technique is more effective and lower in cost than direct observations. In a related study, Levine and McGuire (1970), using similar groups, found that the structuring of oral exams, the standardizing of cases materials and the training of examiners minimizes the problem of evaluation.

Meehl (1954) compared the correctness of prediction for clinical vs statistical prediction. In 19 of 20 studies evaluated, the statistical prediction was equal or superior to the judgmental prediction, regardless of who was the judge (counselor, clinical psychologist or industrial personnel manager). The implication of this research is that the evaluation of aeronautical decision-making skills should be strongly oriented to statistical formulae using multiple criteria.

Although, at first glance, clinical judgments of physicians and in-flight judgments of pilots appear quite different, the underlying demands are reasonably similar. Both have the potential to be life or death situations, both impose enormous information processing demands on the judge, and both require rapid, sequential, interdependent judgments. It seems plausible to assume that techniques effective for evaluating one should be effective for evaluating the other.

In further support of the feasibility of pilot judgmental evaluation, the adaptation of signal detection theory (SDT) to the flight situation has been proposed in Chapter 2. An individual's bias or threshold for response is analogous to the motivative aspect of pilot judgment. In structured situations, this bias before and after training should be measured to determine the effectiveness of the training program.

Jasper Shealy, in a validation study of three tests to predict down-hill skiing accidents (an activity with many of the same characteristics and personal pressures as flying high performance aircraft) found that neither of two generalized tests, the Atkinson 'Need for achievement' and the Rotter 'Locus of Control' could identify people who are high risks for such accidents. However, as shown in Figure 9.1, a situation specific test (SST) developed from the Locus of Control format was highly predictive of these types of people.

Finally, studies in which subjects were taught to be expert intellective judges (Shanteau, 1975) and studies in which subjects were taught 'vigilant' decisional responses (Janis and Mann, 1977) both present successful judgmental evaluation techniques. These evaluation techniques involve the comparison of observed responses in judgmental situations against the expected or correct responses in those situations. These same techniques could be used in pilot judgmental situations provided that proper behavioral responses or evaluation criteria are established.

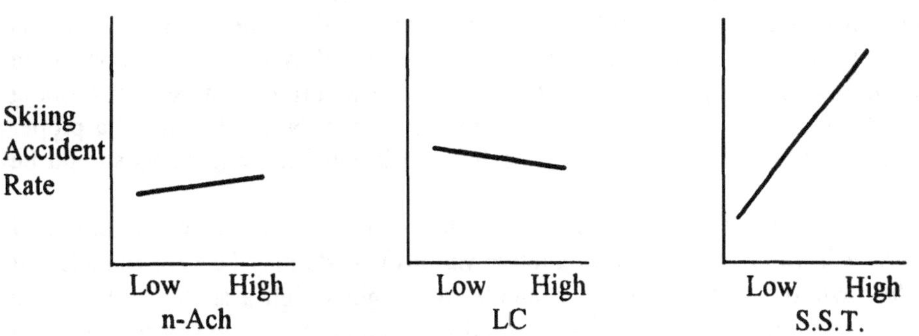

Figure 9.1 Results of three tests to predict down-hill skiing accidents (Shealy, 1974)

Basic psychological testing

The history of formal psychological testing is as old as the history of formal psychology, perhaps beginning in the mid-19th century with the introduction of psychometrics by Sir Francis Galton. The roots of informal psychological evaluation are lost in antiquity and probably began as an answer to 'I wonder why Torg did that'? Both forms of evaluation are still being practised today. The formal evaluation procedures include questionnaires, interviews, objective tests of intelligence, personality, etc., projective tests primarily for personality testing and situational or job sample evaluations. Informal evaluation includes all the less well specified occasions during which we 'size someone up', which may be as much as an FAA examiner can do when approaching a new certification candidate for the first time.

The two conditions essential for any evaluation method are reliability and validity. Reliability always refers to consistency or repeatability while validity indicates how well the evaluation measures what it is supposed to measure. One type of reliability is scorer reliability, or the measure of interscorer differences, often referred to as observer-observer reliability. Stated another way, do two people see it the same way? Efforts must be made to ensure that evaluations made by different evaluators are very nearly equivalent. Another kind of reliability, or more accurately, another source of unreliability, refers to the consistency or inconsistency with which the same pilot performs in the same situation on different occasions, often called ride-ride reliability (Roscoe, 1980).

Validity must be determined with reference to the particular use for which the evaluation method is being considered (Anastasi, 1968). Content validity involves a systematic examination of the test content to determine whether it covers a representative sample of the behavior domain to be measured. This is built into the test and is primarily dependent on the development of appropriate criteria. As an example, the content of a flight check during primary pilot training should sample the content of the previous instruction. Not only should a proper proportion of items be included on such a check, but irrelevant items should be eliminated.

Criterion-related validity indicates the ability of the test to make predictions of behavior in future situations (predictive validity). Tests may be very reliable but useless without validity. As an example, we might weigh a student pilot every day and find his weight remains essentially constant. Our testing instrument (the scale) is consistently accurate, but it is unlikely to yield a valid measure of a pilot's performance. On the other hand, an unreliable score can never be relied upon as an index of validity.

Evaluation might be considered as the end product of a systematic attempt to observe human behavior. However, it is necessary to exert a certain degree of selectivity of observation. It would be both impractical and inefficient to follow an individual day after day recording everything he does. The individual might not spontaneously exhibit the types of behavior in which we have interest, and we would rarely observe the same behavior in other individuals for comparison. One way to control this problem is to provide individuals with one or more specific tasks to perform. The nature of the tasks may be varied to elicit the behavior we wish to observe.

An individual task of this sort is often viewed as the test item and a series of these tasks as a test. To trust results from a test, we need measures of its accuracy and utility (reliability and validity). Generally, the more behavior samples taken, the more accurate the measurement will be (Lord and Novick, 1968). Accuracy is critical when we use measurements to discriminate among test-takers. However, we must weigh the increase in accuracy attainable against the increased cost of taking more samples of behavior. The utility is generally a function of careful selection of observations. Careful selection of our behavior samples will increase the utility or validity of our measurement.

A number of methods have been used by psychologists, educators and personnel managers to evaluate human behavior and personality. These methods include interviews, questionnaires, written tests and situational tests with objective response criteria. Each of these may have some utility in pilot judgmental evaluation prior to the beginning of flight training, during judgmental training and in the measurement of transfer. Those methods useful in initial testing would be particularly valuable in the identification of potential judgmental weaknesses in beginning students, correctable through various instructional techniques.

Interviews. Although the structured interview has limited applicability in judgmental evaluation because it requires trained interviewers, it should be mentioned because of its widespread use in many contexts including airline hiring (Bellows, 1963) and because it may be useful in some pilot training situations. In the formal sense interviews are most commonly used in personnel selection or psychiatric screening, but we all use many of the same techniques informally to evaluate pilot judgment.

Common errors of the interview are likely to occur in both formal and informal assessments. A personal bias in the evaluator may cloud his judgment, and he may base his evaluation on irrelevant personal characteristics of the individual rather than his statements. The way in which the situation is structured may have a profound effect upon the evaluation of the individual's judgment. The individual being evaluated may not recognize the relevant dimensions of the situation because of the format. Despite being highly motivated to render an adequate judgment, he might be unable to do so in the typical interview situation.

Another common problem in interviews is the establishment of rapport between interviewer and interviewee. While the instructor and student pilot may have sufficient time to develop rapport, it is conceivable that changes in evaluators (e.g., check airman) will not enjoy this established rapport. Any deleterious effects will likely be magnified when the situation becomes as important as a flight check. Small changes in the evaluator's approach may produce dramatic changes in performance which would be reflected by changes in evaluations.

In addition to these problems, there is a wealth of data that calls into question the evaluations of highly skilled interviewers. In certain cases, interviewers may escape common pitfalls of interviewing precisely because they don't know enough about the criteria to be biased. The value of the interview as an evaluation method is extremely limited (it may even be negative), but there are justifications for retaining face to face contact between evaluee and evaluator. Strauss and Sayles (1967) indicate that interviews, as performance evaluations, can be considered as a form of coaching and may be useful as part of the overall training program. Interviews are also good because they provide feedback from the evaluee to the evaluator. Evaluees should be encouraged to point out any weaknesses they perceive in the interviewing process.

'Objective' tests. Objective tests may be used to measure many diverse entities. The type of test that most people think of first is likely to be an intelligence or I.Q. test. Anastasi (1968) reviews objective tests and notes that the history of intelligence testing predates Binet by at least 21 years, beginning in 1887. Standardized aptitude and achievement tests are common in our society from early school years through the search for employment.

The typical I.Q. test measures an individual's performance on a wide variety of questions or tasks and normally provides one score called the Intelligence Quotient, or I.Q. Part of what we have defined as judgment resembles part of what is known as intelligence. To this extent, I.Q. tests may give information about the ability to make judgments. As an example of this limited correspondence, an item on the Wechsler-Belluvue (now updated as the Wechsler Adult Intelligence Scale) asks subjects what they would do if they found an addressed, stamped letter in the street (Wechsler, 1939). Responses to this and similar items are scored for the comprehension subscale, but their relationship to judgment, as good sense, appears evident.

The objective personality tests and self-report inventories number in the hundreds. These tests purport to measure emotional, motivational, interpersonal and attitudinal characteristics. Shealy (1974) and Johnson and Davis (1972) have compared risk-taking behavior with several standard tests of this type and have found them to have some value. However, the work on this type of test has not generally been aimed at pilot judgment. Borg (1963) indicated a difficult problem with some self-report inventories. Using these tests may result in subjects' faking replies to create the desired impression. Fortunately, many of the standardized tests are constructed to catch faking.

Verbal protocol. In the last few years the verbal protocol method has been growing in popularity as a device for assessing and analyzing complex cognitive abilities and tasks (Ericsson and Smith, 1991). The procedure used is to ask the subject to perform a form of low fidelity role playing in the form of reacting to a scenario presented by the experimenter. In the aviation context the experimenter presents a flight situation containing cognitive based problems. The subject is asked to state his thought processes and decisions concerning the problem presenting as if he were in the airplane. Verbal data are recorded, transcribed and analyzed later to determine the quality of the thought patterns expressed. In an experimental setting, this method has proven to be a highly effective pilot judgment assessment tool (Jensen, Kochan and Chubb, 1994).

Judgment assessment literature

Definitive evidence of the types of interpersonal factors which can facilitate or inhibit the exercise of good pilot judgment is unfortunately not available in the literature. Shealy (1974) presents what might be described as a disposition toward varying amounts of risk taking. Risk taking appears to transcend specific situations and has been proposed to manifest itself in a variety of situations (e.g. driving, flying, downhill skiing). The critical factor for measuring this disposition is to imbed the test items within a specific situational context.

World War II saw an increase in research relevant to pilot errors or lack of good judgment as a cause of accidents. Kalez and Hovde (1945) reviewed the records of pilots and report what they call a unique psychological group of pilots who wilfully failed to use checklists. They refer to this temporary psychological compulsion as an error in judgment. The pilots who were involved in the resulting accidents presented a long history of nonconformity as evidence from their flight records. Occasionally, they were below-average aviators in every measurable respect. The type of errors in which they were interested were classified as 'pilot errors'.

In their opinion, the 'accident prone' aviators belonged to this below-average group. The accidents resulted from an initial error of judgment which was then coupled with poor technique. They reported 18.4 per cent of the pilots caused 58.2 per cent of the pilot inaptitude accidents. They found five factors valid in the production of repeated accidents. In 23 per cent of the pilots, (1) *a below-average flight training record* was found. The fatal crash rate was 3.4 times that of pilots with average and above-average flight training records. There also appeared to be (2) *below-average adaptability to flight*, but is was not clear whether it was present prior to training or occurred after training began. There was (3*) an excessive grounding history* appeared in 31 per cent of pilots in question. In 9 per cent of the pilots, a history of 'doping off' in the air was present, where (4) *lapses of memory concentrations and attention* initiated a dangerous aviation incident (not an accident). As education standards applied in military pilot selection were lowered, those with (5) *less education tended to be over-represented* in the accident group.

The nature of 'pilot error' was further investigated by Kunkle (1946). The term refers to all the defects which a pilot may exhibit in all aspects of aviation, although it is limited in that it primarily refers to accident causing behavior. He divided the topic into primary or 'non-emergency' error including errors in judgment and secondary or 'emergency' error occurring in a setting of tension and confusion associated with a crisis in some phase of flight. Of course, the secondary error may result from an immediately preceding pilot error. However, true emergencies were responsible for only a very small minority of pilot error accidents. Kunkle reports the general failing of the then current (WORLD WAR II) selection tests for prediction of pilot error accidents and concomitantly an inability to assess an individual's judgment. He further implies that the problems involved in the operation an airplane allow full scope for any of the various manifestations of accidents proneness. Reviewing the accident background of pilots involved in accidents revealed a previous accident pattern in which auto accidents were indeed conspicuously frequent. He concludes that there is a direct, but by no means, rigid, relationship between past performance on the ground and safety record in the air. Kunkle feels this relationship would be greater if the sample were less restricted (only rated pilots were included, therefore considerable selection had already occurred through processes of attrition, suspension, or death).

In the work on printed classification tests for aircrew members, Guilford and Lacey (1947) report that, of all the traits necessary for pilots, judgment stands out as being the most persistent and universal. However, the frequent mention of judgment for the pilot presented a persistent challenge to break it down to manageable components and devise tests for it. In 1940, Kelly (Guilford and Lacey, 1947, p. 124) had flight instructors rate student pilots and found the following items from a 14 item checklist to be related to a judgment factor:

1 How good is his judgment with regard to taking flying risks (weather, stunting, etc.)?
2 Does he show respect for a ship and its motor?
3 How well is he satisfied with his flying ability?
4 Is he inclined to show off while flying a plane?
5 How carefully does he check his plane and engine before taking off?

The research indicated that practical judgment was a part of safe flying suggesting that further investigations of both judgment tests and instructor reports of practical judgment are needed.

Guilford and Lacey (1947) concluded that the judgment factor was best defined by a work-planning type of item. Further, items calling for relatively complicated estimates involving time as well as distance and size were significantly loaded with the judgment factor, while the simpler items were not. The inference which may be drawn is that the judgment factor is a thinking, rather than a perceptual or memory ability. Guilford concludes that judgment was very highly regarded as a factor and received considerable attention (with varying degrees of success). A promising avenue which opened at the end of the war was the hypothesis that thought fluency might be an independent contributor to practical judgment. The facile recall of pertinent experiences for possible use in everyday predicaments would give an individual more potential solutions from which to choose. Apparently the construction of such tests using answer sheets presented some problems, but it was suggested that this hypothesis should have been followed up, if possible. In addition, it might be profitable to collect, from training instances of good and poor judgment, the types of information that would have been pertinent to a successful solution of the problem.

One of the major weaknesses in many evaluations of complex behavior is the frequent use of subjective judgments based on irregular and uncontrolled observations (Abeles, 1973). To avoid these problems, several alternatives are available. When there is no objective measure by which an optimum evaluation can be determined, subjective evaluations from skilled individuals who have the particular competence and experience may be required. These judges will use a limited number of criteria to determine the optimum, but the relative importance or weights for these criteria may vary from one judge to another (Gulliksen in Shelley and Bryan, 1964). Abeles (1973) found that the achieving of a consensus among judges is difficult, particularly when they use impressions. To eliminate this source of variance, guidance should be provided to the evaluators in the particular weights to use.

Assessment of automobile driving judgment. The measurement of the adequacy of judgment is a difficult task. Rockwell (1972) attributes this difficulty to the observer's inability to adequately evaluate the judgment without knowing the alternative courses of action open to the driver and his motivation behind a given decision. According to Rockwell, judgments in automobile driving should be divided three categories:

1 Emergency decisions -- one time situations requiring unique often rapid decision-making and response. Simulation of these events lacks the realism of real-world threats and repeated exposures destroy the uniqueness. Field studies are too dangerous and again limit coverage to a single exposure.
2 Decisions whether or not to engage in driving -- the situation in which the driver is tired, drinking, aware of vehicle defects, etc. These are also unique decisions and not easily researchable, although they may be the keys to good driver judgment.
3 Operation decisions, including routine decisions as to speed selection, passing and merging decision. These decisions are more amenable to observation and research and are important because they dictate the level of subsequent skill required of the driver to executes the decision and maintain a given performance.

Cumming (1964) has suggested that there is an upper limit to the rate at which the operator can make decisions. Obviously, the decisions that an operator must make are many and varied, ranging from minor automatic decisions to highly complex decisions relating to potential accident situations. Not only do decisions vary widely, they also differ by type. Some decisions are dichotomous (e.g., passing, merging, or stopping) while others are continuous, (e.g. vehicle control inputs). It is not apparent from the literature how individual differences in the role of decision-making may account for accidents, but the research in this area is also needed.

In a review of accidents and risk, Taylor (1976) discusses what he calls the intentional aspects of risk. He notes that it is not a logical impossibility to take a risk and not intend to suffer any untoward consequences. Certainly the majority of risks are undertaken with the belief that there is a high probability of success. It is important to remember that the subjective impression of danger, regardless of how minuscule, or whether actual, is a necessary condition to define risk. Tracing the intended risk taking behavior to an accident dictates that, at some point, the events leave the control of the involved individual. Near accidents might be viewed as instances where control was regained prior to the accident. Taylor proposes that maintenance of control is the essence of *safe* behavior. Control is defined as the power to direct the machine in accordance with intention. This removes one ambiguity but leaves the unobservable intention. The driver's opportunity to perceive loss of control and this perception is subjective risk (Taylor, 1976).

The critical aspects of a situation are the perceived risk and the amount of risk an individual is willing to accept. Increases in safety of a system may be achieved by many routes. One possibility is to increase the perceived risk involved with a given course of action. If the acceptable risk level remains constant, increases in perceived risk may cause the decision maker to act in a more cautious manner. Naatanen and Summala (1973) have shown that making the environment safer in both appearances and actuality has eliminated the expected safety gains. For example, better road surfaces allowed increases in both speed and frequency of passing up to the new level of subjective risk, thereby, attenuating the effects of a subjective risk. It is assumed that subjective risk increases are reflected by increased vigilance.

Direct determination of risk taking tendencies would be particularly useful if the demonstrated relationships apply to aircraft operation as well. Gumpper and Smith (1968) attempted to find a relationship between two paper-pencil measures of risk-taking tendencies and accident records in truck drivers. Neither the Life Experience Inventory (designed by Torrance and Ziller, 1957) nor Wallach and Kogan's (1959) Dilemmas of Choice questionnaire were able to discriminate between high and low accident drivers in Gumpper and Smith's sample. Although these results are not encouraging, the possibility exists that risk taking tendencies may be found through other measures. It is also possible that these measures would have greater success when applied to different criterion groups.

Interpersonal groups. Although the evidence is not definitive, it appears that the investigation of interpersonal factors as correlates of poor judgment would be a project with considerable merit. Several very promising approaches to the study of interpersonal factors have not been fully exploited. Perhaps the most promising of these is the situational approach of Shealy (1974). Use of situational tests encompassing the existing risk and decision-making research should have the greatest potential payoff. To the extent that an individual's attitudes will affect the quality of that person's judgments, then personal attitudes would be a useful index. Although not really test data, the biographical history of individuals has frequently been mentioned as a correlate of accident occurrence. The individual with the adverse background may also be shown to hold attitudes consistent with that background and this combination may be sufficient to contraindicate positions of responsibility. Only well-conducted research specific to aviation will resolve the conflicting results reported from the diverse sources.

Judgment evaluation in aviation

The most important initial step in the establishment of any evaluation procedure is the development of criteria against which the student's behavior can be compared. This is particularly true when behavioral evaluations are made subjectively and yet standardized over a national population of evaluators such as flight instructors and examiners. Thus pilot judgment evaluation criteria must be meaningful, they must objective and they must mean the same things to a wide variety of evaluators and those being evaluated so that they may know what is expected of them. Figure 8.1, in the previous chapter, illustrates how criteria and evaluation are integrated into a training system design for pilot judgment.

A crude first step toward the establishment of judgmental evaluation criteria might be through the use of accident investigation reports. A few of the pilot defections frequently reported by the NTSB that we classify as 'decisional activities' are:

- Operating an aircraft with known deficiencies
- Operating beyond own experience or ability qualifications
- Continuing into known adverse weather under visual flight rules (VFR)
- Selecting flight route over unsuitable terrain
- Neglecting adequate preflight planning or preparation
- Flying an unfamiliar aircraft
- Allowing attention to wander from operation of aircraft.

Each of these frequently reported causal factors in accidents represents a qualitative evaluation of a pilot's judgment by the NTSB that could be used as the basis for a set of quantitative criteria in which the pilot's judgment is evaluated on a scale from good to bad, possibly quantitatively.

In pilot training, for each level of pilot experience, certain judgmental hurdles (proficiency levels) could be objectively specified. The instructor or examiner who evaluates the judgments would have a range of acceptable performances, objectively specified. Evaluation of pilot judgment would be a matter of comparing performance against the established criteria. The critical point for judgment evaluation in a national system, such as that within which pilots operate, is the use of the same criteria as well as by the pilots themselves.

While understating the difficulty of constructing these criteria, an approach to this level of standardization is essential to the evaluation of training and transfer. For this purpose, NTSB reports may be used as one basis for judgmental evaluation criteria. As suggested in our definition, the ultimate standard against which pilot judgment is measured is determined by societal norms. Although one might argue with specific findings, the NTSB is officially sanctioned to assess the cause of accidents including those involving faulty pilot judgment. To make any such assessment requires an assumption of what society would have done in that situation for use as the index of desired performance. Therefore, the NTSB findings provide a unique standard for the establishment of pilot judgmental evaluation criteria.

Pretesting for judgment. Pretesting of the judgmental ability of individual candidates prior to their initial pilot training is a potentially useful adjunct to the entire instructional and evaluation process. Pretesting used as a baseline against which to measure the effects of instruction forms an essential component of the classic evaluation of program effectiveness. Ideally pretests would provide an opportunity for the individual to respond to the type of materials that would be used during training. Modification of these responses as a function of training would be taken as evidence of training effectiveness (assuming a similar group not exposed to the specific training manifested no such change). Because most beginning flight students lack the rudimentary knowledge to make appropriate judgments in all but the most obvious aviation situations, pretesting specifically for pilot judgment is likely to be futile.

While training procedures are testable in outcomes and, less frequently, means, pretesting can be accomplished by measuring such personal characteristics as the role concept, values, cognitive structure, personality and experience. Experimental evidence is available to indicate that personal characteristics do bias individuals in the type of judgments they will prefer. For example, Coombs and Pruitt (1960) demonstrated preferences for varying amounts of risk in gambling situations, where some subjects consistently preferred ‘long shots’ and others played it safe. In a laboratory setting, Brody (1963) has shown personality variables of achievement motivation and anxiety to affect risk taking behavior. Taylor and Dunnette (1974) found dogmatism and intelligence to relate to risk taking during decisional behaviors. Additionally, the work during WORLD WAR II on printed classification tests (Guilford and Lacey, 1947) indicates that there are measurable characteristics which the individual has acquired prior to pilot training which can affect the outcome of training.

The question of how to use pretest data must be considered discreetly in our free society. During and following WORLD WAR II, selection tests were used by the military as a screening process to lower the attrition rate from primary flight schools. One major problem when attempting to provide a better fit between man and machine, and hence lowered accident probability, applies to the task of selection. According to Klein (1976), it has been assumed that: (1) a large pool exists from which operators may be selected who embody the proper characteristics; (2) these operators could be forced to behave appropriately, by threats or economic incentives; and (3) performance of the man-machine system could be continuously monitored. Klein admits these assumptions probably hold for the professional operator, e.g. commercial airline pilots or interstate bus drivers. But, Klein quickly indicates that such restrictions may not be applicable to the more casual operator. In the amateur situation, less rigorous selection procedures are in force, economic sanctions are not easily enforced and continuous monitoring is essentially impossible. Although the potential for a rigorous selection approach exists in civil aviation, a more acceptable purpose of pretest data would be to improve pilot training procedures. Individuals might not be screened out, but identified as needing greater attention in judgmental training.

These issues are what Cronbach and Gleser (1965) termed institutional decisions. When an individual considers one person alone (e.g. approving him for private pilot certification), in applying the value system of society, he is making an institutional decision. The decision in this case is made to minimize the loss from an erroneous decision. It will not always be the case that what is best for the individual will be best for society. Any testing authority is forced to consider the ramifications of decisions based on one or a series of tests. A balance between individual freedom and societal protection is needed.

Although the question of how far into an individual's background regulatory agencies should probe is beyond the scope of this report; this question is important if one accepts such statements as 'Man drives as he lives' (Tillman and Hobbs, 1949). It may be said that the entire adjustment complex of the individual reflects the accident record of this person as a vehicle manipulator. The person who makes continual mistakes in facing the demands of the social and personal milieu is also likely to err in the performance of his obligations in vehicle manipulation. This concept has been shown to be basically true, as reflected in the financial, social and employment records of a sample of violators and accident-involved persons (Lucas, 1970). For truly effective implementation of pilot judgment evaluation, a more intensive background investigation would appear to be an essential component to the entire process.

Training assessment. As shown in the judgment training model, following the training a second level of assessment, called training assessment, is needed to show the effect of the training. In this step aviation specific examples or scenarios can be used to test the students provided that the scenarios are written for the level of the training (e.g., private, commercial, instrument, etc.). Tests such as the PDAQ, flight simulator checks or computer assessments are examples of this level of evaluation. Ideally, the training assessment will show both an absolute level of judgment attainment in the student and also it will show the improvement in judgment as a result of the training. At this level, *attitudes* about training, *knowledge* attained, and *behavioral change* can be assessed.

Transfer assessment. The final type of assessment, called the transfer assessment, is designed to determine the amount of the training that has transferred to the operational environment. In aviation, this type of evaluation can only be made after the student leaves training and moves into the real world of aviation operations such as flight instruction, recreational flying, corporate flying, etc. Assessments at this stage can take the form of realistic flight scenarios of the type experienced in that setting and can be a part of six-month checks, in flight proficiency rides, simulator checks, or biennial flight reviews. At this level, the primary objective is to determine if the candidate is *safe* to operate in the context of the system.

Examples of judgment assessment

There are several examples of research in which the investigators developed assessment techniques for evaluating pilot judgment that could be applied to each of these three levels of assessment. Berlin and Holmes (1981) performed a training assessment observing pilots on a flight test prior to their private pilot check ride. Buch (1983) also performed a training assessment. Using an examiner posing as a photographer, she obtained evaluation data by observing pilots after they had obtained their private pilot certificate on an eighteen-item check. As mentioned in the previous chapter, at OSU, we recorded pilots on video tape as they made a 'LOFT' type flight in a Frasca 141 flight simulator until they were forced to turn around (Jensen, Adrion and Maresh, 1986). The real value of the video tapes were in the debriefing that followed the flight. During the debriefing the pilots used the tape to cue their thought patterns as they described what they were thinking when making their decisions to turn around and return to home base. Last, one of the major efforts to develop an assessment tool for pilot judgment was done by our group in Ohio under contract to the US Air Force. Different parts of this tool, described below, can be used at all three levels of the assessment of pilot judgment.

JAMJET

In the mid-1980s my colleagues and I received a contract from the US Air Force to develop a computerized judgment assessment system (Jensen, Adrion and Maresh, 1987). The resulting system, called JAMJET (Jensen, Adrion and Maresh, Judgment Assessment Technique), was geared for the military but has several features that may be useful to civilian judgment assessment. The system was based on the detailed model presented in Chapter 2 and included several tests to determine the abilities of an aviator to perform different functions thought to be necessary for good judgment. The tests were called Vigilance Test, Cognitive Adaptability, Knowledge Test, Pilot Decisional Attribute Questionnaire (PDAQ), Aviator Locus of Control, Situation Discovery and Diagnosis, Alternative Generation and Selection, Aviator Load Scale. The introduction and initial phase of the test consist of instructions for taking JAMJET. After the pilot completes the biographical questionnaire, he proceeds directly to the test battery.

Vigilance. The test is designed to determine the pilot's capability for detecting problems during a mission. In this test the subject/pilot is presented with a display of 25 rows of 80 dots, all of which are colored red. After a random interval no greater than 15 seconds, a dot at a random position changes from red to magenta. When the subject detects the change, he acknowledges by pressing any keyboard key. A mechanism has been employed such that no invalid responses are recorded. Once a valid response is recorded, the dot is returned to red until the next random interval expires. This sequence is repeated a total of 10 times. A score sheet lists the response time for each interval as well as the mean response time for the test items. The same random time intervals and screen positions are used for each subject to ensure consistency in testing.

Cognitive adaptability. An innovative cognitive adaptability test was developed to test for the pilot's ability to change focus of attention at the appropriate time. The test involves two key elements in pilot judgment: cognitive processing and adaptability to requirements for a new strategy. Based on Sternberg (1969) and the auditory selective adaptability test (Gopher and Kahneman, 1971), the test consists of a presentation of three sets of alphabetic characters from which the subject must determine which set contains the highest proportion of characters from a memory set. Each of the three character sets consists of random characters as well as varying proportions of memory set characters. There are a total of 600 characters in each set. The subject can view only three characters from any one set at a time as they move in sequence across a window on the computer monitor, twice each second allowing the test to run for five minutes.

The three character sets are constructed as follows. The first set begins with a probability of memory set characters of 5 per cent and transitions to a probability of 15 per cent at the end of the test. The second set consists of a constant probability of 8 per cent memory set characters, while the third set consists of an initial probability of 15 per cent of memory set characters and ends with a probability of 5 per cent. The objective of the test is to determine which set has the highest initial probability, and which set has the highest final probability.

Prior to the test, the subject is shown two characters which represent the memory set. These characters were chosen at random from the upper case alphabet. The memory set is shown for ten seconds to allow the subject to memorize them. The test begins by presenting the subject with the first window of three characters in the first character set. The remaining windows remain blank. The windows are arranged horizontally across the screen and labeled 1, 2 and 3. Three keypad characters have been re-labeled according to the window numbers. An additional keypad key has been re-labeled 'select' to allow the subject to designate the current window memory set characters. The subject may view either of the two remaining character windows by pressing the appropriate key mentioned above. When a new window is selected, the previous window becomes blank such that the subject may view only one window at a time.

When the subject has determined which character set has the highest probability of memory set characters, the appropriate window is designated by viewing the window and pressing the key labeled 'select'. The time and window are recorded by the computer. As the test progresses, the probability of memory set characters gradually shifts from the initial constant values, to the final values between two and three minutes. Once again, the subject just monitors each of the three windows and selects the character set containing the highest probability of memory set characters. The test is five minutes in duration. One or more tests with differing memory sets can be administered.

Knowledge. The Knowledge Test portion of the series measures the subject/pilots' initial level of knowledge in areas covered in later portions of the battery. If the pilot does not know the facts needed to diagnose the problems offered in the later tests, the usefulness of the other portions of the battery are in question.

The Knowledge Test is a general test of systems and procedures of the applicable aircraft and flight environment. In this case the F-16 aircraft was used in the Air Force environment. Additional subject areas include basic aircraft performance, instrument flying procedures and weather. The test is multiple choice and it is not timed. The knowledge test is computer scored.

The PDAQ. The Pilot Decisional Attribute Questionnaire (PDAQ) is an instrument designed to assess the Air Force pilot's relative tendencies toward five hazardous attitudes that have been postulated and tested in general aviation (Buch and Diehl, 1982, Diehl, 1992). The five hazardous attitudes are Anti-authority, Impulsivity, Macho, Invulnerability and Resignation.

Aviator locus of control. The Aviator Locus of Control test is the second of three tests of pilot vulnerability to pressures found in the Background Problem (Figure 2.1). This instrument is designed after the Rotter Internal-External (I-E) Control scale, Rotter (1966) which is a test of a person's belief that he controls his own fate (Internal Control) versus his belief that his fate is controlled by forces outside of his own power (External Control). It can also be considered a test of a person's reliance upon skill versus chance.

In the aviation context, if a pilot *perceives* a successful outcome of a mission to be contingent upon his own skill he is considered to be *internally controlled*. On the other hand, if he tends to think that the outcome is not entirely contingent upon his own action, and, at least in part, the result of luck, chance, fate, or under the control of others, he is considered to be *externally controlled*. The causal relationship between a pilot's behavior and the outcome is not an all or none situation regarding control. I-E control will vary in a matter of degrees.

The Aviator Locus of Control is a twenty-nine item, forced choice, computer administered instrument. Four of the items are non-scored, filler items designed to conceal the true nature of the evaluation. The questions are stated in such a manner that the pilots' belief about the nature of the world can be evaluated. Though none of the items specifically address the preference for internal or external control, the generalized expectancy measure can be considered to be an estimate of the *value* the pilot puts on internal or external control.

Alternative generation test. The alternative generation test assesses the pilot's capabilities from a highly applied setting. It is designed to determine how a pilot goes about seeking and processing information to generate alternatives to a difficult situation. Four items representing different information seeking and reasoning forms are provided following each scenario for the pilot to consider. Each of these items represents a different approach to the generation of a hypothetical alternative. The pilot is to rank these items in the order he considers best to worst in terms of providing an effective approach for the generation of alternatives to the situation. There are no right or wrong answers, so the rank order of the four items indicates the relative worth of each item to the pilot as an approach to generating alternatives.

The Alternative Generation test provides a form of branching. That is, the scenarios presented are related to each other in that the selection of certain items are considered in following changes in the scenario. The importance, or rank, that a pilot places on each item is expanded upon in subsequent questions taken from earlier responses. The four items to be ranked represent the following reasoning categories: 'hypothesis' directed, 'other' directed, 'data' directed or 'knowledge' directed. The responses to these broad categories will show a trend toward the paths used in processing information as well as the relative importance of the types of information available to the decision maker.

The individual who tends to use 'hypothesis' directed reasoning will tend to select answers which assume a particular outcome to the situation which he believes to be the correct state of the world. Usually these assumptions are directed by one's perception of his own skills and abilities. The information seeking behavior will be an attempt to confirm or disprove the hypothesis. The 'other' directed form of reasoning tends to look to others, especially experts who he respects, for information useful in the generation of alternatives. These individuals may also place a lot of importance upon the interpretation of the information by others. 'Data' directed reasoning indicates an 'open minded' method of looking for more hard data prior to making a decision. For example, the pilot would look around the cockpit, perhaps testing various instruments or controls or he may examine written documentation for more information useful to the solution of a problem or making a decision. Those who tend toward 'knowledge' based reasoning will focus on prior knowledge of a fact or set of facts. This would include the pilot's experience and training.

For example, in one scenario, the pilot is on a mission with a limited amount of fuel. He has several decisions to make regarding the continuation of the mission and where to land. The weather at his home base is currently reported to be too bad to make an instrument approach, though he does have the option to fly the approach and see if he has enough visibility to make a landing. Another option would be to abandon the mission and head for his alternate airport where the weather is slightly better. Some pressure is introduced in the scenario by the fact he has a meeting scheduled with his superiors at his home base

Situation discovery and diagnosis. The Situation Discovery and Diagnosis test investigates the pilot's ability to appropriately detect and diagnose events or problems and satisfies the Diagnosis block of the detailed model (Figure 2.1). This test is also applied in nature and requires a knowledge of F-16A systems and operations. The extraction of the correct information and the solution to each problem go beyond basic knowledge. The pilot must be able to organize his thinking and proceed in a rational manner, given the amount of information available from the scenario and/or graphics and his knowledge. The primary purpose of this exercise is to determine the pilot's general information processing and problem solving capabilities in the context of F-16A operations.

As in the Alternative Generation Test, the scenarios are presented both graphically and in written form. The scenario given above illustrates the cognitive processing the pilot must use in this test. The situation is a combat mission during which a 'surprise attack' is encountered. The pilot first must recognize the threat warnings and then make a determination as to how to handle the situation.

Aviator external stress scale. Major and minor life stressors contribute to an overall level of stress which must be carefully monitored. Stress is one of the greatest factors affecting one's ability to make logical decisions. In piloting an aircraft, the first area to be affected by an overload of stressors is pilot judgment. This could result in 'tunnel vision' or a narrowing of one's point of view. All of the information that is available may not be seen and making choices from the available alternatives may be difficult. In its most insidious form, stress can result in panic during which one may even lose control of motor coordination. The influence of life events on one's ability to cope with short term stressors is an important consideration when evaluating judgment. This questionnaire, designed to evaluate the current stress load on the pilot from life events outside the cockpit, is a computerized version of the Holmes and Rahe (1967) social readjustment scale.

Examples of CRM evaluation

The OSU NASA experiment

In the early days of CRM development we at OSU conducted one of the first studies to determine the effectiveness of communication analysis as a method for CRM cockpit evaluation (Jensen, 1986). In this study communication styles of pilots were determined using a generalized Personal Attributes Questionnaire (as shown in Chapter 6) administered to each member of a volunteer crew of two experienced jet pilots prior to an experimental flight in a T-40 jet simulator. The experimental flight consisted of a realistic LOFT type corporate flight from Milwaukee to Minneapolis with two unforeseen abnormalities (an engine failure while enroute and glideslope failure). Video tape and computer recordings of flight performance during the flights were compared with the results of the questionnaires. The effect of cockpit management style on cockpit communication before and after these occurrences was analyzed. In the analysis of the communication data, a modified Rogers and Farace (1975) communication coding technique was used. In this technique a communication coding scheme designed to show the impact of 'control' being exercised in dyadic communications. The objective is to show when each member is demonstrating 'one-up manship' or 'one-down manship'.

The OSU (Jensen, 1986) coding scheme for cockpit communication (shown in Table 9.1), was designed to identify a similar control factor. It employed some of the particular forces that are present in cockpit communication. The first four terms in the control code, 'command', 'request', 'suggestion' and 'observation', are designed to indicate the level of cockpit communication control being sought on a continuum similar to that suggested by Tannenbaum and Schmidt (1958). Other terms, such as 'disagreement', 'question' and 'initiate new subject' would show different levels of control being sought in a more general way. Finally, our scheme includes two terms that are fairly specific to the aviation communication environment, 'acknowledgment' (usually 'roger') and 'checklist call out' which is a part of standard operating procedures. It should also be pointed out that the terms used in the cockpit context are restricted in meaning because of the technology involved. Usage of some of these terms would be quite different in a classic analysis of dyadic communication.

Table 9.1
OSU communication coding form

1. Speaker	4. Control
a. Captain	a. Command
b. First Officer	b. Request
	c. Suggestion
2. Form	d. Observation
	e. Acknowledgment
a. Statement	f. Checklist item
b. Question	g. Answer
c. Exclamation	h. Disagreement
d. Non-Verbal	i. Initiate-Terminate
	j. None of the above
3. Direction	
a. Other crew	
b. ATC	
c. Self or Aircraft	

Definitions of Control Terms used in Table 9.1:

Command: An authoritative order directing action of the object. Example: 'Gear up'.

Request: A communication asking for something to be given or done, said with less authority than a command. Example: 'Would you please call for weather'?

Suggestion: To mention or introduce an idea for consideration or action. Example: 'Why don't we check the weather at our destination'?

Observation: To offer in communication what one has seen, noticed, or perceived. Example: 'We have just passed the station'.

Acknowledgment: To indicate receiving and understanding a fact. Example: 'Roger, 40RJ'.

Checklist item: Communication read from checklist. Example: 'Flight instruments - Check'.

Answer: To respond to a question, suggestion, checklist item, or communication that is not simply an Acknowledgment, or Agreement. Example: 'Flight instruments, checked'.

Disagreement: To differ in opinion. Example: 'No, I don't think we should try that approach'.

Initiate-terminate: To begin or end a line of thought. Example: 'Where did you learn to fly anyway'?

Table 9.2
Predicted cockpit communication control as a function of form and control type

	FORM			
	Statement	Question	Exclamation	Non-Verbal
a. Command	↑	↑	↑	↑
b. Request	↑	→	↑	↑
c. Suggestion	→	→	↑	---
d. Observation	→	↓	→	→
e. Acknowledgment	↓	---	---	↓
f. Checklist	→	---	---	↓
g. Answer	↓	↓	↑	↓
h. Disagreement	↑	↑	↑	↑
i. Initiate-Terminate	↑	↑	↑	---
j. Other	---	---	---	---

Legend:

↑ One-up-manship
↓ One-down-manship

➔ Across - neither up nor down
--- Unlikely communication or not clearly interpretable

Results. Figure 9.2 shows the overall communication rate for the four control terms that were predicted to indicate decreasing levels of control. As one might expect, a strong interaction is evident along this dimension between the captain and the first officer. Captains make more commands than requests and suggestions. First officers make fewer commands but more requests and suggestions. Both pilots make a lot of observations but the co-pilots make many more.

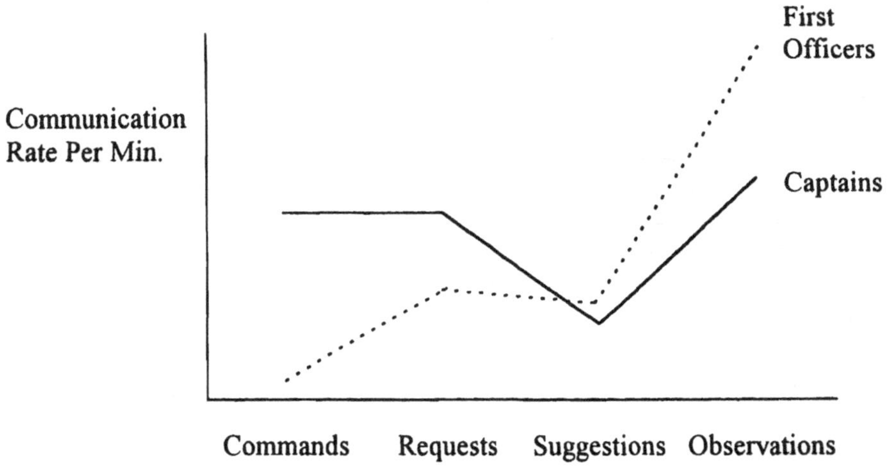

Figure 9.2 Overall communication rate versus control level for captains and first officers

Next let us consider the effects of expressivity on communication rate for captains and first officers. For these results we look only at the communication after the engine failure. Table 9.3 shows the communication rate versus the grammatical form used, for the four combinations of crews based on expressivity as determined by the PAQ. Examining these data one can see that, in combination with High E captains, first officers make fewer command statements and more observation statements than they do when combined with Low E captains. High E captains make about as many commands as Low E captains but they make far fewer suggestions and observations. Low E first officers make more requests than High E first officers whether with High E or Low E captains.

The results of this study indicate that through the use of basic group communication coding techniques modified for the cockpit setting, cockpit communication can be assessed in a reliable, useful and consistent way. However, the method is extremely tedious. Independent study students were used as coders which meant that training for standardization was an important task.

The coders not only put all communication into the designated categories, but they also transcribed the phrases that they coded so they could be easily checked. Thus it was possible to code only about 20 minutes of each flight. Nevertheless, the study provides a useful technique for assessing crew communication behavior and CRM performance.

Table 9.3
Communication rate as a function of grammatical form for the four combinations of captains and first officers by expressivity

	Commands	Requests	Suggestions	Observations
Capt (High E)	0.90	0.57	0.51	0.82
F/O (High E)	0.13	0.26	0.34	1.11
Capt (High E)	0.45	0.50	0.92	1.17
F/O (Low E)	0.00	0.81	0.25	2.20
Capt (Low E)	0.70	0.72	0.47	0.73
F/O (High E)	0.10	0.40	0.83	1.60
Capt (Low E)	0.75	0.58	0.26	0.56
F/O (Low E)	0.19	0.78	0.60	1.30

The KLM communication coding procedure

The KLM CRM course employs an equally effective communication coding technique that is less tedious but places a premium on coder expertise and vigilance. The procedure used requires course facilitators, each to observe and listen to individual and group communication and mark on a checklist every utterance made by each of five people in their group during their discussion exercises. The items on the checklist are shown in Table 9.4 below.

If the observers are very skilled, group conversations can be coded along these 13 categories quite well. In the KLM course this technique is used very effectively. The coding and data acquisition are done in 'real-time' and summarized later. At the end of the course, the facilitators share the summarized results with each participant individually. These results are compared both with the student's own self-perception and with the results of several personality tests taken during the course. These recorded behavioral data are highly effective as a validation of the personality tests.

Table 9.4
KLM Coding checklist

Proposing	Behavior which puts forward a new concept, suggestion or course of action.
Building	Behavior which extends or develops a proposal which has been made by another person.
Supporting	Behavior which involves a conscious and direct declaration of support or agreement with another person or his or her concepts.
Disagreeing	Behavior which involves a conscious, direct and reasoned declaration of difference of opinion, or criticism of another person's concepts.
Defend/Attack	Behavior which attacks another person or defensively strengthens an individual's own position. Defending/attacking behaviors usually involve explicit value judgments and often contain emotional overtones.
Blocking	Behavior which places a difficulty or block in the path of a proposal or concept without offering any alternative proposal and without offering a reasoned statement of disagreement. Blocking behavior therefore tends to be rather bald (e.g., 'It won't work', or 'We couldn't possibly accept that').
Open	Behavior which exposes the individual who makes it to risk of ridicule or loss of status. This behavior may be considered as the opposite of defend/attack, including within this category admissions of mistakes or inadequacies provided that these are made in a non-defensive manner.
Testing Understanding	Behavior which seeks to establish whether or not an earlier contribution has been understood.
Summarizing	Behavior which summarizes, or otherwise restates in a compact form, the content of previous discussions or considerations.

Seeking Information	Behavior which seeks facts, opinions or clarification from another individual or individuals.
Giving Information	Behavior which offers facts, opinions or clarification to another individual.
Shutting out	Behavior which excludes, or attempts to exclude, another group-member (e.g., interrupting, talking over).
Bringing in	Behavior which is a direct and positive attempt to involve another group-member.

NASA/UT/FAA approach to CRM evaluation

One of the primary instruments used to measure CRM is the Helmreich 25-item CMAQ (shown in Table 9.5), which was designed to measure attitude shifts resulting from CRM training. The CMAQ has three scales: (1) communication and interpersonal awareness, (2) the captain's role and responsibility in dealing with the crew in normal and emergency situations and (3) the pilot's reactions to stress. The CMAQ has been given to thousands of pilots to assess their attitudes and changing attitudes about group behavior in the cockpit. It has been validated through performance observations in line flights (Helmreich *et al.* 1986).

Table 9.5
CMAQ -- 1991 version (Helmreich, 1988)

A = Disagree strongly; B = Disagree slightly; C = Neutral; D = Agree slightly; E = Agree strongly

_____ 1. Crew members should avoid disagreeing with others because conflicts create tension and reduce crew effectiveness.

_____ 2. Crew members should feel obligated to mention their own psychological stress or physical problems to other flightcrew personnel before or during a flight.

_____ 3. In order to maintain safety, pilots should avoid disengaging automated systems.

_____ 4. When using cockpit automation, it is less important to double check other crewmembers' inputs and selections.

_____ 5. Increased automation reduces the need for crew communications.

_____ 6. Each crew member should monitor other crew members for signs of stress or fatigue and should discuss the situation with the crew member.

_____ 7. Good communications and crew coordination are as important as technical proficiency for the safety of flight.

_____ 8. Pilots should be aware of and sensitive to the personal problems of other crew members.

_____ 9. The captain should take physical control and fly the aircraft in emergency and non-standard situations.

_____ 10. The pilot flying the aircraft should verbalize plans for procedures or maneuvers and should be sure that the information is understood and acknowledged by the other crew members.

_____ 11. Crew members should not question the decisions or actions of the captain except when they threaten the safety of the flight.

_____ 12. Crew members should alert others to their actual or potential work overloads.

_____ 13. Even when fatigued, I perform effectively during critical flight maneuvers.

_____ 14. Captains should encourage crewmember questions during normal flight operations and in emergencies.

_____ 15. There are no circumstances (except total incapacitation) where the first officer should assume command of the aircraft.

_____ 16. A debriefing and critique of procedures and decisions after each flight is an important part of developing and maintaining effective crew coordination.

_____ 17. My performance is not adversely affected by working with an inexperienced or less capable crew member.

_____ 18. Overall, successful flightdeck management is primarily a function of the captain's flying proficiency.

_____ 19. The effective crewmember always uses the automation tools provided.

_____ 20. Automated protections reduce the requirement for flight crewmembers to monitor aircraft systems and flight status.

_____ 21. The pre-flight crew briefing is important for safety and for effective crew management.

_____ 22. Effective crew coordination requires crew members to take into account the personalities of other crew members.

_____ 23. The captain's responsibilities include coordination between flight and cabin crews.

_____ 24. A truly professional crew member can leave personal problems behind when flying the line.

_____ 25. My decision making ability is as good in emergencies as in routine flying situations.

CMAQ interpretation

As a result of the observation of 114 airline pilots who had taken the management attitude questionnaire, Dr. Helmreich has concluded that the effective pilot/manager is one who (Helmreich, Foushee, Benson and Russini, 1986):

- recognizes his or her own personal limitations
- recognizes his or her diminished decision capability in emergencies
- encourages other crew to question decisions
- is sensitive to personal problems of other crew
- discusses personal limitations
- verbalizes plans while acting as pilot flying

- provides training for other crew
- has a relaxed and harmonious flight deck
- adapts management style to crew and situation
- coordinates cabin crew activities

The expertise approach

Since the mid-1980s, many behavioral scientists have begun to expand the use of a new approach to discover how so-called experts think and how they have acquired their expertise. This approach, which we saw briefly earlier in this chapter (McGuire and Babbott, 1967), is taken in part because of the realization among scientists that complex systems need to be examined as a whole to understand the interactions that occur among the various elements making it up. The scientific method which requires careful control over the smallest element to observe its performance under various conditions does not permit the study of a complex system as a whole. As we saw at the end of Chapter 1, the expertise approach has the potential to greatly expand our understanding of human behavior in complex systems and to advance technology to levels not possible with strict adherence to the scientific method.

The study of expertise seeks to understand and account for what distinguishes outstanding individuals in a domain from less outstanding individuals in that domain, as well as from people in general. The approach is limited to outstanding behavior that can be attributed to relatively stable, learned characteristics of the relevant individuals. The classical expertise literature says that expertise is acquired over decades of experience and practice. The approach includes both a careful analysis of the attained performance and an analysis of how that performance was attained.

Building cognitive models using the expertise approach involves three steps: (1) Identifying representative tasks that capture the essence of superior performance in a specific domain, (2) Detailed analysis of the superior performance through several methods including verbal reports during performance of the tasks (or stimulus materials), (3) Efforts to account for the acquisition of the characteristics and cognitive structures found to mediate superior performances of experts.

The identification of a collection of tasks that can capture superior performance is often not easy. It may be helpful to identify only a small number of such tasks and test them on a larger population of experts and novices in a particular domain. Alternatively, one might identify a large number of tasks to be used on a small number of subjects.

The full range of methods of analysis in cognitive psychology can be applied in the examination of phenomena associated with a particular type of expertise including performance analysis, expert-novice comparisons and extensive studies of single subjects. One cannot, of course, directly observe the mediating steps in information processing but it is possible to specify hypotheses about relations between the internal processing steps and observable behavior. Think-aloud verbalization and retrospective verbal reports are the most common methods of assessing differences in mediating processes as functions of the subject's level of expertise. Subjects are asked to think aloud while carrying out a series of representative tasks.

Although the expertise literature says that aggregation of experience is the single most important factor in the acquisition of expertise (Chase and Simon, 1973), 'meaningful' experiences are at least important if not essential for learning to occur. For example, some tasks never seem to be learned even though they are performed repeatedly. Several studies have demonstrated that, at best, minor improvement is gained with years of experience in comparison with initial training (Ericsson and Smith, 1991). In aviation, pilots know that flight experience does not equate to expertise -- yet, hours of flight time continues to be the criterion for currency, ratings and certificates. Other important mechanisms at work include the teacher, instructional media, motivation and internal adaptation processes.

Compared with statistical procedures, expertise procedures are not likely to yield more reliable results. However, by going through the expertise procedure, we are more likely to encounter and develop a better approach toward interventions than we would using the statistical procedures. This is because much of the expertise approach deals with the acquisition of expertise, not just the explanation of expertise. The statistical approach may retrospectively identify what happened and how often, but it does not suggest the causal factors or remedial actions. The expertise approach actively solicits such information and the results lead more quickly into intervention strategies.

Epilogue: practical judgment evaluation

My first, and still my best experience with pilot judgment occurred at Moody Aviation where first I learned to fly. Moody places a great deal of emphasis on judgment in their selection and training of pilots -- and I was the 'victim' of some of that evaluation, failing two phase checks for that reason. Later, when I was doing my first research on pilot judgment for the FAA in 1976, I asked Dirk E. Van Dam, director of the school (and the most dreaded of Moody's judgment examiners) to give his understanding and application of the subject to the evaluation of his flight students. He provided a brilliant explanation of his (and Moody's) approach to judgment training and evaluation (Van Dam, 1977), parts of which I have used throughout this book. From the following text provided by Mr. Van Dam, one can see some practical lessons in the evaluation of judgment in an aviation school environment. Reference is made in Mr. Van Dam's comments to the Moody flight camp which is a week-long period during which all candidates who have passed a series of prior non-aviation hurdles, are evaluated on numerous parameters including learning ability, attitude, flying ability and many other factors. At the end of the week, half of the candidates are selected to continue in training.

Figure 9.3 Moody Aviation facility at the Elizabethton Municipal Airport as it is today with N4541F the same C-185 in which the author failed his commercial flight check 28 years ago.

COMMENTS FROM MOODY AVIATION
Dirk E. Van Dam, Director

Moody judgment definition

'The ability to perceive various alternatives in a dynamic situation, and to choose the most acceptable in terms of personal, environmental and equipment capabilities, while pursuing the desired or planned objective'.

Judgment as a criterion for trainee selection

It is quite apparent that projection of judgment in flight situations has some parallels in the background life style and record of candidates when we first encounter them. A generally conservative approach to 'borderline' situations seem to be an index of whether there is a regard for margin which is so important in flight operations. The way a person *thinks about* regulations and good operating practices seems to give evidence of this characteristic. In tests that we conduct for personality and behavior patterns, it appears that an *ease in decision-making* is significantly important, particularly for the pilot candidates. The slow deliberator and the person who is threatened by the 'go--no go' type decisions will be often frustrated and sooner yield to other emotional pressures rather than choose the proper alternatives in a situation. I feel quite strongly that a *need to be noticed* is a characteristic which requires critical attention. The 'show-off' tendency may have a decided effect on which option is chosen in flight situations. If this characteristic can be identified and qualified or quantified it may be supportive of the type of judgment which can be expected. The desired emotional nature of the person probably involves a more restrained or less reactionary response to situations. Discretion for pilots often requires being able to distinguish between the quick and easy approach and the *planned and prepared* approach.

It is probably also true that a certain amount of passivity rather than a defensively aggressive response promotes safety consciousness in an individual. I strongly believe that the person who needs to 'prove himself' when faced with questionable situations is a dangerous individual when confronted with hazardous conditions. The person who has a reasonable need to support authority and follow directions, as well as a need for rules and supervision, can be expected to follow standard procedures and good operating practices rather than experiment on the threshold of danger.

Finally, some index of the person's emotional security seems of great importance. The candidate who is 'free to fail' in the sense that an error or mistake is treated as an opportunity to learn, has much going for him. In contrast, the person who in any way prefers not to recognize errors or limitations will probably have difficulty in learning, and is likely to profit little from either his own shortcomings or that of others when they are brought to his attention. A teachable spirit, in my opinion, is a very important index to the learning which will later take place through personal experience or by vicarious exposure to situations which develop judgment. An individual's personal security about his decision-making ability and the credentials which he recognizes in himself for dealing with flight issues, seems to be an obvious characteristic of those who have long records of safe flight operations.

Initial indicators of judgment in preliminary training

During the week long process of flight camp evaluation, some of the factors which figure significantly in the overall evaluation, and particularly those which reflect judgment include:

1. *Obvious effort and attention to instruction.* A casual or careless response is considered negative and usually results in a lack of sensitivity to signals which may affect choices.

2. *Relaxation.* Alertness without being frustrated or given to anxiety about details without consequence.

3. *Division of attention.* An ease in dealing with a context where many things need to be integrated and cannot be done 'when it's convenient'.

4. *Response delays.* The individual who must 'compute' the solution appears to have difficulty in determining alternatives intuitively. Judgment in flight situations seems to require above-average ability in thought processes under stress.

5. *Confidence.* The capacity to realize that the solution *is* available and that it *can* be learned.

6. *Capacity for problem-solving.* The rather obvious awareness and capability that shows even though a procedure has not been prescribed nor instructions specifically given. The situation easily results in a best choice. The 'common sense' approach to weighing alternatives and making a choice in line with objectives and definitely within any perimeters required. An obvious positive trait.

7. *Initiative.* The tendency to take action before a deadline. This quality appears to provide incentive for preparation so that situations which require pre-planning are encountered *with* that benefit. Initiative seems to be the initial step in planning ahead.

Evidence of judgment development

Some of the evidences that practical judgment is being developed through a process of flight training are as follows:

1. *Eagerness to learn or high motivation.* Punctuality, preparedness, attention to details--perhaps even a little tenseness, give evidence that the person will be prepared when a critical decision is to be faced.

2. *Teachability.* The person is a good listener, asks some questions, is inquisitive and applies what he is told, can accept constructive

criticism and suggestions without excuses or defenses. Doesn't *always* require full explanations to be convinced.

3. *Adaptability and flexibility.* The person is able to respond *positively* to change in procedure or technique. Looks for ways to improve and/or develop his performance. Has interest in ideas and concepts as well as in facts.

4. *An intuitive quality in thinking and/or decision-making.* Not overly mechanical or academic in treatment of data. Not afraid to experiment within conservative personal or equipment limitations. Doesn't require explicit details in order to proceed. Can improvise. Not limited to 'computed' answers but can visualize patterns and processes easily. Directs attention to actions which *answer* the problem; not just explore it.

5. *A pattern of good choices.* Decisions *usually* work out to confirm they were the best choice. This person avoids hasty reactions and thinks before speaking or acting. Not impulsive. Generally has alternatives in mind and is not afraid to say 'I don't know' or 'I'd like to think about that some more'. Expects problems but doesn't worry about things he can't control. Relatively disciplined in some typical areas of life style but not so rigid that he can't do it another way.

6. *Applies margins and allowances.* Normally has an answer for the question 'what if----'. Seldom comes up with exact data which is 'guaranteed'. This person seldom 'gets caught' in not having been prepared for assignments, responsibilities and expectations of his peers and supervisors.

Evaluation of judgment in advanced pilot training

In observing many trainees over 20+ years, I'm convinced that a significant evidence of judgment is that the person remains a *learner* throughout his training period. The developing person has a unique balance between confidence through experience, and a discontent because he has not yet fully achieved. It seems apparent that a person who is well aware of his own limitations as well as of his capabilities can make the best judgments in flight situations that are borderline. The learner who continues to have a curiosity about new information on equipment and the environment also continues to be armed with the knowledge that will be necessary. Complacency or overconfidence which become apparent as soon as initial achievements like certificates and ratings occur, is an index of a deterioration in judgment.

The individual who spends much time in 'hangar flying' with even mild exaggerations of his performance, can usually be expected to perform poorly in real-life situations. We observe that the individuals who demonstrate the highest quality of flying skill and judgment are also most often the least demonstrative outside the airplane.

Summary

In general, I think it can be rightly observed that pilot judgment is an important element in *selection* or in *encouragement* of candidates for flight training. Although a very subjective quality, there appear to be valid tools by which it can be identified and measured even before pilot training is begun. It appears also realistic that once started, pilot training gives evidence that a person responds favorably or unfavorably to expectations or criteria by which pilot judgment can be evaluated. The training process itself, also can be arranged to provide the necessary emphasis and standards for development of discriminating abilities and response patterns which are expected for the professional pilot level. Those who teach can be prepared to emphasize by theory, demonstration and directed practice those procedures and guidelines which create a clear and relevant example for trainees. On the negative side, it may be worth consideration that strong indications of behavior or attitudes which identify with poor judgment should result in such candidates being discouraged.

References

Abeles, H.F. (1973). 'A facet-factorial approach to the construction of rating scales to measure complex behaviors'. *Journal of Educational Measurement*, **10**, 145-151.

Anastasi, J.A. (1968). *Psychological Testing*, 3rd edn. New York: McMillan.

Bellows, R. (1963). *Psychology of Personnel in Business and Industry*, (3rd ed.). Englewood Cliffs, N.J.: Prentice Hall.

Berlin, J. and Holmes, C. (1981). 'Developing a Civil Aviation Pilot Judgment Training and Evaluation Manual'. In R.S. Jensen (ed.) *Proceedings of the First Symposium on Aviation Psychology*, Columbus, OH: OSU Department of Aviation.

Borg, W.R. (1963). *Educational Research: An Introduction*. New York: David McKay Co.

Brody, N. (1963). 'N-achievement, test anxiety and subjective probability of success in risk taking behavior'. *Journal of Abnormal and Social Psychology*, **66**, 413-418.

Buch, G. (1983). Pilot judgment training validation experiment'. In R.S. Jensen (ed.) *Proceedings of the Second Symposium on Aviation Psychology*, Columbus, OH: OSU Department of Aviation.

Buch, G.D. and Diehl, A.E. (1982). 'An investigation of effectiveness of pilot judgment training'. *Human Factors*, **26**(5), 557-564.

Chase, W.G. & Simon, H.A. (1973). 'Perception in chess'. *Cognitive Psychology*, **4**, 55-81.

Coombs, C.H. and Pruitt, D.G. (1960). 'Components of risk in decision making: Probability and variance preferences'. *Journal of Experimental Psychology*, **60**, 265-277.

Cronbach, L.J. and Gleser, G.C. (1965). *Psychological Tests and Personnel Decisions*. Urbana, IL: University of Illinois Press.

Cumming, R.W. (1964). 'The analysis of skills in traffic'. *Australia Road Research*. Victoria, Australia.

Diehl, A.E. (1992, January). 'Does cockpit management training reduce aircrew error'? *ISASI Forum*, **24**(4).

Ericsson, K.A. & Smith, J. (Eds) (1991). *Toward a general theory of expertise: Prospects and limits*. Cambridge: Cambridge University Press.

Federal Aviation Administration (1994). *Private Pilot Practical Test Standards* (FAA-S-80811-AS). Washington, DC: Superintendent of Documents, US Government Printing Office.

Gopher, D. and Kahneman, D. (1971). Individual differences in attention and the prediction of flight criteria. *Perceptual and Motor Skills*, **33**, 1335-1342.

Guilford, J.P. and Lacey, J.I. (1947). *Printed Classification Tests*. Washington DC: US Government Printing Office. (AAF Aviation Psychology Program research Report No. 5).

Gumpper, D.C. and Smith, K.R. (1968). The prediction of individual accident liability with an inventory measuring risk-taking tendency. *Traffic Safety Research Review*, **12**(2), 50-55.

Helmreich, R., Wilhelm, J.A. and Gregorech, S.E. (1988). *Revised versions of the cockpit management attitudes questionnaire (CMAQ) and CRM seminar evaluation form* (NASA/The University of Texas Technical Report 88-3-revised 1991). Austin: University of Texas.

Helmreich, R.L., Foushee, H.C., Benson, R. and Russini, W. (1986). 'Cockpit management attitudes: Exploring the attitude-performance linkage'. *Aviation, Space and Environmental Medicine*, **57**, 1198-1200.

Hogarth, R.M. (1980). *Judgment and Choice: The Psychological Decision*. Chichester: Wiley.

Holmes, T.H. and Rahe, R.H. (1967). 'The social readjustment rating scale'. *Journal of Psychosomatic Research*, **2**, 213-218.

Janis, I. and Mann, L. (1977). *Decision Making: A Psychological Analysis Of Conflict, Choice and Commitment*. New York: Free Press.

Jensen, R.S. (1986). *The Effects of Expressivity on Cockpit Communication and Management.* Columbus, OH: The OSU Research Foundation/NASA Final Report.

Jensen, R.S., Adrion, J. and Maresh, J. (1986). *A Preliminary Investigation of the Application of the DECIDE Model to Aeronautical Decision Making.* Columbus, OH: OSU Aviation Psychology Laboratory, Final Report AOPA Air Safety Foundation Grant.

Jensen, R.S., Adrion, J. and Maresh, J. (March, 1987). *Pilot Judgment Skills Test Battery.* Columbus, OH: Aviation Research Associates Final Report to US Air Force.

Jensen, R.S., Kochan, J. and Chubb, G.P. (1994). *A New Approach to Aeronautical Decision Making: The Expertise Method.* Columbus, OH: OSU Department of Aviation.

Johnson, C.D. and Davis, J.H. (1972). 'An equiprobability model for risk-taking'. *Organizational Behavior and Human Performance*, **8**, 159-175.

Kalez, M.M. and Hovde, R.C. (1945). 'Pilots with repeated 'pilot-inaptitude' accidents'. *Journal of Aviation Medicine*, **16**, 370-375.

Klein, D. (1976). Social aspects of exposure to highway crash. *Human Factors*, **18**, 211-219.

Kunkle, E.C. (1946). The psychological background of 'pilot error' in aircraft accidents. *Journal of Aviation Medicine*, **17**, 533-567.

Levine, H.G. and McGuire, C. (1970). 'The validity and reliability of oral examinations in assessing cognitive skills in medicine'. *Journal of Educational Measurement*, **7**, 63-74.

Levine, H.G. and McGuire, C. (1968). 'Role playing as an evaluative technique'. *Journal of Educational Measurement*, **5**, 1-8.

Lord, F.M. and Novick, M.R. (1968). *Statistical Theories of Mental Test Scores.* Reading, MA.: Addison Wesley.

Lucas, R.L. (1970). Attitudes, personal characteristics and driver behavior. In N. W. Heimstra (ed.) *Injury Control in Traffic Safety.* Springfield, IL: Thomas.

McGuire, C. and Babbott, D. (1967). 'Simulation techniques in the measurement of problem solving skills'. *Journal of Educational Measurement*, **4**, 1-10.

Meehl, P.E. (1954). *Clinical vs Statistical Prediction.* Minneapolis: University of Minnesota Press.

Naatanen, R. and Summala, H. (1973). 'Physical and psychological aspects of crash barriers'. *Accident Analysis and Prevention*, **5**, 247-251.

Peck, M.S. (1978). *The Road Less Traveled.* New York: Simon and Schuster.

Rockwell, T. (1972). 'Skills, judgment and information acquisition in driving'. In T. W. Forbes, (ed.). *Human Factors in Highway Traffic Safety Research*, New York: Wiley.

Rogers, L.E. and Farace, R.V. (1975). 'Analysis of relational communication in dyads: New measurement procedures'. *Human Communication Research*, 1, pp 222-239.

Roscoe, S.N. (1980). *Aviation Psychology*. Iowa City: Iowa State University Press.

Rotter, J.B. (1966). 'General expectations for internal versus external control of reinforcement'. *Psychological Monographs*, **80**(1).

Shanteau, J. (1975). Information integration analysis of risky decision making. In M. Kaplan and S. Schwartz (eds.) *Human Judgment and Decision Processes*. New York: Academic Press.

Shealey, J.E. (1974). 'Risk-taking in skilled task performance'. *Proceedings of the Eighteenth Annual Meeting of the Human Factors Society*. Santa Monica, CA: Human factors Society.

Shelley, M.W. and Bryan, G. L. (eds.) (1964). *Human Judgments and Optimality*, New York: Wiley.

Sternberg, S. (1969). 'The discovery of processing stages: Extensions of Donders' method'. *Journal of Experimental Psychology*, **27**, 276-315.

Strauss, G. and Sayles, L.R. (1967). *Personnel: The Human Problems of Management*. Englewood Cliffs, NJ: Prentice Hall.

Tannenbaum, R. and Schmidt, W.H. (1975). 'How to choose a leadership pattern'. *Harvard Business Review*, May-June.

Taylor, D.H. (1976). 'Accidents, risks and models of explanation'. *Human Factors*, **18**, 371-380.

Taylor, R.N. and Dunnette, M.D. (1974). 'Influence of dogmatism, risk-taking propensity and intelligence on decision making strategies for a sample of industrial managers'. *Journal of Applied Psychology*, **59**, 420-423.

Tillman, W.A. and Hobbs G.E. (1949). 'The accident prone automobile driver: A study of the psychiatric and social background'. *American Journal of Psychiatry*, **106**, 321-331.

Torrance, E.P. and ziller, R. C. (1957). *Risk and Life Experience: Development of a Scale for Measuring Risk-Taking Tendencies*. USAF personnel training and research center Research Report 57-23.

Van Dam, D.E. (1977). 'Comments from Moody Aviation'. In R.S. Jensen and R.A. Benel, *Judgment Evaluation and Instruction in Civil Aviation*, Springfield, VA: National Technical Information Service, FAA Report, FAA-RD-78-24.

Wallach, M.A. and Kogan, N. (1959). Sex differences in judgment processes. *Journal of Personality*, **27**, 555-564.

Webb, E.J., Campbell, D.T., Schwartz, R.C. and Sechrest, L. (1966). *Unobtrusive Measures*. Chicago: Rand McNally.

Wechsler, D. (1939). *The Measurement of Adult Intelligence*. Baltimore: Williams and Wilkins.

10 Complacency

Complacency is caused by the very things that should prevent accidents...factors like experience, training, and knowledge contribute to complacency. Complacency makes crews skip hurriedly through checklists, fail to monitor instruments closely or utilize all navigational aids. It can cause a crew to use shortcuts and poor judgment, and to resort to other malpractices that mean the difference between hazardous performance and professional performance.

Continental Airlines *Golden Contrails* (Nov. 1980), Quoted by E. L. Wiener (1981)

When diligence and skepticism fade in the glow of self-satisfaction, a pilot is in for a rude awakening.

Mark Patiky (1991)

One of the most insidious of all human mental states in aviation is complacency. Pilots are well aware of its existence from their own experience. What flight instructor has not seen a student appear content, probably thinking, 'All is well, I've really got this mastered', only to watch him forget to close the cowl flaps, or set the mixture, or activate a flight plan, or extend the landing gear. This state of mind can be referred to as 'FDH: Fat, Dumb and Happy' (Wiener, 1981). Familiarity causes expectations and expectations control our perceptions. Consider the following example from the ASRS data base:

> After preflight, an FAA ramp inspector found a small pin missing from a main door hinge cover that I missed on the preflight on my walk around. I was concentrating on the main gear wheels, the brake debooster and the lower gear doors and, as I walked around the landing gear, I failed to look up at the gear door attach point. If I had, I could have seen the pin missing. This error, I believe, was caused

> by complacency, lack of a positive attitude due to job security and events unfolding at my airline and a short over night (10 o'clock) layover. I had only four hours of sleep.

Pilots tend to blame their errors of omission on complacency. When I asked pilots in the US Forest Service why the most experienced pilots were usually the ones involved in accidents, their almost universal response was 'complacency'. A recent check of the ASRS data base revealed hundreds of complacency-related incidents. Complacency is sometimes mentioned by the NTSB as the cause of accidents. When retired airline captains were recently asked for the leading human frailty causing accidents, over 50 per cent said 'complacency'.

What is complacency? Why does it happen to people? Is it inevitable? Is it caused by automation? Regulations? SOPs? Management? Stress and fatigue? What can we do about it in aviation? The word has become a dumping ground for many disapproved behaviors in aviation. Still, perhaps we can learn from an examination of the word in aviation.

What is complacency?

Webster says that complacency is 'Self-satisfaction accompanied by unawareness of actual danger or deficiencies'. Earl Wiener says that complacency is 'A psychological state characterized by a low index of suspicion' (Wiener, 1981). NASA's ASRS Coding Manual (Billings *et al.* 1976) says that complacency is 'Self-satisfaction which may result in non-vigilance based on an unjustified assumption of satisfactory system state'.

Fahlgren and Hagdahl (1990) asked over 1,000 pilots to define complacency. They concluded that there are many different definitions of the term in aviation but that the best definition was, 'A mental state where a pilot acts, unaware of actual danger or deficiencies. He still has the capacity to act in a competent way -- but for some reason, or another, this capacity is not activated. He has lost his guard without knowing it'. Complacency occurs in a state of low arousal, almost unconsciousness. Deliberate violation of regulations and SOPs without sound justification is *not* complacency. *Unconsciously* following rules, SOPs, automatic systems, management edicts, without question and sound justification is complacency.

Complacency is related to the concept of judgment that we have been discussing. It is found in the Problem Vigil component of our decision model. It is related to affective components found in the Background Problem (Detailed Judgment Model, Chapter 2). Thus complacency may linked to the vigilance decrement described in Chapter 2. As a task is repeated, self-satisfaction with performance develops to the point that it can be performed reasonably well without thinking. At that point, it *is* performed without thinking, vigilance declines and safety is compromised because there is a greater risk of missing important stimuli.

Pilots use the word 'complacency' when describing their reactions to situations such as, receiving the same clearance over and over again, flying into the same airport over and over again, watching the same instruments, engines and automatic systems which work well over and over again. These are some of the 'causes' of complacency in aviation -- according to the pilots. During these repeated normal situations the pilot begins to be less vigilant for safety related stimuli.

Complacency versus boredom

In contrast to complacency, boredom is a heightened state of arousal and a frustration over a need for something to do. Boredom can lead quickly to stress and fatigue. Boredom occurs when talented, highly trained, highly educated people are given tasks that do not require their superior skills which they have worked hard to acquire. Boredom also occurs when people have very little control over their tasks. It is associated with low workload or underload coupled with heightened arousal.

Boredom is a highly distressful condition for the human. Given the choice of being bored or frightened most people will choose to be frightened (consider the types of movies people choose to view). Computer assistance designed to alleviate work overload may aggravate boredom. Boredom usually occurs when a task has been mastered, is being performed well and presents no more challenge. Boredom may induce complacency (Hopkin, 1988). The results of boredom are frustration and high stress which, in turn, lead to early onset of fatigue. Therefore, in designing tasks for pilots, we must realize that the end product of taking tasks away from pilots may be increased stress not reduced stress.

Deci's (1975) theory of intrinsic motivation holds that each person has a certain state of motivational equilibrium on a scale from inactivity (congruity) to hyper activity (extreme incongruity) which is considered comfortable (optimum) and is sought whenever we are forced away from it by our circumstances (see Figure 10.1). When you become overly stressed (or overworked), you seek to get tasks done and avoid new tasks. On the other hand, when you are inactive you become bored, causing frustration which moves you to take on new tasks. Occasionally, animals seem to demonstrate a drive for exploration when they are in a reduced state of activity. Myers and Miller (1954) and Isaac (1962) have suggested that an exploration drive is instigated by boredom in animals and humans.

When a person discovers a new game, he works at it until he masters it and then becomes bored with it. The boredom causes him to seek another activity or problem to solve, and, when it is solved, the new activity become boring. As long as the person is *free* to seek new novel stimulus, he is in a pleasurable state of mental advancement and, as we have seen in Chapter 7, he performs in some optimal way. When new, automatic technology or his organization prevents a person from seeking new problems to work on, he becomes frustrated and boredom is forced upon him.

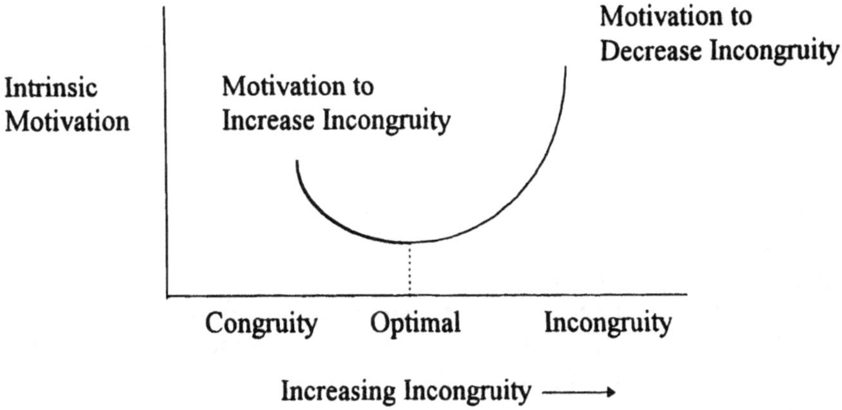

Figure 10.1 Proposed relationship between incongruity and motivation (Deci, 1975)

Boredom can also be related to expectancy. If a person has worked hard to learn to perform a task well and finds it intrinsically motivating to meet ever-changing challenges of the task and then the task is changed through automation, he may become bored and even lose some of the self-esteem he derived from doing the task manually. Task changes could also take the form of organizational limiting factors that do not permit the operator freedom to perform in ways he may think is most effective, thus destroying the satisfaction derived from a job well done.

Countering boredom

Boredom can be countered by changing the factors that cause it. Perhaps, most importantly, human tasks must be designed such that they require the best that the operators can offer from their talents without overloading them. Skilled operators should be given tasks over which they have some measure of control. Within certain limits, variation in the task can be useful in preventing boredom. From the vigilance decrement research it seems clear that increasing the hit rate may be useful in some tasks to prevent boredom. Finally, operators need to be told of the insidious effects of boredom that lead to early stress and fatigue. Mental games can be played using the aircraft systems themselves to counter boredom. Social factors can be introduced including talking with whoever is available, other crew, passengers, or ATC. The pilot can engage in useful activity such as providing reports of flying conditions that might not be reported during heavier workloads.

A word of caution is in order. In all attempts to reduce boredom, one must watch for and guard against the possibilities for greater human error opportunities. Most of the above suggestions place more responsibility on the human operators and carry the added risk of human error. It therefore becomes important that the new designs be tolerant of all human error (Billings, 1991; Wiener and Nagel, 1988). People have different levels of tolerance for boredom as well as different possibilities for error under different levels of workload.

Physiology of complacency

Is there a physiological explanation for human complacency? Repetition and monotony produce habituation of physiological responses, particularly neural ones (Mackworth, 1970). The result is that the person finds it very difficult to pay attention to a monotonous task for extended periods of time. Neurophysiological theories suggest neural inhibition, reductions in the level of arousal and changes in the amplitude and pattern of evoked potential to explain decrements in vigilance. Apparently fewer neural events reach the criterial level after habituation has occurred. At the beginning of a task, arousal responses are prolonged, but they become shorter and may finally disappear as the task proceeds.

Activation refers to the degree of excitation of the lower brain stem reticular formation (BSRF). Sharpless and Jasper (1956) found that generalized cortical arousal was diminished after repeated presentations of auditory stimuli. Once that neural inhibition had been established, they found that increases in sensation were possible only by changing the frequency or after discontinuing the stimulus for an extended period of time. It was as if the body resists stimulation of the same sort if it is repeated for some time.

In the cockpit, performance is best when activation levels are at intermediate levels. As noted in Chapter 7, Yerkes and Dodson (1908) found that if arousal (activation) is too low, performance suffers; if arousal is too high, tunnel vision may occur causing a deterioration of performance and in the extreme, panic ensues with a collapse in performance. Lager (1973) describes this 'day-dream-to-panic' continuum as a situation in which the brain is 'switched out' at low and high levels of arousal. Although it is the panic end of the spectrum that is more spectacular and results in the most study, it is often the other end that is more dangerous, because it is linked to complacency.

What can be done to bring about moderate levels of arousal in the cockpit so that peak performance is achieved from the crew? Fortunately, there is quite a large plateau in the curve at or near the optimum level of arousal. People are able to perform fairly well over a wide range of motivation, workload, stress, fatigue and wellness. However, there are some actions that could be taken to improve the level of arousal and decrease the impact of habituation in flying tasks. First, the cockpit equipment should be designed so as to keep the crew 'in the control loop'. The feedback that results from seeing one's actions is an excellent motivator and it can greatly reduce the vigilance decrement. Second, changing the working environment including the crew or even the airplane can be useful in maintaining a higher arousal and motivation. However, in doing so one must be very careful to avoid introducing factors that increase other errors into the system.

Schemas

When a pilot is in a complacent state, his or her level of mental and sensory arousal is at a low level and artificial signals to raise one's arousal to an optimum level in the cockpit are not easy to find. A common 'complacency' error in aviation is the inadvertent wheels-up landing. It is so common that pilots say, in referring to it, 'There are those that have and there are those that will'. What can be done about it? Consider adding another call to yourself or your passengers that the landing gear is down. Soon this additional call becomes a part of the pilot's 'schema' and the new call is non-arousing.

Greenberg and Baron (1993) identify schemas as, 'Cognitive frameworks for understanding the external world'. Schemas develop with the experience of people and events throughout life. We develop both person schemas and event schemas. Person schemas are internal frameworks representing various traits that others can process. Event schemas are internal knowledge frameworks for various types of situations. These event schemas are also known as 'scripts'. For example, the businessman who attends a routine meeting in which the order of events is the same time after time (call to order, roll call, minutes, etc.), may miss a bit of information if there is a change from the routine. Some pilots have reported errors in similar situations in which the same clearance has been given over and over again, then it is changed and the pilot misses the change. For example, consider the following from ASRS:

> *Narrative:* We had been given a runway change and while taxiing into position on the runway, we were given a new departure frequency. We have a dual head (two frequencies) com radio. I immediately dialed in the new frequency for departure and flipped the transfer switch to the new frequency! After sitting in position for several (2-3) minutes, departure control called us on the new frequency and told us to contact tower. It was then that I realized I had switched the radio erroneously to the new frequency. After returning to tower frequency, received takeoff clearance and remainder of flight was normal. This was a case of pilot error and complacency. I am in the habit of 'flipping' the transfer switch after putting in a new frequency, like I normally do enroute when changing controller frequencies, and I 'flipped' it at the wrong time and didn't notice it. A heads up tower controller, was able to regain communication by figuring out what frequency I may have changed to. Technology is great, but there is still no substitute for

> 'attention to detail'. Sitting on the active runway at a busy commercial airport is no place to be 'off frequency'!

These schemas have powerful effects on our cognitive processes once they are formed. Figure 10.2 shows how schemas fit into the cognitive framework. Schemas affect our information processing, attention, memory retrieval and encoding. They can compromise our focus of attention for the immediate environment and leave us in a complacent state of low arousal.

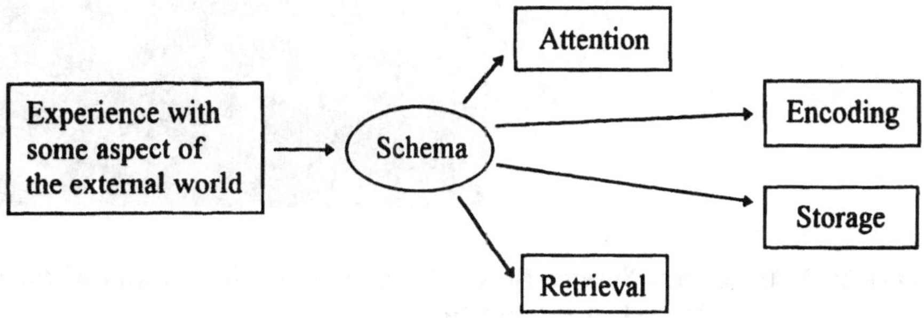

Figure 10.2 Model of schema development and impact on mental processing (Greenberg and Baron, 1993)

Task/job induced complacency

Pilots are sometimes placed into flying situations that require them to use all of the skill they have to accomplish the flight. Some pilots, such as those in the US Forest Service who fly 'leadplanes' in 'formation' with 'tankers' over fires in mountainous regions in poor visibility are faced with difficult, risky flying tasks all during the 'fire season'. It is interesting to look at their accident statistics which show that the most accidents occur after the fire season is over and they usually occur during what most pilots would consider routine, point-to-point flights. Some examples of accidents that have happened during the 'off' season include: running out of fuel during point-to-point transportation of passengers, wheels-up landings (the pilot said, 'because of extreme anger with a supervisor') and making a takeoff with the nosewheel toeing tug still attached.

These pilots complain that complacency is causing these accidents. They say that during the fire season, they are 'keyed up', always on their guard. After the high risk fire is over they relax and let down their guard and become complacent. That is when accidents happen. This is a form of task induced complacency.

Figure 10.3 US Forest Service Beech Baron as lead plane ahead of tanker dropping retardent on a fire

Another form of task induced complacency occurs when a flight or segment of a flight is repeated over and over again. Expectations build in the pilot's mind about how the flight will go, what clearances to expect, etc. These expectations cause pilots to reduce their vigilance for novel stimuli and, when they come, they are not perceived or, even worse, the expectation is perceived. For example, consider the following incident from ASRS:

> *Narrative:* During preflight preparation I usually load the ACARS and in so doing send a Miscellaneous 4 entry which automatically uplinks the ATC enroute clearance. However, the captain loaded the ACARS and the clearance request was never sent. We ran the before starting engines checklist which includes transponder. This is where both of us should have caught our omission. The captain replied, 'set and checked', I looked at the transponder, but I was looking to see if a TCAS was installed which is part of the transponder call on the checklist. I did not notice 1200 set in the transponder. This was our eighth trip DFW to JAC this month and we already knew our route, departure plan and departure frequencies, so we didn't really miss the ATC clearance until we were airborne and departure control gave us our transponder squawk. I pulled up our ATC clearance on the ACARS and verified our route. I guess this whole experience shows how easy it is to become complacent and the importance to verify checklist responses.

Another curious statistic in US Forest Service aviation accidents is that most involve highly experienced pilots. When I asked them about this, they said that, with experience, pilots become complacent. Consider the following incident from ASRS:

> *Narrative:* At push time, load planning was still waiting to hear from ramp regarding cargo weight. All three bell doors were open; lots of mail was being loaded. Company operations manual states that a two-person crew aircraft should wait at gate until load close-out is received. However, at the discretion of the captain, one can begin to taxi, but should be stopped while receiving the numbers. (The first officer must handwrite ten lines and read back three.) The norm is to taxi without the numbers if the distance to taxi and the first officer's experience level will not cause haste. In this case there was airport construction with several closed taxiways and the ground control's clearance was confusing. I was half listening to the first officer when he said, 'boy, the payload sure dropped off' (clue #1). I was half listening because the taxiway I was supposed to hold short of, I couldn't get to because of construction. I forgot to get the payload from first officer to get a rough check of gross weight. This is not a company requirement, but good sense. I asked the first officer for the maximum takeoff weight for the use of flex power. He said 7000 lbs. We used flex power. He made the takeoff and remarked how nose heavy the stabilizer trim was (clue #2). The computer gross weight had been entered in the PMS. During our climb to fl310, we encountered engine/airframe ice. The climb was a little slow, but we attributed it to bleed air losses from ice protection (clue #3?). The PMS computer indicated climb to fl370 OK; flight plan, based on expected weight said not for another hour (clue #4?). Company operations requires new flight plan if weight is heavier by a certain amount, not lighter. To request a new flight plan at push time would be rare, indeed. Anyhow, autopilot/PMS (vertical performance management system) flying aircraft in attempted climb to fl370. Speed low; crew figures it was because of power loss from ice protection at high alt (clue #5?). aircraft couldn't get above fl360. Down to almost minimum holding (mach). Wings started to rock back and forth (big clue #6 -- crew started to wake up to situation). I told first officer to turn off autopilot. He did, the rocking ceased and the aircraft started to lose altitude. I didn't know if it was his input or not. My question was answered when he grabbed the mike and requested a block altitude from ATC of fl350-370. They could not comply because of crossing traffic. Crew now aware that aircraft was losing altitude no matter what. I requested clearance back to fl330 and it was approved. With sweaty hands the weight manifest was reexamined -- and there it was:

> payload 11344 lbs. and 15 passengers. No way, unless they were all midgets. I'll bet the experienced load planner left out the passenger weight or made a 10000 lb. error. Twenty years ago, the freighters I flew had a means to measure the weight and per cent of MAC. One can make a case for such a system on a freighter. I wonder if a similar case could be made for passenger aircraft? Who said experience is the best teacher? He should have spoken with the person who said, 'with experience comes complacency'. I liked what the first officer said: 'you gotta fly the aircraft with the idea that everybody out there it trying to kill you'. Amen. It is obvious that we should not have attempted to fly the aircraft to fl370, but that is not all. Because of the weight difference, flex power might not have been used, or at least adjusted up. V speeds were wrong by about 10 knots. Stabilizer trim setting is wrong and the flap setting is wrong. The second segment climb would most likely have been a descent.

Organization induced complacency

In an article published in *The Daily Reporter* (January 26, 1990), Harry Franken quotes NTSB Board member John Lauber as saying that 'There is no accident that has ever taken place which involves individual or crew error which does not also illustrate errors in management and in the organizational side of the effort. Any time we have a failure, it is never a failure that stops at the cockpit door. It extends all the way to the top corporate offices. It is quite clear that in those airlines which have exemplary safety records, the commitment to safety begins at the top'.

Complacency can be a direct result of poor management. Failure to hold people accountable for their actions or recognize their accomplishments can cause them to be complacent just as quickly as automatic systems that direct your steering without providing feedback. Organizations that allow superiors to break rules and penalize employees who show justified creativity beyond the rules cause complacency in their pilots. Organizations that permit non-productive people to hold their jobs while others are asked to take up the slack of the non-productive employees cause complacency in their productive employees. Negative thoughts and feelings replace creative thoughts and soon the employees are acting without conscious thought. Consider the following incident report from an airline environment:

> *Narrative:* During preflight (last 5 minutes before push back) the first officer was extremely busy with a weight and balance problem due to an excessive number of changes in cargo weights and passenger counts. The captain completed through flight (before push back checklist). He missed fuel on board checklist item. During his completing of the checklists, there was an excessive number of interruptions, changes and

> disturbances. The effect of all this was that the aircraft took off with only 1 hour fuel rather than required FAR 121 domestic fuel for a flight from PHL to MDW. I think that if the weight and balance was done off the aircraft by ground personnel, that the first officer could then participate in the cockpit preparation for departure and complete a viable check. I also think, by not doing the weight and balance in the cockpit, it would significantly decrease the interruptions to normal crew operations. I also think required IFR fuel should be located not only on a before start checklist but also on an after start or taxi checklist. Both the captain and first officer should respond to fuel on board.
>
> A *Callback* conversation with reporter revealed the following information: This was the fourth time this airline has had an air turnback because of insufficient fuel. The reporter, who was just a cockpit rider, discovered the error and brought it to the flight crew's attention. They returned to PHL for more fuel. The reporter was very critical of the airline's director of training. The procedure for working the weight and balance in the cockpit is his baby and he does not see that it causes any problems. The association is trying to build a case against this procedure. The principle operating inspector is a good friend of the director of training so the flight crews cannot get any relief through this part of the FAA. Another procedure developed by the director of training and approved by this same operating inspector is to have only one pilot read and respond to the checklist if they stay on the same aircraft through a stop over and this removes the crosscheck factor. The airline had a new chief pilot and there is hope he will be more responsive to flight crew's recommendation. The director of training is ex-military and is influential in the hiring of new pilots. He has many of his old 'buddies' in supervisory positions. The company is very economy minded and had cut any extra help to the point that it is always a scramble to get 'out on time'. This is an important factor in the operation. The company policy is to accept any passenger up to the time the gate is closed and sometimes even after the aircraft door is secured thus increasing the cockpit workload to the point that it is a one man show with the first officer disturbed by last minute weight and balance changes.

Fatigue/stress/frustration induced complacency

Pilots report that stress, frustration and anger are sometimes causes of complacency. They say that when they become stressed, they pay less attention to details of flying and they become complacent. The sources of complacency causing stress can be outside of the cockpit including one's family or other relationships or the organization for which they work. For example, a US Forest

Service pilot went out for a flight after a major argument with his superior which made him furious. He was so mad that he could not stand the sound of warning horns in the cockpit and pulled the circuit breaker on the gear warning horn. He then proceeded to forget to lower the gear before landing. He attributed the accident to complacency due to the anger that he felt for his superior.

Fatigue can result from almost any activity that is overdone. Because fatigue can increase the chances for pilot error, it is essential that pilots choose their activities outside of aviation in a responsible way providing enough recovery time before flight is planned. There are certain physical and mental aspects of flight activity that are particularly fatiguing including reduced levels of oxygen, loud noise, turbulence, uncertainty, risk, requirements for precise aircraft control sometimes repeatedly or for an extended period of time, flight at all hours of the day, loss of sleep, movement through several time zones, etc. The particular combination of these factors is unique to aviation and thus fatigue is a particular concern to aviators. The following incident illustrates how fatigue can result in complacency and error.

> *Narrative:* The first officer was flying and we had a straight out departure from ORD to the south with a slight tailwind. We needed to make 4000' by 8 DME for the ORD 9 SID. I had misread the SID. On climbout, ATC appeared to give us a heading to go behind another aircraft, which was at 3000'. If we had made the SID restriction, there would not have been a problem, I don't think. We were between 3700 and 4000' at 8 DME (we were at 3000' at 6 DME, climbing about 1000 feet per minute at 160 knots). I always monitor the SID until 4000', then do climb checks. But this time for some reason I did climb checks as soon as I had time. I think I let my guard down, probably because I was very tired. I have no problem with the way the trip was scheduled (early morning departure). I just was not used to that early hour. One thing I know now is that taking off on runway 22L, we were already 2.0 DME from VOR at rotation. I will take that into account in the future. The captain and I noticed that we were at 3000' and 6 DME. That would leave 2 mile to climb 1000'. That wouldn't normally be a problem, but the tailwind made it close. When we were almost at 4000', departure control turned us to 15 degree heading for traffic, either departing or going into Midway. Later when the captain and I had thought about it, we wondered if the controller had to turn us because we had not been at 4000' by 8 DME. The captain hadn't flown in 6 weeks and I'd flown 6 out of last 7 days, both of which could have been factors. Also being a little complacent might have contributed, since we fly in/out of ORD every day and always comply with the ORD 9 departure.

Dependency and complacency

Sometimes pilots report that the person with which they are flying can cause them to be complacent. For example, if one is flying with a very dependable captain performing non-flying pilot duties one may become complacent thinking that the other person will not make a mistake. Consequently precise checks of their tasks may not be made because you may think he or she to always make sure everything is done right. Some US Forest Service pilots have even reported to me that their maintenance people are so dependable that they become complacent in checking for mechanical problems during preflight. This form of complacency is especially true if one is flying with an instructor in the right seat, but it can also be a factor in routine flying. Consider the following example from ASRS incident reports:

> *Narrative:* It was the first officer's leg and he flew J-86 off of BLD VOR (094 degree radial) Via Mead 8 departure as assigned (BLD 108 degree radial to intercept PGS 265 degree radial). Both routes go to PGS and he misinterpreted the clearance. I didn't notice until 5-8 miles off course. He then overshot PGS radial by 3-5 mile when correcting after discovery of deviation. I obviously wasn't paying enough attention...This first officer is very experienced (he's already passed captain upgrade) so unlike a new guy who you always watch out for...I was simply complacent.

Automation and complacency

Since the evolution of the airplane, cockpits have become a beacon for millions of bits of information that must somehow all be able to be displayed and controlled in a very small area. This has been accomplished mainly through automation; advanced technology which has gone so far as to change the pilot's role from that of controlling the that of monitoring. This can lead to complacent, bored and/or unmotivated pilots who are not satisfied with their new roles. Figure 10.4 shows a Wiener and Curry (1982) conceptualization of the impact of various alternatives to automation and task design. At the far end of the automation scale is boredom, complacency and erosion of competence.

Technology has been called upon to improve the information and control functions in the cockpit. When control tasks are automated, the pilot's role changes from operator/controller to monitor and supervisor. There is a growing concern in aviation that automation induced changes are leading to new types of errors, error recognition failures, takeover transients, complacency, rigidity of expectations and boredom (Jensen, 1989, p. 42-43).

Hurst and Hurst (1982, p. 75) in discussing the problem of takeover transients suggest that there is a significant 'warm up' delay when the human changes from a passive monitor to an active controller. Pilots may be lulled into a complacent or bored state in which they may be easily distracted from the monitoring task by unimportant events.

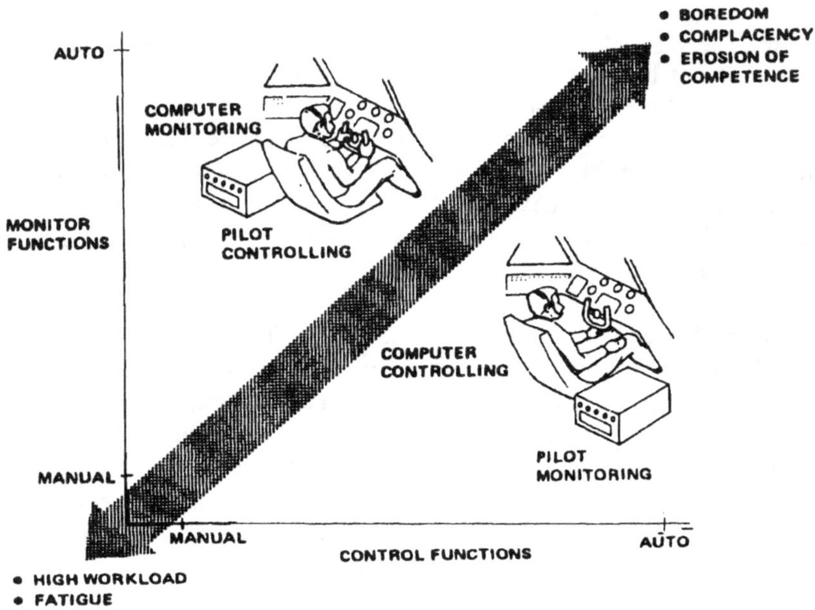

Figure 10.4 Conceptualization of different levels of automation on controlling and monitoring tasks (after Wiener and Curry, 1982)

> *Narrative:* The departure station (BWI) was inadvertently entered as the departure and arrival stations in ACARS (BWI/BWI). Clearance for the trip was received through ACARS. Since I had flown this trip sequence several times with the same first officer and the routing was always the same I only checked the transponder code and altitude. I did not question why we were cleared to 10000 ft instead of the usual 4000 ft. No check of our clearance was made with BWI since none is required (usually at least the reading back of the transponder code is required at some airports). We received the correct load closeout during taxi and then took off. The first clue that something was amiss was when departure control wouldn't answer us. A quick return to tower got us on the right frequency. I called departure control and said that we were climbing to 10000 ft. He didn't question why we were climbing to 10000 ft instead of 4000 ft (the real clearance altitude), but instead cleared us to 17000 ft and told us to change our squawk (clue #2). Clue #3 came when we requested our flight plan from ACARS and it gave us a flight plan for a DFW direct BWI trip that was flown the

> day before. Now it all came into focus. I checked our ACARS clearance and it was indeed for flight #1 DFW to BWI flown the day before instead of flight #2 BWI to ORD! Factors that I feel contributed to this include: 1) fatigue -- it was an early morning go plus I had flown 7 out of the previous 9 days. 2) Complacency -- I had flown the same trip with the same first officer several times before. 3) Procedures -- if we were required to at least read back the transponder code to clearance delivery we would have found out something was wrong. 4) Computer programming -- bias the computer to prevent it from giving out clearances that are 12 hours old. We did get the correct load closeout so why not the correct clearance?

Singh, Molloy and Parasuraman (1993) developed a 'Complacency-potential rating scale' for measuring attitudes toward commonly encountered automated devices that could potentially lead to complacency in the cockpit. Four factors were used: Confidence (in the reliability of the device), Reliance (may be used in place of manual systems), Trust (the automatic device can do the task better than I can manually) and Safety (I feel safer using the automatic device than I do when operating manually). In a test of validity these factors clearly showed their potential in revealing attitudes that could lead to complacency. However, the authors conclude that more research is needed to prove its value in aviation.

Designing jobs to reduce complacency

Organizational psychologists have a set of tools designed to improve employee morale, self-esteem, motivation and performance. These tools have filled several textbooks and most are only marginally useful to combat complacency in aviation. However, one model that may be useful in aviation to reduce complacency through job redesign is the Job Characteristics Model (Hackman and Oldham, 1980).

Figure 10.5 shows the five core job dimensions of the Job Characteristics Model: skill variety, task identity, task significance, autonomy and feedback. The first three dimensions are grouped together because they contribute to an employee's interpretation of the task's meaningfulness. It is up to management to provide autonomy and feedback to the employee. When workers are given autonomy they feel more personally responsible and accountable for the work they perform because they have freedom to succeed or fail based on their own decisions. Personal goals become identified with company goals. Feedback gives workers knowledge of the results of their work. Autonomy and feedback can result in 'critical psychological states', which in turn can lead to the achievement of both personal and company goals. The authors suggest that the model holds best for workers who have relatively high 'growth need strength'.

Most pilots are high on growth need strength. Therefore, this model should apply well. The model speaks to the need for control in the cockpit. Most would agree that because of automation, pilots have lost a great deal of their autonomy. Now computers do much of the control functions. Wiener says that pilots complain of 'being along for the ride'. Aircraft designers and regulators need to reconsider their decisions in terms of their impact on motivation and complacency. We have seen that complacency is not totally the fault of the flight crew; much of the blame can be attributed to the design of the cockpit, the pilot's organization and the FAA regulations.

Many pilots chose flying as a career because it is an individual job in which the pilot has full responsibility for a very large and expensive aircraft system. Now changes in ATC technology, ACARS, TCAS, etc. have reduced some of the pilots responsibility for navigation and flight control. It is no longer the pilot's sky; autonomy has largely disappeared; he or she is also being watched carefully by ATC for deviations that could mean reprimand and even threaten his or her job. Therefore, the job is not what was expected. Thought needs to be given to redesigning the characteristics of the pilot's job to avoid these factors.

What can we do as pilots to counter complacency?

No one would suggest that the way to counter complacency and make flying safer is to give less training, or provide less reliable equipment because it will 'keep you on your toes'. It is one thing to define the word and even set down why complacency happens. It is quite another to provide ideas for dealing with it. Complacency is a factor that everyone who operates hazardous machinery seeks to avoid. It can be seen in the posters, signs and warning labels found around such activities. Let us look at a few suggestions about awareness and attitudes that may be useful in our attempts to avoid complacency in aviation.

Awareness

Making everyone aware of the problem we humans have with complacency is a step toward countering it. If we rehearse in our minds the consequences of complacency through reading incident and accident reports, we can counter its affects. Vigilance researchers have found that occasional stimulating activity, either within the task or separate from it, can improve performance on the task. This can be done in aviation by varying your approach to flying within the safety envelope. Interact with other pilots concerning better approaches to your flying tasks. Develop a curiosity for why things are done the way they are done. Don't accept standardization as the only reason.

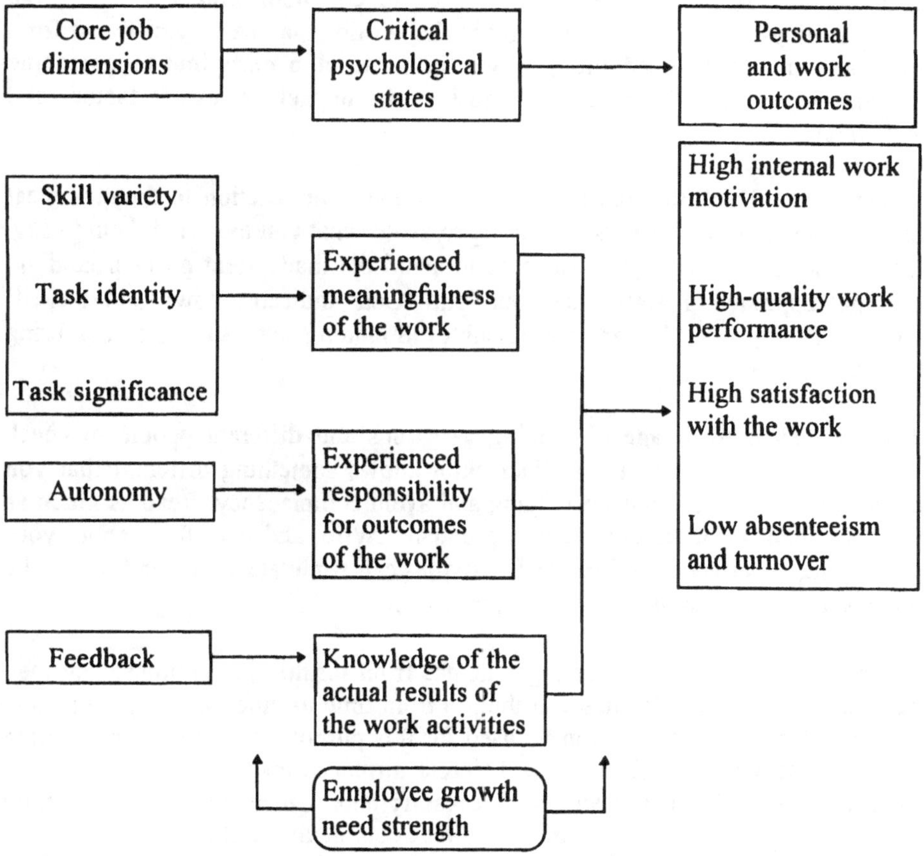

Figure 10.5 The Job Characteristics Model (adapted from Hackman and Oldham, 1980)

Attitudes

As we have seen, there is a very significant attitude component in complacency and it is in this component where the greatest potential for change exists. There are three parts to attitudes that are important as we consider what we can do:

- Affective: Your feelings for the object.
- Cognitive: Your knowledge about the object.
- Conative: Your volition or predisposition to respond in some way toward the object

The first two parts are methods we can use to communicate through to our subconscious where the third resides. The third is where the real power lies in changing our behavior. Care must be used in communicating because if the wrong feelings are evoked, the changes to the volition may be very different from what was intended. In addition, as we try to develop discipline to overcome complacency it may be helpful to consider the impact of several factors and techniques.

Expectations. Think through system failures and your reaction to them. Repeat back consciously what others say, trying to avoid what you expected them to say. Use verbalization when you can (e.g., to avoid the inadvertent gear-up landing, on final approach, I always say out loud 'Gear indicates down and locked' whether flying a fixed gear or retractable). Sounding out can sometimes bring back a conscious state.

Training. Take advantage of training variations and different people to check your flying whenever you can. Each pilot knows something different that you could use to improve your own flying and avoid complacency. Read as much as you can about aviation and aviation problems. Write and tell others about your own flying experiences. All of these activities can stimulate your mind out of the unconscious state that is known as complacency.

Job redesign. A key to preventing schemas from resulting in complacency may be to make meaningful changes to the job from time to time. In flying, changing airplanes from time to time can be used for this purpose. Pilots who must apply their knowledge (and schemas) to different airplanes may be less prone to let them cause complacency. Caution is necessary here. Changing planes frequently can also lead to confusion and increased task complexity to the extent that errors of other forms occur. Another method would be to change flight crews. Working with a new person may require explaining one's schemas to a new person can serve to add novelty and cause one to think through why the schemas are being used.

Ask, 'what if'? To avoid getting into the complacent unconscious state, continually ask yourself questions such as, 'What if an engine fails on this takeoff'? 'What if the weather closes behind me'? 'What if a bird nested in the cowling over night'? 'What if the weather forecast is wrong'? 'What if the vacuum pump fails'? 'What if I am asked to hold'? 'What if the runway is closed'? 'What if I lose communication with ATC'?

Critique yourself. Keep your head in the cockpit continually critiquing yourself. Think through anticipated situations. Consider whether you may have forgotten something. Consider how you would do this task if an examiner were sitting next to you. Consider the consequences of your actions. Ask yourself if there might be a better way. Plan ahead to avoid getting yourself into situations that will cause you to sweat.

References

Billings, C.E. (1991). Toward a Human-Centered Aircraft Automation Philosophy. *The International Journal of Aviation Psychology*, **1**(4).

Billings, C.E., Lauber, J.K., Funkhouser, H., Lyman, G., and Huff, E.M. (1976). *NASA aviation safety reporting system* (Tech. Memo. No. TM-X-3445). Moffett Field, CA: NASA Ames Research Center.

Deci, E.L. (1975). *Intrinsic Motivation*. New York: Plenum Press.

Fahlgren, G. and Hagdahl, R. (1990). 'Complacency'. *Proceedings of the 43rd Annual International Air Safety Seminar*, Rome, Italy: Flight Safety Foundation.

Franken, H. (1990). *The Daily Reporter*. January.

Greenberg, J. and Baron, R.A. (1993). *Behavior in organizations*. Needham Heights, MA: Simon and Schuster.

Hackman, J. and Oldham, G. (1980). *Work redesign*. Reading, MA: Addison-Wesley.

Hopkin, V.D. (1988). 'Air traffic control'. In E.L. Wiener and D.C. *Nagel, Human Factors in Aviation*, New York: Academic Press.

Hurst, R. and Hurst. L.R. (1982). *Pilot Error: The Human Factors*. New York: Jason Aronson.

Isaac, W. (1962). Evidence for a sensory drive in monkeys. *Psychological Reports*, **11**, 175-181.

Jensen, R.S. (1989) *Aviation Psychology*. Brookfield, VT: Gower Press.

Lager, C. (1973). 'Human factors'. In *Proceedings of the Aviation Accident Prevention Seminar*, Stockholm: Royal Institute of Technology.

Mackworth, J.F. (1970). *Vigilance and Attention*. Middlesex, England: Penguin Books, Ltd.

Myers, A.K. and Miller, N.E. (1954). Failure to find a learned drive based on hunger: Evidence for learning motivated by 'exploration'. *Journal of Comparative and Physiological Psychology*, **47**, 428-436.

Patiky, M. (1991). 'Complacency in the cockpit'. *Aviation Safety*, **11**(18), Sept 15.

Sharpless, S. and Jasper, H. (1956). Habituation of the arousal reaction. *Brain*, **79**, 655-680.

Singh, I.L., Molloy, R., and Parasurman, R. (1993). 'Automation-induced 'complacency': Development of the Complacency-Potential Rating Scale'. *The International Journal of Aviation Psychology*, **3**(2).

Wiener, E.L. (1981). 'Complacency: Is the term useful for air safety'? In *Proceedings of the 26th Corporate Aviation Safety Seminar* (pp. 116-125). Denver, CO: Flight Safety Foundation.

Wiener, E.L. and Curry, R.E. (1982). 'Flight deck automation: Promises and problems'. In R. Hurst and L.R. Hurst, *Pilot Error*, London: Granada, pp 67-86.

Wiener, E.L. and Nagel, D.C. (1988). *Human Factors in Aviation*. San Diego: Academic Press, Inc.

Yerkes, R.M. and Dodson, J.D. (1908). 'The relation of strength of stimulus to rapidity of habit-formation'. *Journal of Comparative Neurology and Psychology*, p. 459.

11 Risk management in aviation

Risks are a necessary part of a worthwhile progressive life. The risk should not be greater than the value of the undertaking. Proper supervision increases the value and decreases the risk. Accident prevention is a worthwhile goal that is intimately connected with efficient management and intelligent supervision.

We need not treat safety as a separate part of our planning or our accomplishments, neither should we separate the principles of safe driving from safe flying, safe recreation from safe working, or safe living from ethical thinking. In the final analysis, accident-free days are the pleasant by-product of efficiency.

Moody, Elementary Safety Practices (1968)

Is flying a 'risky business'? Absolutely! Does this suggest that it should be strictly avoided? Absolutely not! Risk is something that we live with everyday. The question is not whether to engage in risky activities, but instead, when and why to do so. All of us can be pressured into taking risks by our circumstances. What factors cause you to take risks that you might not otherwise take? When you are in a long line of cars at a busy intersection, how long does the gap in traffic need to be before you will risk turning left across and into the traffic? Does the risk gap get shorter as you wait longer? If the driver behind you is honking? If you are in a hurry to catch a plane? If you have impatient passengers? We take risks in flight for many very legitimate reasons (e.g., transportation, convenience, economics). Sometimes, we make risky choices based on less legitimate reasons as well. After a close call, many people rationalize, 'I had no choice, I had to do it'. Safety in our aviation system depends, to a great extent, upon the amount of control we exercise over our choices to take risks.

Risk assessment in the absolute sense in every aviation decision is not possible. However, it is possible to assess the relative risk comparing one alternative with another. This form of risk assessment requires a thorough knowledge of the aviation system. We use this procedure in aviation to manage our risks. This chapter is offered to pilots as a guide to recognizing, assessing and managing risks.

Risk management in aviation must be considered from two perspectives: the organization and the individual. As we will see in the next chapter, the organization for which the pilot flies is responsible for setting the risk management culture around which the individual pilot operates. You as the individual pilot also must develop your own risk management plan both to fit the requirements of the organization and the safety philosophy you have established for yourself.

However, while flying for personal reasons, there is another factor that you must consider. Recall the two-part model of good judgment given in Chapter 2:

> *The ability to discover and establish the relevance of all available information relating to problems of flight, to diagnose these problems, to specify alternative courses of action and to assess the risk associated with each alternative.*
>
> *The motivation to choose and execute a suitable course of action within the available time frame.*

Where:

a. *The choice could be either action or no action and,*

b. *'Suitable' is a choice consistent with societal norms.*

The term 'societal norms' refers to the level of risk that your passengers expect you to accept on their behalf. Unfortunately, in most situations, this factor is unknown to the pilot. In the context of a company, these standards reflecting societal norms are (or should be) set by the top people. When you are flying on your own, you will use your own standards but you would be well advised to consider standards that you could reasonably assume your passengers would expect. In many cases, you will not know these standards until you get to court after an accident! Later in this chapter, I have presented an example of how personal standards can be set for takeoff go/no go decisions.

Risk definition

In the insurance industry, where risk is a central factor of business, the following definition is used:

$$\text{Risk} = P \times C \times E$$

Where:

P = Probability of a loss
C = Cost of a loss
E = Exposure

In the dictionary risk is defined as an exposure to the possibility of harm, loss, danger, or hazard (Funk and Wagnalls, 1968). In the insurance world and in most aviation applications, this is a term with a decidedly negative connotation. In the financial world, risk can have both a positive or a negative connotation. It can be positive to the investor willing to risk some of his resources on the chance for a large financial gain. In fact, this is often true in the aviation world as well. In virtually every pilot decision there is a potential gain (economic, convenience, enjoyment, etc.). We will look at these values in a minute. In examining the definition to gain an understanding of risk, each part of the definition must be considered.

The terms harm, loss, danger and hazard are not as simple as they may first appear to be. The denotative definition of harm is 'physical or psychological damage'. By expanding the definition to the consideration of psychological well-being, the risk environment becomes more complex. This raises the specter of such concepts as the risks of being late, being reprimanded, failing to follow through on a commitment, being embarrassed, having additional expenses, not being promoted, being inconvenienced and others too numerous to mention. Obviously, these risks are real and do affect the personal decision process of every pilot.

Risks may be categorized as, physical, financial, legal, egocentric, sociological, aspirational and temporal. Each or all of these categories could be a part of one's thought process in many decisions. The weighting of each of these factors in the judgment process is very personal and is situationally dependent. Each factor can be a psychological motivator, as perceived and can bias decision making. Your judgment can sometimes affect the risk exposure of others. Depersonalized or displaced risk has a different significance than immediate or personal risk. Risk is not a highly specific or localized concept. It is a matter of perspective and is relative in its implication. The ramifications of risk can be many and varied in our highly technical, litigious, communicative and financially volatile society.

The second aspect of the definition of risk is the 'exposure to the possibility' of harm. This says that there must be exposure to the situation and that there must be some probability of harm. If there is no exposure, there is no risk. If there is exposure, but no uncertainty, the situation is not probabilistic. It is instead deterministic. A deterministic situation does not involve risk; there is only certainty. In such an instance, simple deductive logic is sufficient. That is, if 'A' occurs, then 'B' must occur. Realistically, there may be very few encounters with purely deterministic events. There may be only situations that are virtual certainties by virtue of their overwhelming odds. Simple, straightforward problems are not necessarily deterministic. There is almost always the chance that something may go awry. Because there is always the possibility of a negative outcome, we are trapped in a risky environment.

There is a simplistic corollary that must be examined. If risk is the exposure to harm, should it not be minimized if it cannot be avoided altogether? The answer is no. Every decision has a negative or risky side. Minimizing or attempting to avoid risks would mean that nothing would ever get done in this world. Refusing to accept risks is equivalent to refusing to seize opportunities. Minimizing risks is tantamount to minimizing costs with no concern for, or appreciation of, value. Instead of simply trying to minimize risks, the appropriate course of action is to optimize risks in view of the payoffs. Mathematically, this involves balancing the expected value of the payoffs against the expected cost of the risks for a given alternative. In the actual flying environment, it is not usually practical to perform such a quantitative analysis. However, an understanding of the general concept will provide a useful decision making tool.

Risk acceptance

The process of familiarization with risk acceptance is the 'SMART' program: 'Systematic Management of Acceptable Risks Training'. After gaining a proper understanding of the risk generation phenomenon, it is possible to approach risk-acceptance decisions more objectively. In choosing between two or more alternatives using expected values, the alternative with the greatest expected value is chosen. The expected value for each alternative is the sum of the probability for each outcome, times the value of the expected outcome within the alternative.

$$EV = \sum P_{OC} \times V$$

Some of the outcomes will be negative in nature and some will be positive. Negative and positive can be thought of as desirable or undesirable, respectively. The choice of negative or positive is subjective and is based on personal and societal norms. The negative represent risks, by definition, and the positives represent payoffs or benefits. This is especially true for psychological factors since it is very difficult to attach any meaningful value to them, though they are no less real or less influential than some less abstract factors.

Suppose one were able to step away from the actual flight and compute risks. Some observations can be readily made. Ignoring the positive side of an outcome for the moment, a high risk would be generated by a situation with a high probability of a highly negative outcome. A low risk would be generated by a situation with a low probability of a slightly negative outcome. The continuum of risk levels would be defined as, 'The probability and severity of negative outcomes over the range of these extremes'. If all decision alternatives could be reduced to measurable risks and payoffs, a potential benefit/cost quotient could be defined for each alternative and a simple choice could be made. One would obviously opt for the most favorable alternative which maximizes the quotient and, therefore, optimizes the risks that are accepted. This is more easily said than done. Even if it were feasible to attach values to these abstract concepts, it would be too cumbersome to do so in the flight environment.

Consider the decision every multi-engine pilot makes during takeoff from an airport that is above the single-engine service ceiling for that airplane. Because the pilot knows that an engine failure means he must put it into the trees, some pilots may choose to leave a passenger or two behind to avoid taking on that risk. Other pilots might choose to accept that risk knowing that the probability for that outcome is very small and that he knows that pilots accept that risk all the time in a single-engine aircraft. What some pilots fail to consider in that scenario is that, the probability for an engine-failure in a twin-engined aircraft is twice that of the single making the outcome twice as likely.

Of the many factors that may be at risk, both physical and psychological, safety *must* dominate your analysis and decisions. Although strictly speaking, safety is defined as the absence of risk, safety is customarily thought of as referring to physical well-being. Safety is a relative term, especially since there is no escape from risk. However, proper training can help to avoid undue and unnecessary risks.

The systematic management of acceptable risks is largely procedural. First, review the decision alternatives and estimate the positive and negative outcomes for each alternative and their respective probabilities. Determine the expected values for risk and for benefit for each choice. If the benefit/risk quotients are all unacceptable, find ways to reduce the risks. This may include the creative development of new alternatives. Risk reduction can be accomplished in preparation for a flight and as a long-term policy or life-style. It can also be performed within the decision environment or in-flight as a compensatory measure. An example program to prepare you for risk management decisions prior to takeoff is provided at the end of this chapter.

Perceived risk and risk expectancy

Real or actual risk may be removed from and irrelevant to antecedent decision making. This fact gives rise to the saying 'hindsight is better than foresight'. Such is the case with judgment. Before the consequences of a decision are observed, the decision can only be based on the perceived consequences. Therefore, judgment is fueled by perceived risk exclusively, not actual risk (O'Hare, 1990; Layton and McCoy, 1989). It is not reality but the perception of reality that is the basis of judgment. There are inherent problems with this. Judgment is subjective in nature and operates in an environment of uncertainty. Improving judgment requires improving the veracity of perceiving risk. One approach to this is to review the reliability history of systems to obtain an idea of the probability of system failures. The same approach can be used to make more accurate estimates of risk expectancies for eventualities other than system failures. Likewise, accident scenarios should be reviewed to assist in the prediction of the consequences of encounters with risks of all types. It has been said that 'experience is the best teacher'. Sometimes, the price of experience is too dear. It is seldom sufficient and it always takes too long to accumulate. Experience can be gained much faster and more efficiently if it is gleaned deliberately through training. Experience is a very valuable and important asset that can form a vital component of a thorough knowledge base. Whether acquired personally or vicariously from others, it can be a powerful weapon in the arsenal of wisdom.

It is informative to analyze the nature of errors in perceiving risk. Taking the extreme cases only, real and actual risk may be either high or low. These can be used to define a two by two matrix as shown in Figure 11.1. The one cell in the matrix that is the most dangerous is where the actual risk is high and the perceived risk is low. This would lead to accepting what was thought to be a low risk and getting into a high risk situation. Two of the cells provide perfect information. The actual and the perceived risks are the same, an ideal combination. The last cell should produce a conservative and safe basis for judgment, but not be optimal since it is incorrect. This is the cell that combines low actual risk with high perceived risk. If there is a perceptual error, it is preferable that it be in this direction.

		Actual Risk	
		Low	High
Perceived Risk	High	Undue Concern (Excess Caution)	Preparedness (Caution)
	Low	Preparedness (Safety)	Unpreparedness (Danger)

Figure 11.1 Perceived risk and actual risk

Knowledge base

The absolute importance of a good knowledge base cannot be overemphasized. Knowledge is essential for preparing and coping with emergencies. Actually, more than knowledge is required; there must also be understanding. Though it is possible to fly without understanding, it is not possible to fly safely without it. It is necessary for improving skills, coping ability, creating, recognizing and evaluating alternatives, estimating expectancies, foreseeing outcomes and applying logic. Acquiring knowledge is not always easy, fast or inexpensive, but it is the key to risk management in aviation.

Coping ability and risk sensitivity

In some situations we may have little or no control over the outcome once we become involved. In others, we may have the opportunity directly to affect the outcome. In the former instance, all that can be done to manage the risk is to avoid the situation, if possible. In the latter case, there may be an opportunity to match our abilities against the demands of the situation. The measure of the match is our coping ability. Coping ability is a function of confidence, skill, knowledge and understanding, strength and stamina, emotional set or attitudes, physiological and psychological state, assistance from others and equipment. The synergistic combination of these determines our ability to cope with the demands with which we may be confronted.

The degree to which our coping ability exceeds the demands of the situation is a measure of risk sensitivity. Low sensitivity would indicate relatively low demands while high sensitivity would indicate relatively high demands. High sensitivity would mean that there is little chance of our being able to prevent a negative outcome. Coping ability is one's capacity successfully to handle a situation. Since we do have considerable control over coping ability, we can reduce the net risk by maintaining the highest practical level of coping ability. In any case, it is critical to know your own limits. It is also necessary to foresee the demands of a decision alternative in order to estimate risk sensitivity. This is a matter of training, anticipation, experience and understanding.

Errors in the perception of coping ability also create a problem. Defining a matrix of extremes, actual and perceived coping ability can be either high or low. The resulting matrix shown in Figure 11.2 reveals two cells which show correct perceptions of coping ability where the low's and high's are matched. The greatest danger occurs when the actual coping ability is low and the perceived coping ability is high. This is a matter of overconfidence and could lead to perilous behavior. The other mismatch occurs in the cell where actual coping ability is high but the perceived coping ability is low. This is an instance of insecurity. It will probably produce conservative behavior, but the individual may not live up to their potential and may forego opportunities that are within reach.

If a risk situation does develop and you are unable to cope, the consequences default to a negative outcome. This negative outcome is the basis of the risk penalty. It is the cost of getting into something that you cannot handle. A low penalty would be an outcome which is only slightly undesirable. In contrast, a high penalty would be a catastrophic outcome. Risk penalty is a measure of the value of the most likely outcome if you are unable to cope with the situation resulting from the development of a problem.

Risk assessment

Risk level is assessed by the estimation and combination of the risk components expectancy, sensitivity and penalty. It is an extrapolation of the expected value technique previously described. High risk is generated by high expectancy, sensitivity and penalty. There are three basic questions to be answered for risk assessment: (1) What is the probability of the occurrence (risk expectancy)? (2) What is the likelihood of not being able to cope (risk sensitivity)? (3) What is the cost of failing to cope (risk penalty)? The product of these three factors determines your personal risk level for one alternative.

		Actual Coping Ability: Low	Actual Coping Ability: High
Perceived Coping Ability	High	Over confidence (Danger)	Realistic esteem
	Low	Realistic Esteem	Insecurity (Excess caution)

Figure 11-2 Perceived coping ability and actual coping ability

It is only after risks have been assessed as accurately and objectively as possible that they can be dealt with rationally. The logical acceptance of risk must be predicated upon sound risk assessment. Once assessed, risks must be balanced against the payoffs associated with accepting those risks. If the SMART principle for risk acceptance mentioned earlier is forgotten, consider the 'CAREFUL' process. CAREFUL means Consciously Accept Risks Evaluated with Forethought, Understanding and Logic.

Utility valuation

Utility is a measure of what the payoff is worth to you. Utility means usefulness, which can be misleading. It is easier to think of utility as meaning desirability. This viewpoint removes the burden of practicality from the concept of utility. Then it is simply the value that you place on something. Utility is based on personal and societal values, is subjective, situationally dependent and may change with time. The objective here is not to specify a value system for balancing risk, but rather to point out that risks should only be accepted deliberately and with justification.

Unnecessary risks, those without sufficient associated utilities, are unacceptable and frivolous. This is the difference between investment and gambling.

Realizing the difficulties inherent in estimating risk expectancy, sensitivity and penalty and also for estimating expected payoffs and utilities, an acceptable risk is simply one that has a benefit to risk ratio of greater than one. That is, the benefits outweigh and, therefore, balance the scale in favor of accepting the risk. This is graphically portrayed in the figure below.

In the discussion of risk penalty and utility, there was no measurement scale specified. There are two reasons for this. First, a quantitative analysis in the cockpit would be distracting, add to the workload and be a risk in itself. Second, it is difficult to assign specific values to the worth of such factors as ego, peer pressure, etc. As we have said, in aviation situations that call for risk assessment, the relative values of the risk factors will be such that quantification will be unnecessary.

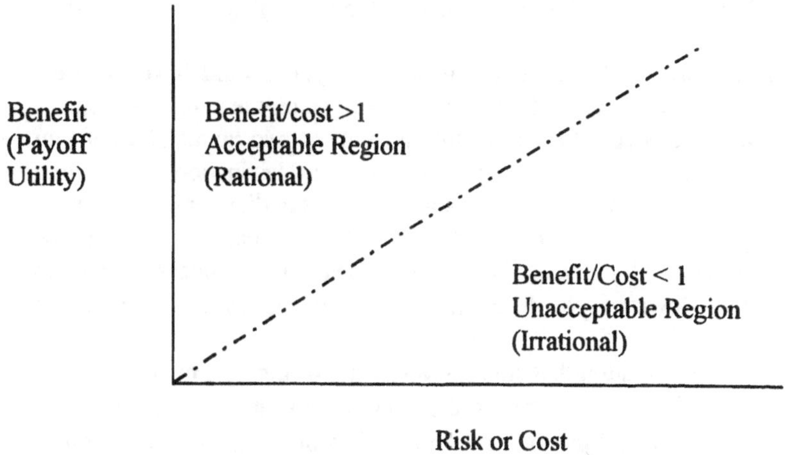

Figure 11.3 Benefit versus cost matrix

Judgment evaluation

Judgment evaluation should always be based upon this criterion of balancing risks and utilities. A common pitfall in both learning from experience and in the general retrospective evaluation of judgment, is to use a successful outcome as the sole benchmark. This is in itself poor judgment. It is also a great danger in aviation where a pilot may be applauded for the successful completion of a flight, even if it was accomplished against the odds and with great risk (see for example Wilkinson, 1994). A good decision is not one that simply results in a successful outcome. It is instead a decision that is arrived at by CAREFUL means, that systematically manages risks. For example, consider the following scenario in terms of risk vs utility:

> You are a 300-hour commercial pilot with an instrument rating. You have some recent actual instrument time and your experience is somewhat more than adequate for being legally current on instruments. You are flying alone and after having covered four hundred miles of the seven hundred on your trip home after a weekend flying holiday, you land for fuel at dusk on Sunday evening. Most of your flight has been VFR, but after having your aircraft refueled, you get another weather briefing which indicates no adverse weather enroute except for moderate turbulence and cloud cover. The clouds should be solid with ceilings at 1500 AGL and tops at 15,000 MSL. They are expected to clear along the route by 0400 local time tomorrow morning. You are flying a single engine high performance retractable which is IFR equipped. The autopilot is out for service. Two hours before you landed here, your electric turn coordinator failed and you have been unable to fix it or to get it serviced at this late hour. This is an ideal time for assessing the risks of the flight ahead of you.
>
> You are somewhat tired but you feel that you should have no great difficulty with the last leg of the trip. You will be on the gauges at night in moderate turbulence. This trip should take about two hours. Even if you have to make an instrument approach at your destination, the ceiling should be well above minimums and you are familiar with the airport. Subjectively, the flight appears fairly routine. Though there may be several aspects of the trip that are not ideal, none appear to be too demanding. However, you decide that an objective assessment of risk is in order.
>
> There is one situation that could develop that would be a major problem. Since you will be hand flying on the gauges, your attitude reference will be critical. With no functioning turn coordinator, your only real attitude reference will be your vacuum powered artificial horizon. In the two years that you have owned this aircraft, it has failed twice while you were VFR. If it fails tonight, you will be without an adequate attitude reference, in moderate turbulence, at night, for an extended period. This is the major risk that you envisage. You estimate the expectancy of the failure to occur tonight to be about one per cent. That is, you think there is about one chance in a hundred that the vacuum gyro will fail during the two hour flight home. This is the risk expectancy.

If the failure does occur with essentially no back-up, in turbulence, at night, in the clouds, you are in a very difficult situation. Realistically, you probably could not maintain control of the aircraft for long enough to reach a visual horizon reference. Therefore, the risk sensitivity is high. Your coping ability is likely to be inadequate for the situation. If this outcome does occur and you are unable to cope with it, the consequences could be catastrophic. Losing control of a high performance single at altitude, at night, would almost certainly mean the loss of the aircraft and your life. Therefore, the risk penalty is very high.

Combining risk expectancy, risk sensitivity and risk penalty, the risk level for the proposed flight can be assessed. Stating it simply, there is a small but real and significant chance that continuing this flight tonight would mean absolute catastrophe. Now is the time to ask if the risk is justified.

In order to answer this question appropriately, it is necessary to look at the available decision tree. The basic reasonable alternatives are: (1) continue the flight tonight as planned or (2) stay where you are until morning, then continue after the clouds have cleared. Another way to view this problem is to consider the postponement of the flight to be a means of reducing the risk that you are confronting. To summarize the attributes of each alternative:

1 Continue the flight tonight.

Benefits: The trip would be concluded tonight. You could stay in your own home tonight with all its conveniences and without additional expense. You could start the new week normally and with plenty of rest.

Risks: There is some chance of having a catastrophic accident.

2 Postpone the flight.

Benefits: The risks would be no higher than for any other normal flight. You would be flying in clear weather and probably without turbulence.

Risks: There is a chance that the weather may not clear as soon as forecast and you could be late to work Monday morning. It is inconvenient to stay overnight away from home, especially since you had planned to be back by this evening. You will probably have a motel expense and possibly a rental car or taxi expense. Besides, you really want to get home tonight.

Risk assessment

Evaluating the risks (and benefits) is now in order. This is a classic example of 'get-homeitis'. The additional expense of staying where you are tonight is probably not critical. Essentially then, it is a matter of balancing the risk of death against some subjective inconvenience. Would you be willing to play 'Russian roulette' using a gun with a 100 shot cylinder and only one live round just to have a chance of getting home tonight?

Risk acceptance

Logic suggests that the prudent course of action is to reduce the risk by choosing Alternative 2, and continuing your flight tomorrow morning. The risk exposure is easily preventable; not to do so would be foolish.

Guidelines for risk management

Compensatory risk reducers:

1. Always be CAREFUL in judgment. Consciously Accept Risks Evaluated with Forethought, Understanding and Logic.

2. Avoid hazardous attitudes, emotions and illogic that can adversely bias judgment. Be as objective as possible. Step back away from the situation that you are in and analyze it as if you were not personally involved. Then take your own advice. Remember that the risk itself could be a motivator. Avoid being a 'thrill seeker'.

3. Estimate expectancies conservatively. Estimate benefit expectancies on the low side and cost (risk) expectancies on the high side. Remember that *it can happen to you!*

4. Anticipate negative, especially catastrophic outcomes. Frequently, they will dominate other risk factors. Evaluate them carefully.

5. Always plan ahead. Never stop planning and always be ready to adjust your plans as necessary to manage risks. Always have an alternative plan and leave yourself strategic escape alternatives (an 'out'). Remember, 'Proper prior planning prevents peril'. Box canyons are for playing cowboys and Indians, not for flying.

6. Never compromise safety, yours or someone else's. Safety *must* dominate all of your decisions.

7. Anticipate situational demands and avoid those which may jeopardize your coping ability.

8. *Never* accept unjustifiable risks. If it is not worth doing safely, it is not worth doing.

9. Do not resign your fate. Take command of the situation.

10 Take the time to make deliberate decisions.

11 Allow adequate safety margins in all phases of flight.

12 Avoid the 'white cane' syndrome - never use the aircraft to 'feel' for the ground, especially on approach.

13 In assessing risks, remember that they are generated by the product of risk expectancy, risk sensitivity (the expected adequacy of coping ability) and risk penalty. Ask the questions:

a. What can happen?
b. Will I be able to deal with it?
c. What will be the consequences if I cannot cope?

14 Remember the definition of fool: 'One who is deficient in judgment, sense, or understanding; one who can easily be deceived; to take unawares, surprise'. Do not be so categorized.

15 Remember that risks are increased by a chain of bad decisions. If action is taken before it is too late, this chain can be broken.

16 Never panic; remain or return to being calm. Always respond with objectivity, analysis and logic, not with emotion.

17 Remember operational priorities: fly the aircraft, navigate, communicate, etc.

18 Manage your workload. Be aware of the onset of excessive stress and unload your workload to satisfy priorities.

19 If you must fly at night, be especially cautious, widen your safety margins, and anticipate special night flying risks. It may be easier to spot properly lighted aircraft and obstructions in visual conditions, however, all other hazards may be more difficult to see. Also, if you have a survivable accident at night, it may take much longer to find you.

20 If the risk level is unacceptable for a decision alternative, either find ways to reduce the risk and choose another alternative. This may require the creative development of a new alternative.

Preparatory risk reducers:

1 Review the SMART process (Systematic Management of Risk Acceptance through Training).

2 Try to anticipate and prepare for any foreseeable emergency. This is especially important if the preparation is simple or if the risk is great.

3 Maximize your coping ability:

a. Keep yourself in good physiological and psychological condition.
b. Keep your equipment in good condition. Inspect it thoroughly and frequently. Correct any anomalies promptly.
c. Know your limitations and those of your equipment.
d. Plan your flights carefully.
e. Always get a good weather briefing, understand its limitations and update it frequently.
f. Always file a flight plan.
g. Always carry spares for critical items: flashlights, radios, fuel strainers, microphones, tools, fire extinguishers, first aid supplies, etc.
h. Learn as much as possible about your aircraft. A good knowledge base is essential for effectively planning your flight and for dealing with emergencies.
i. Do not let your judgment of risk be falsely biased by successful completions of dangerous flights made with poor judgment.

Weather predicting and analysis

Decision making in the weather arena begins with the development of a sound knowledge base concerning meteorological phenomena affecting aviation and receiving a proper weather briefing. On particularly bad days/nights for weather, it is well worth dropping by the Flight Service Station (if at all possible in your locale). The added information gained will help you in your decisions regarding go, no-go, routing, fuel load and perhaps even determine the type of equipment you will fly if you have a choice. An example of a decision process regarding weather is told by a pilot flying from Chicago's Midway to Cleveland's Burke Lakefront. As he tells the story:

> Dispatch did not advise me that the weather observer (at BKL) was on vacation. Due to other jobs the weather observer does at BKL, I can never find him to get current weather before take-off as prescribed by FAR Part 135. So we have an agreement that he will call me with the weather whenever it is IFR. On this night, there had been no IFR call and CLE, eight miles away was VFR. So I headed for Burke Lakefront VFR. Twenty minutes out from the destination, I called on Unicom for weather and was told the weather man was on vacation and the weather was poor -- about 900 overcast and 1 1/2 miles visibility. There were nine airplanes on the ground all waiting for me (they had landed despite the weather and without

> the required observation). So, I felt pressured into quickly turning onto the approach outside the FAF and giving it a try. I asked for and was granted a Contact approach to stay legal, but I probably wasn't very safe because I was barely clear-of-clouds and 1 mile visibility. I hope I won't let someone else's actions influence me again.

Is this starting to sound familiar? If you are not familiar with the area around this particular airport, the obstruction environment is worth a look on an approach chart. The environment requires constant and concentrated watch, that is weather in general, but especially, the anomalies associated with thunderstorms, icing, fog and high winds. The more experience you have in making decisions regarding these factors, the better you will be in making future decisions. One pitfall however, of the experienced aviator is in thinking 'I made it through one like this last time...'. Recognizing those times which you would not and should not repeat is a key to making better decisions given the same situation in the future.

Your understanding of the limitations in the predictive capabilities in any weather reporting will greatly aid in assessing the risks involved in a given flight. Predicting the weather is not an exact science, although it has become much better over the years. Still, some pilots think that the weather information received is never right and this can lead to error in judgment as well. The following statistics regarding the probability of a correct forecast will help to clarify what really should be expected.

General limitations of forecast science

Forecast accuracy stems from what is known and what can be measured. These two limitations determine the reliability factor of various weather situations, how much detail is available and changes over time. An important point to remember is that forecast accuracy decreases with the passage of time since forecast issuance. Since the forecaster must consider a complex combination of many factors, information on many of these factors may be partially or completely lacking. Studies of aviation forecasts (Moody, 1968) indicate the following:

1 Up to 12 hours and even beyond, a forecast of good weather (ceiling 3,000 feet or more and visibility 3 miles or greater) is much more likely to be correct than a forecast of conditions below 1,000 feet or below 1 mile.
2 However, for 3 to 4 hours in advance, the probability that below VFR conditions will occur is more than 80 per cent correct if below VFR is forecast.
3 Forecasts of single reportable values of ceiling or visibility instead of a range of values imply an accuracy that the present forecasting system does not possess beyond the first 2 or 3 hours of the forecast period.
4 Forecasts of poor flying conditions during the first few hours of the forecast period are most reliable when there is a distinct weather system such as a front, a trough, precipitation, etc., which can be tracked and forecast, although there is a general tendency to forecast too little bad weather in such circumstances.

5 The weather associated with fast-moving cold fronts and squall lines is the most difficult to forecast accurately.
6 Errors in forecasting the time of occurrence of bad weather are more prevalent than errors in forecasting whether it will occur or will not occur within a span of time.
7 Surface visibility is more difficult to forecast than ceiling height. Visibility in snow is the most difficult of all visibility forecasts. Skill in these forecasts leaves much to be desired.

Forecasters *can* predict the following at least 75 per cent of the time:

1 The passage of fast-moving cold fronts or squall lines within plus or minus 2 hours as much as 10 hours in advance.
2 The passage of warm fronts or slow-moving cold fronts within plus or minus 5 hours up to 12 hours in advance.
3 The rapid lowering of ceilings below 1,000 feet in pre-warm front conditions within plus or minus 200 feet and within plus or minus 4 hours.
4 The onset of a thunderstorm 1 to 2 hours in advance if radar is available.
5 The time rain or snow will begin within plus or minus 5 hours.

Forecasters *cannot* predict the following with an accuracy which satisfies present aviation operational requirements:

1 The time freezing rain will begin.
2 The location and occurrence of severe or extreme turbulence.
3 The location and occurrence of heavy icing.
4 The location and occurrence of a tornado.
5 Ceilings of 100 feet or zero before they exist.
6 The onset of thunderstorm which has not yet formed.
7 The position of a hurricane center to nearer than 80 miles for more than 24 hours in advance.
8 The occurrence of ice fog.

Occurrences of both icing and turbulence are local in extent and transient in character. Once a pilot receives a forecast of these hazards, he usually plans his flight to avoid them. Because actual passage of aircraft through the weather system is the only way to verify these phenomena, they usually go unverified. With the present state of the science, forecasts of icing and turbulence specify volumes of airspace which are quite small when compared to the total volume used by aviation, but relatively large compared to the localized extent of the hazards. Existence of the hazards in these relatively large forecast volumes is thought to be fairly accurate - occurring perhaps 50 to 75 per cent of the time, but intensity forecasts typically are less reliable.

One other important weather hazard with which pilots have to deal is windshear and the even more well-known 'microburst'. Some avionics manufacturers are experimenting with a specially developed windshear detection device. The cockpit information presented is computed from accelerometers constantly and instantaneously measuring vertical and horizontal speed changes. This may become a very worthwhile safety device in the future, but in the meantime awareness of the potential dangers of windshear is essential if one is to remain clear of these extremely hazardous conditions.

Personal minimums as a risk management tool

The Ohio State University aviation research team has developed a preliminary training program to help pilots develop personal minimums for preflight go/no-go decisions. As we have seen, risk management is an important part of the over all task of flying airplanes. All pilots recognize that there are risks in their flying activities and that some activities are riskier than others. Many pilots will not perform certain kinds of flying activities under certain circumstances but may do those same flying activities under other circumstances. For example, some pilots will fly a single-engine aircraft at night over mountains with a student in training who needs such experience but may be reluctant to do so with their own family (unless they have an important business meeting the next day at their destination). Although most pilots have not established guidelines for these types of decisions, these are, in fact, risk management decisions.

In well-established flying organizations such as airlines and corporations some of the risk management is governed by the establishment of clearly defined limits or minimums (SOPs) for flying activity. On the other hand, most general aviation pilots do not have a large organization that sets down the limits of their flying choices. Risk management for these pilots is left almost entirely to their discretion under the scrutiny of the FAA. In the area of preflight decision making, the Federal Aviation Regulations (FARs) require a pilot to 'become familiar with ALL available information concerning the flight' (FAR 91.103) and then the pilot is 'directly responsible for' and 'the final authority as to the operation of the aircraft' (FAR 91.3a). Furthermore, FAR 91.13 requires that the aircraft be operated in a way that does not endanger the life or property of others. Nevertheless, in setting these limits, the FAA still places most of the responsibility for risk management in the hands of the individual pilot. The FAA limits are based on the capabilities of ideal pilots and equipment because, in addition to regulating for safety, they are tasked to promote aviation. Before each flight the individual pilot must determine whether his or her currency and aircraft meets this ideal level of skill and capability. If ideal levels are not met, pilot judgment is expected to produce more conservative operational minimums for that situation.

The importance of pre-takeoff decisions

The importance of decisions made prior to flight is underscored by studies that have shown that a significant number of pilot decisions (or lack of decisions) made during the preflight phase led directly to accidents and incidents much later in flight (McElhatton and Drew, 1993). Of the 125 ASRS incident reports reviewed in their study, 90 per cent of all time-related human errors occurred in the preflight or taxi-out phase of operation (See Figure 11.4). General aviation examples of such errors might include:

- Knowingly overloading the airplane and failing to clear an obstacle at the end of the runway on takeoff.
- Running out of fuel enroute because of poor flight planning (e.g. insufficient fuel reserve).
- Deciding that the high ranking passenger (boss or family member) must get there despite the weather, resulting in premature contact with the ground on approach.
- Flying with a known medical condition then becoming disoriented (e.g. in haze).

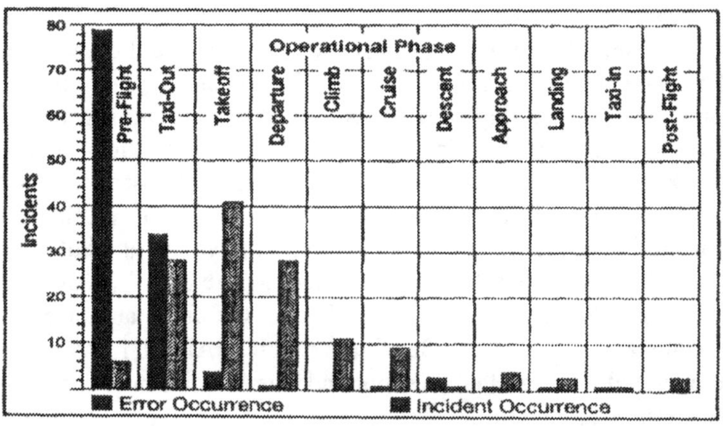

Figure 11.4 Operational base vs error and incident occurrence (after McElhatton and Drew, 1993)

Incident and accident data such as these are not surprising. A flight is a sequence of events that are related and influenced by prior events. Once the commitment to flight is made, a strong psychological force exists in the pilot's mind to continue to the intended destination. Also, social or peer pressure may make admitting an error and turning around difficult or seemingly impossible. Furthermore, once airborne, certain options disappear, including the option to add more fuel. Pilots need to recognize the extent of the forces that will be applied to them once they make the decision to fly and factor them into their preflight go/no-go decision.

New preflight personal minimums training

In its preliminary form the proposed training program has four major parts, (1) an introduction to preflight decision making, (2) assigning risk factors to categories, (3) establishing guidelines for personal minimums and (4) evaluation. It is expected that the program can be completed within the 90-minute time frame of most FAA seminars. It is also expected that, with the helps provided, most FAA ASPMs will be able to facilitate the program with only a brief training session. We believe that, because of the nature of the training, the program is best conducted by a facilitator who interacts with the students. It is not the kind of training that can be accomplished simply by playing a video tape.

Introduction to preflight decision making. In the first part of the session students are given the basic concepts and motivated to learn and establish their own personal minimums through an examination of why preflight decisions are so important to flight safety. Preflight decisions are defined as: 'Any and all decisions made prior to taxiing the airplane onto a runway with the intent to take off'. In aircraft other than airplanes, preflight decisions are any and all decisions made prior to the first task to make the aircraft airborne (i.e., gliders--giving the initiate-launch signal; balloons--giving the command to launch; helicopters--advancing the throttle). Under this broad definition, preflight starts with the first idea to make a flight until the pilot initiates the first action to take the aircraft into the air or until that possibility no longer exists. This definition includes normal preflight decisions, but emphasizes the pilot's complete situation, including all of the assessments and commitments pilots make.

Personal minimums are defined as an individualized set of decision criteria (standards) to which the pilot is committed as an aid to preflight decisions. Because each pilot manifests a unique set of skills and personality, he or she is the best judge of the risks involved in the contemplated flight. Psychological theory suggests that, when they know they are not being watched, people are much more likely to follow standards that they have made themselves than those imposed upon them by someone else. Therefore, the emphasis in this program is placed on 'personal' minimums.

In contrast with imposed minimums (organizationally or FAA established limits or SOPs) personal minimums are self-generated. We believe that this difference establishes the key to the commitment of each pilot to follow the guidelines. If the pilot has determined and established his or her own minimums, there is a strong likelihood that they will be followed when faced with pressures to the contrary. We believe that, in situations where no one is watching, that commitment will be stronger for personal minimums than it is for imposed minimums. It is understood that, in most cases, personal minimums are (and will be) more conservative than the FARs. The FAA, aircraft passengers, the legal system and the general public expect pilots to use more stringent minimums than those mandated by the FARs.

Flying is not unique as an enterprise of trust. In automobile driving, the actions of the driver are not observed by the police at every corner. Instead, the safety of the driver mostly depends upon voluntary compliance with the rules of the road and setting some personal standards depending on the situation. While some pilot operations take place under the direct supervision of regulators, management, supervisors, or others in authority, most flights in general aviation, remain unsupervised and are seldom observed.

In business aviation, commercial flight operations, flying clubs and other situations, operational limitations or sets of rules define operational minimums. In these operations, pilots with less experience must operate with higher ceilings, higher approach minimums, lighter winds and lower crosswind components. In airline and military operations, an extensive set of SOPs provides tools to assist pilots in making critical preflight go/no-go decisions. Flights may be canceled by dispatchers, chief pilots, or supervisory personnel, even before the pilot has an opportunity to address the go/no-go decision. Such supervision has proven to be effective in improving safety in these organizations. At least in part, this may be the reason for the very good safety record of these organizations.

How can we use the experience of these highly supervised operations to improve safety in general aviation? Application of personal minimums offers an approach to improving risk assessment and management by self-supervision. Personal assignment of minimums requires an awareness of the risk factors involved before one can attempt to manage them. Furthermore, because some pilots may think that their responsibility ends with understanding their own safety risk, personal minimums training teaches them to acknowledge and consider the level of risk that is acceptable to passengers, the company, or others potentially affected by any flight consequences.

Risk factors. The second part of the training program is an individual or group activity in which the students are required to generate and classify a list of preflight risk factors into an organized taxonomy. Research has shown that when people make decisions, they usually consider only a very small number of factors. The primary purpose of this activity is to expand the pilot's knowledge base of risk factors that go into preflight decision making and to classify them into categories that are easily recalled when they are needed. Risk awareness is an effective tool to increase safety, but requires experience and knowledge.

To stimulate thinking about risk factors and how they apply to preflight decisions the instructor can offer one or more case studies in the form of 'trigger tapes' of situations developed in the FAA's 'Back to Basics' program or stories found elsewhere. One excellent example was published in *Flying Magazine*'s, 'I learned about Flying From That' (McCutcheon, 1991). In this case a pilot feels pressure to make a medical evacuation flight in a Cessna 210 with inoperative radios, in questionable weather with night approaching. After arriving for the pick up, he finds that there is both child and a mental patient to transport plus families and other cargo causing the plane to be overloaded as well. Despite the child's condition, because of the numerous risk factors, he decides not to takeoff.

As students read this story, they are asked to identify the risk factors facing the pilot including the subtle psychological factors such as the condition of the child. When all have listed these factors individually, the class is opened for discussion and all are invited to share their discoveries. The same could be done with any aviation preflight scenario including those from individuals in the class either in front of the whole class or in small groups.

After listing risk factors, the students are to organize them into a taxonomy with six suggested categories: pilot, aircraft, environment, operation/mission, organization/social and miscellaneous. The first three categories are those normally used to represent the pilot's world. The operation/mission category is added as a place to put factors regarding personal pressures to complete a mission. The organization/social category is included as a place to put risk factors that organizations can add to the pilot's decision process including subtle pressures to complete flights on schedule. Finally, the miscellaneous category is included to underscore the emphasis on personal freedom in the construction of this tool in anticipation that some may not wish to identify a particular risk factor with one of the given categories. The structure presented is offered as a starting point, not a required set of categories. What is placed under any category will be a function of each pilot's mental model of how these factors are classified.

Guidelines. Guidelines are simple acronyms, rules of thumb, sayings, if-then statements, forms, or memory aids representing what the pilot intends to do about the risk factors developed in the previous exercise. The development of guidelines is more personal and should be done individually with direction from the facilitator and handout materials. In this activity, pilots will write examples of personal minimum guidelines organized into an easy-to-use checklist format. Some examples are provided to assist in getting started.

Table 11.1 is a checklist form on which the pilots write their initial guidelines. The headings on the checklist provide an initial organizing structure following the risk factor categories. The pilots are encouraged to consider changes to the checklist that may better match their own ways of organizing the guidelines.

Evaluation. The final exercise consists of a review and evaluation of items included in the personal minimums checklist. Pilots are encouraged to have peers review and evaluate completed guidelines and to sign a statement of commitment to use them. This exercise can be highly rewarding because it offers opportunities for verbal expression and public commitment. However, the instructor/facilitator must be very sensitive to students who may not wish to share their ideas; some might consider them too personal to share. The activity includes presentation of suggested evaluation criteria and evaluation of a sample personal minimums checklist. Suggested steps in this evaluation include:

- Check for missing, unclear and duplicate items.
- Ask general questions such as:

  -- Are the personal minimums flexible?
  -- Do they cover the range of intended activities?
  -- Do they address flying currency questions?
  -- Do they cover the range of equipment and aircraft routinely used?

- Will the pilot understand the intended purposes for each item when used at a later date?
- Is a general statement included about how the pilot will approach 'non-routine' activities not covered by this set of minimums?

To avoid the possibility of embarrassment, the instructor should have available a completed sample personal minimums checklist for students to use and discuss. Samples of factors and guidelines should match the certification levels, flight activities and other variables of the students in the class. Particular items on the checklist can be used or changed to reinforce subjects from other training efforts or the newest 'hot' topic in aviation safety.

The instructor uses the discussion summary to reinforce the requirement for review and revision of any personal minimums checklist, and to check regularly one's checklist during biennial flight reviews, instrument currency checks, safety program participation, or other recurrent training. A personal minimums checklist should be considered a flexible document that develops and changes with time and experience.

Conclusion. In closing, the instructor reminds each pilot that the personal minimums checklist is unique to her or him. Other pilots' minimums will be different. The set of risk factors and checklist provided is never complete. They must be reviewed continuously and changed as required to accommodate the changing pilot activities and capabilities. To provide each pilot with further sources of information, they are referred to advisory circulars, magazines, books, newsletter (IFR Refresher, FAA Safety Review, etc.) and standard pilot references (Airman's Information Manual, FAR's, etc.).

Table 11.1
Personal Minimums Checklist.

PILOT	**AIRCRAFT**	**ENVIRONMENT**
Recent experience in type:___ landings within ___ days	Number of NAV/COM radio(s)___	VFR Ceiling ___
Number of meals at normal intervals ___	Auto pilot functioning properly	Visibility ___
Hours of sleep in past 24	________________	Crosswind ___
________________	________________	Terrain ___
________________	________________	________________
________________	________________	________________
	________________	________________
		________________
OPERATIONAL	**ORGANIZATIONAL**	**MISCELLANEOUS**
Runway length ___ Obstructions ___	Alternate plans available for later than expected arrival or not being able to reach destination	These include: items to be added at next revision, limited use items, one-time items, etc.
Altitude ___ Time ___ for no oxygen	All passengers briefed on personal minimum operating procedures	Time since review of emergency gear extend ___
________________	________________	________________
________________	________________	________________
________________	________________	________________

References

Funk and Wagnals (1968). *Standard College Dictionary*. New York.

Jensen, R.S. Adrion, J. and Lawton, R. S. (1987). *Aeronautical Decision Making for Instrument Pilots*. Springfield, VA: National Technical Information Service, DOT/FAA/PM-86/43.

Layton, C.F. and McCoy, C.E. (1989). 'General aviation pilot perceptions of deteriorating weather conditions'. In R.S. Jensen (ed.), *Proceedings of the Fifth International Symposium on Aviation Psychology* (pp.377-383). Columbus, OH: The Ohio State University.

McCutcheon, D. (1991). I Learned About Flying From That: Better Late Than Never. *Flying*, May, pages 118-119.

McElhatton, J. and Drew, C. (1993). 'Time pressure as a causal factor in aviation safety incidents the 'hurry-up' syndrome'. In R.S. Jensen and D. Neumeister (eds.), *Proceedings of the Seventh International Symposium on Aviation Psychology* (pp. 269-274). Columbus, OH: Ohio State University.

Moody Institute of Aviation (1968). *Elementary Safety Practices* (Class Notes). Elizabethton, TN: Moody Institute of Aviation.

O'Hare, D. (1990). 'Pilot's perception of risks and hazards in general aviation'. *Journal of Aviation, Space and Environmental Medicine, 61*, 599-603.

Wilkinson, S. (1994). 'The November Oscar incident'. *Air and Space*, February/March.

12 Management factors in pilot judgment

NTSB member John Lauber once said, every pilot error accident is a management accident.

(Franken, 1990)

Up to this point, we have focused our attention on the pilot or flight crew in aviator judgment. However, the flight crew operates within an organizational framework that sets the tone for safety and risk management. More than any other factor, the attitude of the management of the aviation organization, as expressed in its safety practice, sets the tone for the judgment used by its pilots to manage their risks. The specific standard operating procedures set down by management are not what determines pilot judgment. Judgment is determined by the way management is seen to practise its own standard operating procedures. If management is seen to depart from standard operating procedures without providing reasonable explanation to the pilots, the pilot's attitude toward safety will be eroded. Therefore, the corporate culture is paramount in setting safety standards by example.

Although some airlines advertise that safety is their number one goal, this cannot be the case. Airlines are primarily in business to make a profit for their owners. Safety is, at best, a supporting objective to transporting people and goods and providing a service to the public (ICARUS Committee, 1994). Safety can provide a great assist in the productivity of the company. Not only does it prevent loss due to damage or injury but it can serve to promote the public image and profits of the company. Therefore, companies in the aviation business find it sound fiscal policy to invest heavily in safety at all levels of the company.

Organizational risk management

As we have seen, the management of risks is primarily a task of preventing or compensating for human error. Until recently, responsibility for human error was placed almost exclusively on the pilot. We are finding that the organizational environment may be even more responsible for these pilot errors than the pilot. Some unscrupulous operators have been known to use job-related motivators to force pilot compliance saying, 'If you won't make this flight, I'll find someone who will'. A cartoon by Brickman called 'The Small Society' once showed two passengers walking down an airport terminal corridor. One says to the other, 'My plane came in late because the pilot didn't like the sound of the engine during warm-up. We had to taxi back to change pilots'.

Although this is a humorous story that does not seem plausible in the real world, in fact, it happens too often. A macho pilot-supervisor who I once knew got a taste of his own application of this phrase. The following is his story:

> A pilot for a large university transportation service was asked to make a flight to a large midwestern city. Drawing upon his considerable experience and good judgment, he determined that the weather was not sufficiently safe for such a trip in a single-engine aircraft because there was moderate icing reported in the clouds. His chief pilot, a personal friend of the passengers who had persuaded them to fly instead of drive, was angry with the pilot's decision to cancel the flight, considering that the passengers had already arrived at the airport. The chief pilot himself hastily loaded the passengers into the aircraft and took off on the original flight plan for the destination. Nearing the destination uneventfully, he was asked to hold for an approach clearance; at that time, an almost routine occurrence at this busy airport in adverse weather.
>
> It was at this point that he realized that he had not fueled the aircraft prior to takeoff and was quite uncertain about the amount of fuel remaining. Not wanting to appear concerned however (he had the 'Right Stuff'), and feeling that there is 'always' more fuel left than the gauges show, he sought no special handling until the engine quit due to fuel starvation. Luckily, he managed to crash-land the airplane in a field near a gravel quarry. He came away unhurt, but tragically, one of his passengers was injured and disabled for life.

We do not know what this supervisor-pilot was thinking -- he certainly did not offer it in the investigation, but from his decisions we can see that there was not much rational thought. Again, this is anecdotal evidence, but we can see the effect of motivation in blocking out rational decision making. It might be argued that such irrational thought is the result of defects in the personality of the pilot involved.

Listening to pilots who work in professional organizations, I have gained the impression that many of these organizations have worked against the best interests of safety. There are common expressions heard among pilots in these organizations who describe their departures from the 'norm' for safety reasons that sound like, 'Facing the front office' or 'Spending three days filling out paper work', or, as two Swedish researchers found, 'His reward is three slaps in the face from a system without a brain' (Fahlgren and Hagdahl, 1990). Pilots often say that they wish their management could face the reality of the life or death situations into which they push their pilots. On the positive side, some organizations, including the US Forest Service, support pilot decisions and have fostered a better attitude about judgment in their pilots.

A major part of the blame for these corporate attitudes must be ascribed to the FAA and to local communities because they too create the system in which we fly. At Washington National Airport, for example, the last message every pilot hears is a sign on the taxiway at the departure end of the runway from the local community that reads, to wit, 'Pilots are expected to observe noise abatement procedures'. Combined with the message from the manufacturer and his accounting department sent through his training department about potential damage to engines for setting the EPR too high on takeoff, the pilot has a mind set to avoid pushing the throttles all the way forward even when special conditions such as ice, snow and an activated stick shaker arise (e.g., the Air Florida accident in 1982).

The pilot's organization

What does this organization that can cause pilot errors consist of? It varies depending on its size but in a large airline it consists of:

- Senior management
- Line management
- Inspectors and quality control people
- Dispatchers
- Operational people

In addition, the aviation organization operates within the framework of national and international aviation including:

- The civil aviation authority
- The accident investigation agency
- Safety reporting systems
- Local airport authorities

Finally, there are numerous other organizations that can have influence if not direct control of risk management including:

- Aircraft manufacturers
- Pilot organizations
- Aeronautical chart makers

It is the duty of management to provide clear, challenging direction for their pilots to avoid the two extremes that can develop: the 'abdicrat', the pilot who abdicates leadership responsibilities on the flight deck, and the autocrat, the pilot who claims all responsibility to do as he wishes without consideration for the company (Hackman, 1993). Management also must be clear about outer-limit constraints on what can be done to achieve the desired ends and then let pilots do what is needed to achieve company goals within the constraints.

Hackman (1993) illustrates the tradeoffs of these management principles. There was a manager who stated his goals for the organization: You are to maximize simultaneously (1) safety, (2) efficiency and (3) passenger service. At first, all of the pilots nodded their approval, but as they thought about it, realized its impossibility. In fact, the manager's point was that they were to make constant tradeoffs among these three, often opposing goals and these decisions should be made within the constraints of the organization.

A recent accident (October 26, 1993) of an FAA Beech Super KingAir near Front Royal, Virginia is an example of a 'management accident'. The accident happened while the crew was conducting flight tests of navigation transmitter systems. After departure from Winchester Regional Airport in an area of low clouds and mountainous terrain, the crew requested an IFR clearance. They were told by ATC to 'Maintain VFR for right now, it's going to be about five minutes before I can get to you, I'm extremely busy at the moment'. The crew made a request for an IFR clearance to a higher altitude to another controller and received a response to 'Standby, I have traffic just over the VOR right now', followed by a suggestion to change to a third controller. No more communications were heard and the aircraft crashed into the mountains about that time.

The NTSB found that this PIC had two previous DUI convictions, numerous complaints were communicated to the FAA management regarding him, he had required three attempts before passing his type rating in the BE-300/F, eight of the eleven SICs had previously requested scheduling preference to avoid flying with him and the list goes on. There were numerous errors in pilot judgment in this accident, but the NTSB cited the following management factors (NTSB Reporter, 1994):

1 The crew was under pressure to complete the mission and the PIC requested and received approval for one hour overtime to complete it.
2 Standardizing PIC flight checks from AVN headquarters;
3 Maintaining and monitoring a central repository of pilot training and performance records;
4 Implementing meaningful crew resource management programs;
5 Rewarding PICs for adhering to standards of safety equal to rewards given for mission efficiency and accomplishment;
6 Providing management continuity, and encouraging oversight of operations and maintenance by entities outside of the AVN.

The NTSB further stated the need to:

1 Establish effective and accountable leadership and oversight of flying operations;
2 Establish minimum mission and operational performance standards;
3 Recognize and address performance-related problems among the organization's pilots; and
4 Remove from flight operations duty pilots who were not performing to standards.

Small charter operations

The commercial pilot for a small charter operator, in particular, is faced with some of the most difficult decisions in aviation today. As in all commercial enterprises, the bottom line is to make money. Both pilots and flight operations managers have decided that, in the larger sense, to make money is worth the risk that flying poses. Unfortunately, pilots flying for hire are often faced with the need to balance financial reward with safety on a day-to-day basis. In these situations they may be faced with pressures by superiors, peers, and/or passengers who are more in tune with the financial or convenience side of the balance than with the safety side. These people usually assume that the pilot will withstand their strong pressure to press on to the destination if there were any 'real' danger. The pilot, conversely, wonders whether a decision against the wishes of his/her superiors will cost his/her job.

Service industries are the toughest of all businesses to operate. Aviation is probably the most difficult of all service industries. The difficulty refers to the ability to maintain a viable corporation, earn a living, keep the employees happy, satisfy the requirements of all the government agencies (FAA, State, Airport Authority, etc.) and service the customers. Servicing the customer is the name of the game and management is often the buffer between all of the opposing forces. How does decision making come into play from the managers' standpoint? Educating your crews in aeronautical decision making is well worth the time and effort invested. There is no doubt that the benefits of such training will, over time, show up where it counts to you -- on your bottom line.

In the meantime, your pilots and training personnel will need your support and cooperation to make the program a success. Decision making skills for your pilots are a life and death subject. Open communication and the sharing of new ideas and the discussion of grievances are the key to successful businesses. Your attitudes towards your employees will contribute to the life of your business, as well as to the lives of your pilots.

Regulations concerning air-taxi operations were created to accommodate generally smaller aircraft and less load capability than air carrier operators. These regulations opened a whole new era of air commerce. Operators of air taxi operations must adhere to the many regulations regarding the qualifications of pilots, certification of aircraft, operations specifications, maintenance and record keeping (paperwork). The managers in many air-taxi operations often double as director of operations, chief pilot, director of training, etc. This puts additional pressure on you as 'one of you' and 'one of them'.

Commuter management

The following is another account of a commercial pilot who heard the words 'fly that airplane, or I'll find someone who will!' Those who have worked their way up the aviation ladder will no doubt relate to it. The recommendation put forth by the pilot in this situation would probably be one of the best solutions. The story took place in a small transport aircraft operated by a commuter airline in Florida.

> After a successful career with Air Carrier 'A', I found myself in dire need of a job when they went out of business. The first opportunity was with Commuter 'B'. One of the aircraft had a sick right engine, however the other pilots (all inexperienced in my opinion) were flying it. I departed Miami enroute to Sarasota IFR (Autopilot was inoperative) and had a communication failure at 12,000 feet over LBU (due apparently to no static wicks on aircraft or other effects of precipitation). Communication was restored descending through 5000 feet. On final at Sarasota, the right engine quit. During ensuing phone conversation with the boss, he begged me to continue and promised the engine would be repaired that night. I refused to carry passengers, but stupidly succumbed to ferry the aircraft to Ft. Lauderdale. Descending into Ft. Lauderdale, the right engine quit again.
>
> After performing some form of aeronautical voodoo on the engine, I was supposed to continue flying trips in this aircraft. After an unsuccessful run-up, I returned to the hanger refusing to fly the aircraft. Within the hour the 'Hero' of the operation (his first flying job) was airborne with passengers. Several days later, I was fired for refusing to fly another aircraft which previously had 19 write-ups, all of which were signed off 'ground checked normal' and none were fixed. Three other pilots quit shortly thereafter. The other 'more experienced' pilots were replaced soon with more 'time builders'.
>
> These are not isolated instances, everyday was more of the same. Noteworthy comments to me were: 'This isn't the airlines' (even though they are operating as air carrier 'X' and 75% of the passengers believed they were on air carrier 'X'). 'If you won't fly this airplane we'll find someone who will' (they did). 'If you think this is bad, you should see so and so' (from the pilots).

The pilot summarized his feelings about the situation. 'Yes, I am bitter about the demise of air carrier 'A'. Out of thirteen civilian flying jobs to date, they were the only company who attempted to do things by the book. Maybe we would still be around if we had falsified training and 'ground checked normal'.

This pilot's suggestions for a resolution to this situation were two-fold. First, more government supervision is needed to force compliance with the regulations. Second, he suggested that management should fly with the pilots on every flight. Do you think that may change their behavior?

Owner/operator management

The second account addresses the same type of management issue, but this time even closer to home. This pilot had a disagreement with his employer (who owned the aircraft) just prior to departing on a trip from Casper, Wyoming to Oklahoma City, Oklahoma. As he puts it 'I was not able to suppress my desires to be out of this man's (his boss's) reach any longer'. He admits to flying the small jet after consuming alcoholic beverages five hours before the flight and flying single pilot although the aircraft was not certified single pilot. He attributes this behavior to lack of rest due to a heavy flying schedule imposed upon him by the boss over the previous week.

This pilot's recommendations included mandatory courses for the owners of high performance and turbine aircraft on the subjects of stress, crew rest and pilot judgment. He feels this would aid corporate pilots greatly by relieving a great deal of stress and tension. As it stands, if the corporate pilot says 'no go' for a legitimate reason, there is a good chance the owner/boss will find another (probably less experienced) pilot to go. Unfortunately, the new pilot may eventually be a name on an accident report, along with the boss.

Understanding management's point of view is not too difficult, but usually when it comes to piloting an unsafe aircraft or making a trip in unsafe conditions, usually the management person is on the ground. Consider the following incident reported to ASRS:

> *Narrative:* While departing SJC for BEL on the SJC 6 departure, the captain, who was also the pilot flying, began his turn to a heading of 110 degrees after reaching approximately 700', rather than waiting until 1.8 DME as depicted on the SID. I attribute this event to 2 factors: pilot complacency, on my part as well as the captain's. The second factor, in my opinion, is much more important: The rapidly deteriorating morale at ABC airlines. Since acquiring ABC 2 1/2 years ago, the XYZ corporation has become increasingly confrontational, condescending and oppressive toward the ABC pilot group. XYZ has made it clear that substantial wage and benefit cuts are in our future. This kind of atmosphere is fatiguing, numbing, unhealthy and downright dangerous to work in. Flying the SID and the trip were not on my mind, or the captain's today -- our deteriorating working conditions were.

'Downeast minimums'

In his book, *Blind Trust*, John Nance (1986) relates an almost unbelievable scenario of control and pressure exercised by Robert Stenger, owner and operator of Downeast Airlines, on his pilots. The pressure led directly to the crash of one of its Twin Otter airplanes killing 17 people in May 1979. At one of Downeast's primary destinations, Rockland, Maine, the published FAA instrument approach minimums were 440 feet. Nance said the following about that approach and the pressure from the owner:

> Nowhere could it be found in the Downeast manuals, nowhere could it be posted on a bulletin board and by no one could it be clearly defined, but Downeast minimums were anywhere from 200 feet to 350 feet above ground level, depending on whom you talked to -- when you talked to them (anything under 440 feet, of course, was illegal). If the FAA was around, for instance, there was no such thing as Downeast minimums.
>
> Over the years a system of psychological reward and punishment had instilled the dangerous idea that a Downeast pilot could safely grope through the fog, descending lower and lower over the trees leading to the runway and be able to see the tree tops before running into them. If you could see the tree tops, you could, so the belief went, stay just above them until the runway appeared. A good pilot could do that on a regular basis because he would know the Rockland airport and its instrument approach with great intimacy.

Stenger believed that he could not make a profit operating on FAA minimums because the ceiling and visibility in that section of the country was low most of the time. He instilled in his pilots a sense that to be a real pilot (and a real man) you had to be able to make it into airports such as Rockland at lower than FAA minimums and many of his pilots promoted that belief within the organization.

Commercial bush flying

The pressures to compromise safety in 'bush flying' are known to most pilots who have heard any 'hanger tales'. Such flying illustrates to the extreme the pressures that we as commercial pilots can face. Passengers unfamiliar with flying safety, and particularly, those macho passengers who love to hunt and fish, are fond of playing on the 'macho' attitude of the pilot. The following untrue story (from Russ Lawton, 1986) illustrates:

> A pilot flew into the bush to pick up some hunters who he had left the week before and found that they had two moose to take out. He said to them, 'I told you when we departed that this airplane will carry only one moose!' His macho passengers responded with, 'The pilot who flew us in last year flew out with two moose!' This pilot, feeling responsible for the maintenance of the macho pilot image said, 'OK, I'll give it a try'. So he packed everything, started the engine of his float plane, taxied around to stir up the glassy water and opened the throttle for takeoff. He hung for quite some time in the ground effect before gaining some altitude. He barely cleared the tree tops at the water's edge before he realized that he

> just could not beat the rise in the terrain. He mushed it into the trees at full power about a mile from the edge of the lake. Fortunately, none of his passengers was hurt. He turned to his passengers and said, 'I guess I just don't have what the other pilot did to get out of here'. One of them responded, 'Oh, you did very well. The last pilot only made it a half mile!'

A safety investigator in Alaska, Michael Mitchel (1983), has made an extensive study of the factors that may cause the higher accident rate in commercial operations in that state compared with the rest of the United States. He found that pilots are being paid upon successful completion of their flights and some are even threatened with the loss of their jobs if they refuse to take a flight that they think would be unsafe. This practice is so blatant that, in some cases, Mitchel said that the blame for an accident should be shared with the operators and/or managers. Mitchel suggested that, if it is found that a pilot is being subjected to strong pressures from superiors or economics (remuneration by commission) to fly in conditions beyond his skill level or the capability of his equipment, at least four forms of action could be taken to reduce the effect of these pressures:

1 Change the form of remuneration for the pilots.
2 Change the training program to reduce the effect of such pressure.
3 Evaluate pilots considering their ability to resist such pressure.
4 Transfer some of the negligence normally assigned to the pilot for poor judgment to management or others causing the pressure.

The thrust of this chapter is on training to reduce the effect of such pressure. However, if you are a commercial pilot, you should recognize such pressure and avoid it if you feel the manager will not be sensitive to your concerns for safety. Recall the example presented at the beginning of this chapter of the university transportation service manager who took the flight when his pilot thought it too risky? An aircraft was lost and one passenger was injured for life. The accident situation contains many hazardous attitudes fostered by economics and personal commitment. The pilot who decided not to make the flight was vindicated for his decision, but the price was very high! This accident shows that managers need to be especially aware of the pressure that they can place on pilots to make bad decisions.

Another example of management becoming involved in pilot decisions has occurred recently in the United States involving decisions to fly ATR-42 and ATR-72 aircraft into icing conditions. Following the crash in Indiana of an ATR-72, many pilots came forward to report that they had experienced control problems with the aircraft in icing conditions and some made the decision not to fly in those conditions. As a result, one of the more vocal pilots who refused to fly was dismissed from his flying position with the airline. Initially, the FAA also supported the airline decision against the pilots. However, later the FAA changed its position and ruled that these ATR aircraft could not fly in these conditions. At this point, the aircarrier grounded the aircraft in these conditions.

The manager, who is faced with the decision as to just how hard to push that pilot to 'go', becomes a party to the risk management process. It is understandable from an economic point of view that the 'mail, checks, boss, passenger, whatever, must get through'. The question 'When is the success of the task not worth the risk', must be kept in mind during all decision making and dispatching of aircraft and crew.

What can the organization do to manage risks?

Just as manufacturers can reduce pilot errors and provide cockpit systems that are error tolerant (Trollip and Jensen, 1991), company management has many tools at its disposal to effect the same desirable result. The first is to have and to communicate to all parts of the organization that safety is a deep concern and vitally important to the company. Safety concerns can and must be stated directly but must also be demonstrated in the way top people are seen to behave and in the policies they develop for the company (ICARUS, 1994). As we saw in Chapter 10 for the individual pilot, it is critically important that company standards exceed those provided by the FAA.

Allocation of resources

It is important that pilots see that decisions are made to provide equipment that is safe and error tolerant. Periodic safety courses (e.g., CRM training) should be made available to all people at levels of the organization. Refresher training should be offered to pilots at their request. For those managers who may think that pilots will take advantage of this policy, my experience with the US Forest Service tells me that pilots will not ask for more than they need.

Safety officer

There must be a safety officer in the organization who is responsible to the top officer -- not to anyone else. The safety officer must be someone who is seen as approachable by the pilots so that they can communicate their own weaknesses, errors and perception of hazards without fear of reprisals. The safety officer should provide feedback through an internal newsletter with contributions from company personnel and articles from other publications that may be relevant to the aircraft being flown.

Safety literature

The company should subscribe to the appropriate safety literature and encourage management and pilots to be members and attend safety organizational meeting (e.g., The International Symposium on Aviation Psychology). Any and every notice of a safety nature should be distributed to the pilots by the safety officer as quickly as it is determined to be relevant.

Incident reporting and feedback

The safety officer should run an internal incident reporting program that rewards pilots for reporting hazards and incidents that could have implications for safety. Pilots should be encouraged to report incidents to national incident reporting systems as well (ASRS in the USA, CHIRP in England). Safety managers should seek and distribute relevant feedback from these programs frequently.

Standard operating procedures

Standard operating procedures reflecting the environment in which the company operates should be developed and used by everyone in the company. These amount to the primary means the management has to set the corporate policy for safety. As mentioned above, it is important that consideration be given to setting these standards higher than those of the FAA.

Risk management is vitally important to any flying organization. It requires that management know the hazards faced by their pilots and to determine the risks to the company associated with those hazards. Management must make a deliberate decision to accept or reject (avoid) those risks and they should communicate those decisions to the pilots. Reducing the risks can be accomplished through any of the means mentioned in Chapter 11 including reducing the exposure, obtaining better equipment or providing better training permitting the pilots to cope better with the hazards.

It is recommended that every company obtain periodic safety audits. The Flight Safety Foundation, Inc (of Arlington, Virginia) is one of the better places to look for such a service. Safety audits might also be available from civil aviation authorities. In the USA the FAA's Advanced Qualification Program is an excellent organization from which to obtain high standards for safety and pilot training.

What can you, the pilot do?

In Chapter 11 I have offered many suggestions for developing your own risk management standards but for the most part those standards are aimed at the private operator. As you work your way up the aviation ladder, you will experience many forms of 'management push' to extend your risk management limits. Generally, as you gain experience, you will be flying higher performance aircraft and most often multi-engine aircraft. However you managed to obtain the minimum required hours, you probably started your first commercial job as green as could be. At that point, you were most vulnerable because you were faced with your own limitations and well as those from your equipment, the weather and other factors. When assigned a trip, you *will* go as expected. How can you admit that perhaps the situation is worse than you can handle? How would your management react to requesting a co-pilot in single-pilot IFR operations? How would they react to a 'no-go' decision?

In learning to make better decisions regarding your aviation endeavors, whether they are immediate (in-flight) or long-term career decisions, take into consideration all of the risk factors that have been discussed. Preflight discussion regarding personal minimums and standards need to be discussed before the situation comes up so that you will be prepared. The reality of the situation is that the decision you make on the ground, will follow you into the skies.

Flight planning

Flight planning in the commercial and air-taxi business, as in all aeronautical trips, involves much more than just planning the route and phoning (or taking) the flight plan to the Flight Service Station specialist. The successful result of each leg of each trip embarked upon depends on a multitude of factors. A cursory look at the accident statistics for Part 135 and other commercial operations, leads one to the realization that a large proportion of accidents and incidents are, in part, the result of 'inadequate pre-flight planning'.

The biggest culprit in the area of flight planning comes with increased experience. Unfortunately, as we saw in Chapter 10, when the pilot becomes more familiar with his aircraft and the operation, he often becomes more *complacent.* Thorough weight and balance calculations, exact performance numbers and complete weather briefings seem increasingly tedious and unnecessary as one's skills and familiarity with the system increases. Adequate time is usually programmed into duty times for all of the above chores to be completed in a professional manner. Unfortunately, some pilots believe that the extra time is for coffee, cigarettes and other priority habits.

There are indeed some different circumstances that affect the pre-flight preparation in the commercial aviation environment. When flying for personal business, for pleasure or even when flight instructing, the decisions as to whether to go, where to go and when to return are at the option of the pilot. In the commercial aviation business, you are providing a transportation service, either for individuals or cargo, from point A to point B. The wishes of the customer play a major role in your decision making process.

These same customers are also in the position to make last minute changes to your careful flight planning. It is not uncommon, for example, for the passengers to show up and weigh significantly more (rarely less) than had been communicated when making the charter booking. When carrying cargo, you will seldom know the exact weight of the load until it arrives for loading. A series of tables and calculations for your mission worked out ahead of time will help in these variable situations. If the total weight of passengers and/or cargo does not allow flight within the limits of the aircraft, it is *your* responsibility to handle the situation. It is not an easy one, but the decision must be made. Another related situation occurs when your passengers show up for the flight intoxicated. The FARs are explicit regarding this, you are the decision maker in this difficult situation.

In-flight decision making

A favorite practice of the charter or corporate customer is the change of destination -- often mid-flight. You had carefully mapped out the entire flight, only to have the customer (whom you are serving) change his/her mind. If in your estimation the change can be handled safely and efficiently, then you can comply. If not, other options need to be investigated.

The practice of using standard routings in air-taxi operations affords many pilots the luxury of having 'canned' flight plans filed with the controlling facility which eliminates having to replan and file for each flight when the same routes are routinely flown. A few precautions about 'canned' flight plans are in order. Occasionally, the routes are changed by the controlling authority. If this become necessary, your company will be notified. Make sure the communication patterns in your operation are such that you, the pilot, will also be informed. Also, 'canned' plans may not list 'alternates', because you would only choose an alternate after having been given the destination weather and that of the potential alternates. It is your responsibility to add alternates to the plan when they are required.

Pre-flight planning and pre-flighting the aircraft are items which, if performed properly can alleviate many problems before they occur. Maintenance of the equipment is a major part of flight planning. A minimum equipment list should not be bypassed for any reason. Also, the type of cargo being carried can be a factors in commercial decision making. Consider the dilemma faced by this commercial pilot:

> Six nights a week, the air-taxi pilot flew the same route (Ft. Wayne - Chicago - Indianapolis - Pittsburgh - Ft. Wayne) carrying 'hazardous materials' used in medical diagnosis. The aircraft and cargo were authorized to land only at designated airports. The point had been made many times that the pilot's career may end in paper work if he were to put down at an unauthorized airport. The pilot was 'haz-mat' qualified and the aircraft was equipped to carry the radio-active materials.
>
> A line of Level V and VI thunderstorms lay on a north-south orientation from Chicago to Indianapolis. The materials that were picked up in Chicago were time critical, having a useful life of only three hours and a critical patient waiting for them. The single-pilot of the light multi-engine aircraft had already navigated around the weather to the north on his way from Ft. Wayne to Chicago. Now, the trick was to get back around it to the south. Due to the weight of the cargo, minimum fuel plus reserves was all the plane could carry.
>
> As the pilot with his particularly 'hot and heavy' load departed the Chicago area, it was obvious that he would have to deviate around the weather quite soon. He had decided to try to fly west around the storm and then south and approach Indianapolis from the west. Deviations were approved by Chicago Center, though it was mentioned by the controller that several of the commuter airline flights enroute from Chicago to

Indianapolis were able to get around the heavier activity by going east rather than west and then approaching Indianapolis from the east. Since neither the aircraft's nor the controller's radar range covered far enough to the southwest to see the backside of the squall line, the pilot decided to take the already tested route hoping to out-run the weather.

After heading eastbound nearly 50 miles, it became obvious to the pilot, both from his radar, his storm-scope and listening to the deviations of the other aircraft in the area, that there was more than one line of thunderstorms. A second line had developed out in front of the one he was trying to get around. His choices seemed to be either to punch south between the lines or to continue further east around the newly developed line. To make matters worse, he was then informed that the Indianapolis airport, where he had to drop of some off his cargo and get refueled, was closed due to a severe thunderstorm over the field with wind gusts to 65 knots. The second line of storms (reported to be a solid line) was now approximately 30 miles west of Dayton.

The weather in the entire midwest was deteriorating rapidly, fuel was being burned and the turbulence had increased from moderate to severe. Fortunately, the cargo could not shift because it was packed tightly in the cabin. The thought process this pilot was going through went something like this: 'If I don't get headed south pretty soon, I won't have enough fuel to make it at all. But, that storm scope looks like a laser light show and the radar shows thunderstorms everywhere. I wonder what really would happen to me if I had to put down at an unauthorized airport'?

At this time (he is approximately 60 miles to the north-west of Dayton), ATC inquires as to his intentions. The controller indicated that he would soon have to amend his clearance. 'Well, maybe I could have the controller try to call company operations to see about landing at another airport, maybe Dayton. I better tell him something pretty quick, he's got enough to do without me sitting out here going nowhere'!

Fuel and time (the materials would need to be in Indianapolis within the next hour) have now become critical, and the thinking process increasingly complex. 'I guess I'm just going to have to suffer the consequences and put down until this weather breaks. Now, where do I go? Maybe a small strip where no one would know. Yes, but there wouldn't be any fuel this time of night. I guess I'm stuck going to Dayton. This controller is going to think I'm a real jerk'!

The pilot proceeded to land at Dayton. Five minutes after landing, the airport was closed due to severe thunderstorms and high winds. Indianapolis was still closed.

The above story is an example of 'in-flight decision making'. What is the situation? How do we go about deciding if something *needs* to be done? How do we select from among the alternative actions? How do we decide when to act? There have been numerous studies involving decision making during actual flight. Most often the notion of a 'critical in-flight event' (CIFE) and the pilots' response to it is studied. The results of these studies show pilots' reactions and decisions regarding these CIFEs vary tremendously with training, experience and attitudes towards aviation.

The key for the commercial pilot in making good decisions during flight is based on good decisions on the ground. Consider again our discussion of flight planning, where many of the 'problems' and answers to them could be made prior to ever leaving the ground. Another 'lifesaving' technique used often in training is the game of 'what if'? It is possible to practise making certain decisions while still on the ground. This gives the decision maker *time* (the aircraft is not moving), *information* (an instructor pilot, or the reference material, or the bosses' answers) and the *opportunity* to make a poor decision and learn from it. In other words, it offers a 'safe' environment in which all of the options can be explored.

'What if'? is not just an exercise in story telling, but rather the serious practice in decision making. Even though you may not get all of the answers on the ground, you will become better at knowing how to ask yourself the questions. Looking back at the above pilot's situation, it is interesting to note how many of his questions could have been answered by 'what if'? prior to making this flight. Many of the hazardous attitudes exhibited in his thinking process could have been eliminated by prior thought and planning.

The 'option' factor

As pilot in command, the option is always yours. Whether it is to go-around on landing, miss the approach, or divert to an alternate, the option is yours. Most important to recognize, is the option to say 'no'. Sometimes in our quest for 'heavier iron' we lose sight of that option. When your situation calls for a 'go' or 'no-go' decision, remember, it is not the company's limits, the aircraft limits or ATC's limits, but your own limits that must govern your answer. Don't let a bad decision be your last one!

References

Fahlgren, G. and Hagdahl, R. (1990). 'Complacency'. *Proceedings of the 43rd Annual International Air Safety Seminar,* Rome, Italy: Flight Safety Foundation.

Franken, H. (1990). *The Daily Reporter.* January.

Hackman, J.R. (1993). 'Teams, leaders and organizations: New directions for crew-oriented flight training'. In: E.L. Wiener, B.G. Kanki and R.L. Helmreich, *Cockpit Resource Management,* New York: Academic Press, Inc.

ICARUS Committee (1994). The dollars and sense of risk management and airline safety. *Flight Safety Digest,* **13**(12).

Lawton, R. (1986). Presonal Communication.

Mitchell, M.K. (1983). The bush pilot syndrome: A critical incident analysis. In R.S. Jensen Editor, *Proceedings of the Second Symposium on Aviation Psychology*, Columbus, OH: OSU Department of Aviation.

Nance, J.J. (1986). *Blind Trust.* New York: William Morrow and Company, Inc.

NTSB Reporter (1994). 'Controlled flight into terrain: FAA Beech Super King Air 300/F Front Royal, Virginia'. **12**(8), August.

Trollip, S. and Jensen, R.S. (1991). *Human Factors for General Aviation.* Engelwood, CO: Jeppesen Sanderson, Inc.

13 Special applications: ballooning and gliding

Seul le silence est grand (Silence alone is great)

Alfred de Vigny
(as quoted in Bernard Desestres, 1988)

Although the discussion previous to this chapter applies to all pilots and aircraft, there are certain aircraft, including balloons and gliders, that have special concerns. In contrast to airplanes, both of these aircraft require careful attention to vertical speed and have variometers as required equipment. Both ballooning and gliding are team activities usually associated with recreation. Several people are needed to inflate the balloon. A 'chase crew' is needed to assist in deflating and reloading onto a trailer following the landing. Communication is needed between the pilot and the chase crew to assist in making decisions concerning the landing site. In gliding a tow plane or winch is needed usually requiring the assistance of a crew for takeoff. Ballooning and gliding tend to form a different organizational cultures among enthusiasts that can cause pressures to take certain risks that may be somewhat different from other types of pilots. In this chapter we will look at how certain aspects of these operations may differ from those of the fixed wing pilot and we will point out a few scenarios from these operations to illustrate the unique decision variables and pressures from these operations.

Ballooning

Balloons, also known as lighter-than-air vehicles, take two forms: free air balloons, which have no form of horizontal guidance except the wind, and airships which are motorized and can maneuver from place to place. The original successful aircraft was the hot air balloon dating from June 1783, when the Montgolfier brothers of Annonay, France inflated a large paper bag with hot air (and smoke) sending it sailing upward 6,000 feet (1,800 meters) into the air. Ballooning is the oldest of aviation activities.

Balloons have been used in many applications since that day in France including transportation of people and cargo, observation and reconnaissance. Today, balloons are commonly used as camera platforms for major sporting events and for recreation. Many of these are operated commercially carrying passengers on sightseeing trips, but the majority of the hot air balloons are owned and operated by private individuals who fly them for sport. The premier exhibition and sporting event for hot air balloons in the USA is held in Albuquerque, New Mexico. Because of the orientation of the mountains and prevailing winds, this location often has an unusual weather phenomenon which permits balloonists to launch to a certain altitude, ride the wind for some distance, climb to a higher altitude and find a wind in the opposite direction thus permitting landing at or near the launch site. Balloons can often be seen at this popular site any time of the year, but in the Fall (October) at the annual balloonist convention, over 600 balloons can be seen in the air at the same time.

A first balloon flight

I have one experience as a passenger in a balloon which gave to me an impression of the various factors involved in balloon pilot decisions. On this occasion, I was invited for a balloon flight as payment for giving a talk to a local group of balloonists on decision making. It was to be an early morning flight and, over the phone, the pilot agreed to let me take my 15-year-old son as well as a good friend. We all arrived at the appointed launch site with great excitement and anticipation for a thrilling new flying adventure together. At sun up, the weather was beautiful with no clouds and no wind. However, after arriving, we were told that the temperature was too high for four people and that one of us would not be able to go -- maybe there would be a second flight if the weather held. It was decided that for the first flight, we would leave my friend behind -- much to her disappointment. The crew had no difficulty inflating the balloon's envelope in the calm conditions and we climbed into the basket for pictures.

As we lifted off, we found the wind at just above tree top level to be surprisingly strong and it began to carry us along fairly quickly. My son and I loved the ride. We went by farm houses and people came out to see us and follow us. I could feel the pilot growing concerned over the fact that the wind was carrying us right over the top of a high-tension power line which did not change direction as far as we could see. Also, we were moving at a fairly fast rate. The chase crew was below us following us on country roads.

Very soon after launch, it seemed as though nearly all thoughts in the mind of the pilot were focused on finding a proper landing site. He talked with the crew about the power line that stayed directly under us and the fact that our speed was increasing. Soon, the wind changed slightly and the pilot decided that he had an opportunity to set down. He chose a small field with lower-level powerlines on all four sides. We needed to get down quickly because the increasing wind speed would make landing more difficult and dangerous. From the air, the field seemed very small considering our speed, but I was only an airplane pilot. The pilot, told my son and me to squat down as low as we could on the bottom of the basket, brought it down nicely touching first in the middle of the field with a pretty good thud. We then lifted off slightly and came back down as the pilot pulled the cords to dump the hot air from the envelope. It did not release fast enough and the basket flipped over on its side and the wind began to drag us along the ground toward the powerline at the end of the field. It was a rough ride and we could easily have fallen out of the basket if we had not held on tightly to handles at the bottom of the basket. The pilot was yelling at the chase crew, who had parked at the edge of the field, to come and help stop the balloon but before they got there, the pilot had dumped enough air bringing the envelope down to the ground. We crawled out of the basket with only slight bruises to our knees and an impression that ballooning is thrilling but can be quite dangerous.

Judgment factors in ballooning

This experience point to several factors in ballooning that are present in all aviator decisions but have special meaning in balloon pilot decisions. These include, the weather, the landing site, the launch site, landowners, terrain, the temperature, the fuel, the weight, the crew and the passengers. Because these factors are especially important to hot air ballooning, we will consider them in that application.

Control factors. Compared with other forms of piloting, balloons have a very long or delayed reaction to control action by the pilot. This large delay means that the pilot must plan ahead, several seconds and sometimes minutes for 'burns' to provide lift that may be needed to avoid obstacles and properly reduce vertical speed for landing.

Weather. In contrast to other forms of flight, the most important weather phenomenon to balloonists is the wind. Over 10 knots of wind can make landing very hazardous. The complexity of the balloon pilot's is more complex, because the FAA Flight Service Stations are not equipped to provide the kind of precise wind data, either present or forecast, needed for their operations. Therefore, the balloon pilot must rely on his or her own expertise and judgment for these decisions. Usually, they will choose to fly in the early morning when the surface wind is light and hope it doesn't pick up too quickly, or they will fly at dusk and hope the wind does not go dead leaving them becalmed over a mountain or city with no suitable landing sight while darkness is coming on and fuel is running low.

Landing site. Next to the wind, the landing site is the biggest uncertainty in ballooning. The landing site should be a large field with no obstructions and no crops growing that could be damaged. It should also be located at the proper place so as to satisfy pilot, passenger, chase crew and fuel requirements. Unfortunately, these criteria are hard to meet in the real world where the wind may force you to compromise many of these criteria. Adaptability is the key to making decisions about the landing site. As one landing site alternative is lost, the balloon pilot must seek another and optimize the situation as best he or she can.

Launch site. In ballooning, the launch site must be chosen based on the wind, the desired landing site and the goals for that flight. The pilot decides where he would like to land, considers the wind direction and picks a launch location from where he expects the wind to take him over the landing site. It is far from an exact science. Winds at different altitudes differ considerably in direction and some adjustments can be made in flight to compensate for errors in picking the launch site and estimation of winds. If the pilot misses the chosen landing site because of the winds, he must choose an alternate site based on the terrain, wind, fuel and all other factors involved in ballooning.

Fuel. In hot air ballooning the fuel is critically important because it sets a time limit on flight and determines decisions about the landing site. In pre-takeoff decisions, fuel can be traded against the weight or space allowances for passengers in the basket. The fuel burn will be higher for a heavier load. In flight, fuel becomes critical in decisions concerning whether to extend the flight at passenger request or to pick a better landing site.

Landowners. It is frequently not possible to land in a predetermined place where the landowner is friendly. Balloonists are usually aware of landowners who are likely to give an unfriendly reception to anyone landing on their property -- the pilots often have known unfriendly land marked on their charts. In some cases, balloon landings can cause damage to crops in the field that can cost a considerable amount to the owner. Therefore, the value of the crop in the field being contemplated for landing as well as the nature of the owner becomes another factor in the balloonist's decision, both before flight and during flight.

Terrain. The terrain over which a balloon flies is sometimes the primary incentive to fly for some passengers and pilots because it is stunningly beautiful and can be viewed from a slow-moving platform allowing passengers slowly to drink in the beauty. However, such terrain places a premium on the pilot's decision making ability. In mountainous terrain and where there are rivers and lakes, the beauty is captivating but the landing sites are few and far between. Major decision tradeoffs are necessary between the enjoyment of the scenery and choosing a safe landing site.

Weight. In ballooning, the ability to carry weight depends upon the volume of the envelope, the outside air temperature and the temperature inside the envelope. Different balloons have different tolerances for higher temperatures. Care must be taken not to exceed the temperature tolerances offered by the manufacturer. A frustration offered by many balloon pilots who carry passengers for hire is that the weight of the passengers sometimes exceeds what they were told while making plans for the flight. Decisions concerning the amount of weight to carry are sometimes influenced by the force of the demands of passengers.

Temperature. As mentioned above, the outside air temperature (OAT) has a direct affect on the buoyancy of the balloon. As the OAT cools, the buoyancy increases and the same balloon can lift more weight. Unfortunately, the opposite is also true. In higher temperatures, the balloon is less buoyant and less able to lift weight.

Crew. Ballooning is not practical to experience alone. A crew is required for each flight to inflate the balloon and to assist in deflating, repacking and loading the balloon and basket onto a vehicle. The crew becomes a factor in many pilot decisions, both before and during the flight. The crew is on the ground doing much less enjoyable duty than those in the air, in the hope that they will be rewarded with a later flight. Therefore, the time and interests of the crew are important to the pilot. The terrain, roads and landowners are also considerations because the crew must deal with each of these on the ground. The crew assists the pilot in decisions about landing via two-way radio communications.

Passengers. The major motives for balloon flight often comes from the passengers and the balloon pilot's interest in showing them a good time. These motives can be very powerful, especially, if there is money involved. The balloon pilot must weigh his or her motives for flight against all of the decision factors mentioned above.

Hazards of ballooning

Ballooning is a fairly safe form of recreation but accidents do happen and sometimes they are fatal. The single most hazardous ground object to balloonists is the power line because it has great *potential* to be fatal. Balloonists put a great deal of effort into warning each other about running into power lines. On the other hand, the most common accident in ballooning, a hard landing, usually is not fatal. Equipment failures are fairly rare as the cause of an accident.

As in other forms of flying, statistics show that most balloon accidents are the result of pilot error (Hamilton, 1989). Figure 13.1 shows the percentages of accidents that have occurred as a result of the various types of errors that pilots have made. These statistics show that most pilot errors resulting in accidents are decisional in nature. The balloon pilot usually has enough time to think through situations although, at times, as situations develop, the pilot sees danger coming but has no control to avoid it.

One of the worst balloon accidents in ballooning history occurred in Columbus, Ohio in December 1990, an accident that killed four people. The NTSB gave as the cause, 'improper planning/decision making by the pilot'. In this accident scenario, the balloon was launched and flown by experienced balloonists from a site near downtown Columbus so that the passengers could get a close look at the office buildings in the morning. The winds on the surface were 4 knots and at 3,000 feet, 13 knots. The wind took the balloon toward a 1,029 foot high television tower. When the pilots saw the tower, they tried to climb over it but they were faced with a temperature inversion that slowed their ascent. They struck the tower near its top ripping open the envelope. The balloon immediately lost all lift and the basket fell to the ground. These balloonists were tempted by the lure of floating past office buildings at an altitude lower than their tops (witnesses said that the passengers talked with people in the buildings from the balloon). Although they should have been aware of the change in wind speed and the temperature inversion, they did not make proper allowance for either in choosing their launch site and their altitude to fly once aloft. After the accident, local balloonists said that the closest weather aloft forecast station located in Dayton, Ohio, nearly 70 miles distance, probably did not provide the correct wind speed aloft in Columbus.

Table 13.1
Balloon pilot error accidents by cause (Hamilton, 1989)

Cause	%
Misjudged clearance	11.86%
In-flight decisions	28.81%
Pre-flight decisions	18.31%
Failure to see obstacles	6.78%
Unsuitable terrain	5.76%
Other	28.47%

Incident reports from balloonists

In safety meetings balloonists are generous in sharing their experiences that they believe could help others. Some send reports to the Aviation Safety Reporting System. The following is an article published in the ASRS *Callback* (No. 125, Nov 1989) Called 'On the horns of a dilemma':

> On a flight 7 minutes after sundown time without lights, we became nearly becalmed over a railroad yard with numerous high tension lines. We had to fly low over a plant to find safe landing spot. We landed in parking lot shortly after sundown. All were OK -- no damage. In hindsight, we should have landed in a field with cows 15 minutes earlier. We should not have been so afraid of cattle late in the day.

From the same ASRS Callback the following was presented under the title, 'Downed, but not out':

> After flying low level, 200-300 feet AGL and attempting to land three times previously after making an initial approach from 1200 feet AGL (the winds shifted each time, making landing impractical due to the proximity of power lines), I was making another attempt to land on the east side of road in a large field, when the winds not only shifted again back to the

west and south, but I also encountered an unusual 'downdraft'. This resulted in a descent from the 75-100 feet I expected to be at, to about 40 feet. The shift also took the balloon on a path that was at first head-on into the oncoming traffic on the road, and then back toward the power lines on the west side of the road. Landing in the middle of the road not being an option, and knowing I could clear the power lines, I ascended to 1,300 feet AGL and landed a short time later a mile west.

I had checked the winds below me several times during the approach...they were consistent, blowing us to the east of the road where I planned to land...However, the unexpected quick descent of the balloon left me little room for error, and caused traffic on the highway to slow with the concern that I might land on the road.

...It is difficult to suggest corrective action other than a more thorough study of micro-meteorology. I am already aware of the causes of 'false lift' and this did not appear to be the cause of this descent. The same type of descent also occurred on my final landing over an open field where it was not as much of a concern, and I was prepared for it.

'False lift' reminds us of the ASRS old eagle (fixed wing type) who quipped, 'You must be either a fatalist or optimist to fly hot air balloons and not expect to find yourself in trouble -- frequently'. We don't know whether this reporter was either, but skill and experience provided ample ballast for the occasion.

Taking a dip. The following experience was offered by a balloon pilot on a solo training flight while preparing for commercial license flight test. It clearly shows how the fascination with balloon flight, the beauty of the scenery and the adventure of trying something new can lead one to make a series of irrational decisions. The pilot had 41 hours in balloons, 32 assents. He was flying a Barnes AX-6.

I Left Norwich High School and within 8 miles reached 8000' MSL. I did my first terminal descent, reaching 900 FPM and went back up to 7,000' MSL and did second terminal, reaching 800 FPM. I had told my crew at takeoff that I would try to reach Cooperstown. By this time, I was flying over totally unfamiliar territory -- searching for landmarks that would lead me to that village. By the time I spotted Cooperstown, I realized that it was too far to my right to get a successful line into it. As I approached Otsego Lake, which ran north out of the village I realized that I was going to cross the lake at right angles and that by dropping down into the lake valley I might be able to veer 90 degrees to the left and run right up the middle of the lake. By this time, I was almost two hours into the flight, fuel was getting low and I was going beyond the two hours after sunrise self-imposed limit.

Nevertheless, it was a beautiful day. I had been having a fantastic flight, and I judged that I had enough fuel to run the lake and land at its north end.

I therefore aborted an approach to land just short of the lake and decided to continue over and down upon it.

As I made this maneuver, the balloon in fact, turned 90 degrees to my left and I was approximately 150' above the lake surface heading up its length. I decided to do a touch and go. By this time a small boat had come out and was running alongside the balloon about 50 feet to my right. My first approach to the lake was too steep so I burned -- burned hard and in fact, overburned which put the balloon into a steep ascent. Frustrated with this I vented. The venting was obviously too long and the balloon once again began a steep descent. I was unable tc stop the descent with another burn (low pressure?) whereupon the basket hit the water.

Throughout this segment of the flight, I was in the middle of this long, thin lake, with the shore about half a mile to either side. The waves beneath the basket were moving with the balloon and appeared to be stationary. The boat likewise showed no motion relative to the balloon. The sense from the basket was of a calm, slow ride up the lake when in fact I was going close to 20 mph.

When the basket hit the water, there was a second or two of calm as the water flowed through the wicker. The balloon, however, continued up the lake at 20 mph. Suddenly, there was a quick jolt and the basket was yanked sideways up the lake -- completely filling it with water. Two tanks were under water and items in the basket began to float about and out. The balloon itself had turned to a spinnaker and I was in full sail up the lake. By this time, fuel pressure had begun to drop and I had much reduced output from the burner. I was barely able to get the flame into the balloon. I was able to maintain one long continuous burn into the balloon and gradually she filled out and lifted me and the basket full of water out of the lake. The water drained out of the wicker steadily and I gained altitude and quickly looked for a landing spot. Finally, at the north end of the lake, I was able to put down at a state park, but not before searching desperately for a field large enough to take me in a strong wind.

The total flight lasted 2.3 hours and I had no fuel left upon landing. There was no damage to balloon, basket, pilot, or landing site. I would suggest the following flaws in my decision making:

1 The major flaw was not to land short of the lake. My fuel was low, the flight had been long and good, it was two hours after sunrise.
2 The second flaw was to attempt a difficult maneuver in these conditions. There was little margin for safety should anything go wrong.

Why did this pilot use such poor judgment? The pilot offered the following contributing factors in making his decisions:

1 His relative inexperience.

2 The desire to keep a good flight going forever.
3 The irresistible urge to fly the length of the lake.
4 The lulling sensation created by beautiful weather, a fantastic landscape and a great flight.

The pilot added that the winds aloft and over all weather was as predicted. In fact, he said that it would have been possible to land in Cooperstown as predicted had he not mistaken another village for it as he made his approach into the area.

IFR in a balloon. This next incredible incident happened to a pilot with 188 hours of PIC time in balloons and 154 ascents. This flight was also made in a Barnes AX-6. The pilot's objective on this flight was to try for distance on a solo flight. He wanted to practise for a 100-mile flight.

> This was a cool March morning with high overcast and a slight breeze. Launch from Norwich HS took me to the southeast -- toward the Catskill Mountains. I would not be able to make any 100 miles today but would be able to give a try at a good distance run.
>
> When I got aloft, the cloud ceiling was much lower than predicted. I think I hit the bottom of the deck at about 5000' MSL with ground below me at about 1000'. I gave this a thought but had 360 degrees of visibility with no lowering in sight and no precipitation in any direction. Soon after takeoff, my crown line untied from the basket and swung free. I made a mental note to be careful upon landing not to snag it. About an hour into the flight, I noticed that the free end of the crown line went straight up into the air (an American flag, not visible from the basket was attached half way up the crown line). I was completely puzzled as to why the line should go straight up. I checked my altimeter and variometer and confirmed that the balloon was in level flight. I felt no wind or turbulence in the basket or sensed it in the balloon. I was left with a feeling that the weather was rather strange and should have made the decision to abort the flight at that point. Again, the sky was clear and I had plenty of room ahead before the Catskill Mountains. Fuel was no problem and I had now passed into completely new territory.
>
> My intention was to land in the Walton area just short of the mountains. As I approached Walton, I descended to about 1000 AGL and began looking for a landing field. Unfortunately, during this time I was unable to find one as the area was heavily wooded and my speed now made it necessary for a good sized field. I approached a hill region that had a gradual upslope on the windward side with a sharp drop off into the valley just before the mountain range. It was my plan to do a quick descent into the valley and land in what I call a 'wind shadow'. It was an ideal plan but could not be carried out.

> As soon as I began my descent at the point of sharp drop off, the balloon hit a violent wind shear which completely distorted the shape of the envelope. Chase crew said they had never seen anything like it. After successfully breaking through the shear zone, I indeed found stable slow moving air but was now located directly over a river with trees on either bank. I tried to ascend above the river to get off to either side but couldn't. The river ran right to the base of the mountain and veered 90 degrees to the right. Unfortunately, the balloon could not turn and I proceeded up the mountain slope. At this point, the ceiling had descended to below the mountain tops and I estimate that I entered the cloud deck approximately half way up the slope. At this point, I was flying blind -- totally reliant on my instruments. I estimated what my ground speed must have been and how long it would take me to fly over the mountain ridge. I had a good idea of the mountain's altitude and made sure that I had enough to clear. I had lost sight of any trees soon after entering the cloud. I began my descent after the time I thought it would take to clear the crest and eventually broke out below. I was in a small valley completely bounded by mountain walls with the exception of a few small houses and an open field. As the balloon approached the field I ripped out and did a rapid descent to catch the field. The balloon and I hit hard but we were down. The total flight lasted 1.8 hours and covered 35 miles. There was no damage to balloon, basket, pilot, or landing site.

The primary flaw in this pilot's decision making was to continue flying after it was apparent that weather conditions were not as forecast and that they were 'in flux'. Many fields were available to land in prior to reaching Walton. Contributing factors were:

1 Desire to make the goal of Walton.
2 Sense of wanting to push to the limit.
3 Not sensing that cloud deck would descend to hide mountains.

Stranded by no wind. The following incident happened to a pilot with 236 total hours and 197 assents. It occurred during a pleasure flight with two people in mountainous terrain, also in a Barnes AX-6.

> Our launch was from family hillside summer farm in the Green Mountain National Forest of Vermont. Two other balloonists were present but decided not to fly. Mountain flying is tough in the best of times and a weather system was very slowly moving in from the southwest. In fact, once aloft, we could see precipitation about 40 miles off to the southwest. While overhead was cloudy, slightly to our north was clear blue sky all the way into Canada.

We went to 7500' where we had good motion to our east. This would have carried us over the eastern ridge of the Green Mountains and would have caused our chase crew to drive many miles on to retrieve. We decided to descend into the valley (1000' MSL) and to land short of the ridge. Upon descending, we turned 90 degrees to the left and proceeded up the main valley. Upon reaching the end of the valley the balloon stopped on a small ridge. The crew waited in the valley below. First we would descend down one side of the ridge toward an open field but then stop and move up the ridge, over the other side and down toward the village side of the small ridge. We kept this up for nearly two hours -- never being able to get off the ridge. We had loaded extra tanks of fuel on board as we always do for mountain -- so in the beginning there was no problem with fuel.

Smoke from the plywood mill down the valley went up about 100' and just hung -- spreading out like a tablecloth. I can remember our emotions moving from slight amusement, to frustration, to anger, to fear -- when it became apparent that we had to land in the trees.

We eventually went up (not out of the valley because by the time the decision was made, we didn't have enough fuel to carry us over the main ridge) and headed north into even deeper forest. When it was apparent that we had to ditch in the trees, we were able to throw our line to the crew and they stopped the forward motion of the balloon. They attempted to pull us into a small clearing but the tree branches stopped the line and we vented into a very small opening between trees. The balloon dropped rapidly to the ground and we were finally down.

Our total flight was 2.5 hours and covered 5 miles. There was no damage to balloon, basket, pilot, or landing site. One chase crew member was slightly injured when the basket landed on him upon final descent.

The pilot offered that, although the initial decision to go into the valley to stop the crew from going over the main ridge was reasonable as it occurred within about 15 minutes of the beginning of the flight, the decision to remain in the valley when it became obvious that there was little wind for maneuvering was the fatal flaw in our line of reasoning. The pilot offered these contributing factors:

1 Desire to save chase crew from a long trip around the mountain ridge.
2 Lack of clarity on who was the PIC. My partner and I shared the flight. I took the first hour and he the remaining 1.5 hours until I took over for the landing. Although I wanted to go high, I did not express this forcefully enough nor did I ask to resume command. By the time it was obvious that going up was the only way out, we did not have sufficient fuel to chance it.
3 At times we were only a thousand feet or so from the fields we needed to land in. We sensed being so close it was difficult to believe, with the

amount of fuel we had, that we would burn it all before we had enough motion to land.

4 The weather conditions were pretty much as forecast. Perhaps the movement of the front became a factor for my partner in not wanting to go high -- I really can't say.

Soaring

Soaring (or gliding) is the 'purist' aviation approach to flight that seeks to capture the efforts of birds. Traditionally, soaring consists of motorless flight, but a recent trend is to use small engines to launch and retrieve the glider. A similar progression occurred in early aviation. Early pioneers used gliders to understand aerodynamics and to teach themselves to fly. Most notable were Otto Lilienthal and the Wright brothers, Orville and Wilbur. Lilienthal conducted over 2000 glider flights in Germany during the 1890s in a series of aircraft that resemble present day hang gliders (Boyne, 1987). His work provided the basis for further aircraft development by others. The Wright brothers, after studying the work of those before them, used gliders in 1901 to teach themselves to fly as well as to test their design concepts prior to the first powered flight. Such flights must have held a special place for them, because after their successful powered flights of 1903 they returned to Kitty Hawk in 1911 to set world gliding records that lasted for many years (Schweizer, 1988).

Soaring is primarily a recreational activity and has little commercial value. The various craft used in soaring are classified as sailplanes or gliders, ultralight gliders, hang gliders and paragliders. Each of these has a range of performance and operational standards that characterize the meteorological conditions and pilot knowledge required. Table 13.2 shows typical performance values for the various classes of gliders. Soaring has become quite popular in the USA as well as in other countries. Due to hang gliding's lower cost and easier accessibility, participation in this sport has increased rapidly in the last several years.

Like ballooning, soaring is less regulated in the USA by the FAA than conventional flying. Sailplane (glider) pilots must possess FAA issued certificates and undergo recurrent training (biennial flight reviews), but an airman's medical is not required. Ultralight glider, hang glider and paraglider pilots possess neither certificates nor medicals, but are primarily self-regulated by the individual pilots and the national organizations, such as the US Hang Gliding Association (USHGA). In all cases, the primary source of safety training takes place through local groups with assistance from the national associations such as the USHGA, the Soaring Society of America (SSA), etc. The flight vehicle certification requirements also vary along a range from full FAA certification for sailplanes to manufacturer's group standards for ultralight gliders and hang gliders, or less. Again, the national associations play a major role in promoting safe equipment.

Table 13.2
General performance characteristics of soaring vehicles: typical glide ratio/empty weight in gliders (World Meteorological Organization, 1993)

Glider Type	Glide Ratio	Speed/Empty Weight
Paraglider	6:1	30 km/h 6 kg
Hang glider	11:1	50 km/h 30 kg
Ultralight gliders Sailplane Trainer	27:1	70 km/h 400 kg
Racing	45:1	100 km/h 300 kg
Open	60:1	110 km/h 500 kg

Like ballooning, soaring often takes more than one person to participate and thus tends to take place in small groups or clubs. Typically, at least one other person is required to help put the glider together, to move it on the ground, to launch or tow it into the air and to retrieve it at the end of the flight. The amount of difficulty in the assembly and ground handling tasks is related to the weight of the glider. Weight also influences launching or towing techniques. In the USA, single engine airplanes provide most of the tows for sailplanes, while ultralight aircraft sometimes tow ultralight gliders and hang gliders into the air. Powered winches and automobiles can also launch sailplanes, ultralight gliders and hang gliders. Hang gliders and paragliders may be foot launched by the pilot running from a raised platform or hill as Lilienthal did a hundred years ago. Retrieving can be as simple as packing the paraglider into a bag, loading a hang glider onto the top of a car, or as difficult as removing a large sailplane from a remote field into a 35 foot trailer.

Competition is also a part of gliding and world or international competitions are held for each glider type. The spectator appeal for soaring is less than that for ballooning because non-participants cannot easily see and follow the activities. Typical competitions consist of cross-country flying for distance and/or speed. Sailplane competition can which take place over several hours (typically 3-5 hr), at altitudes up to 18,000 feet (typically 3000-7000 AGL) and at distances over 600 miles (typically 100-300 miles).

Soaring as adventure

Many soaring pilots find sheer joy in the sport because it offers a challenge and freedom unknown to the airplane pilot. The challenge comes from the combining of piloting skills, an understanding of weather and the lure of the uncertain. The freedom emanates from the sound of the wind in the ears and the lack of a controllers voice. If you choose your lift carefully, for example, along a ridge where a dependable uplifting wind is present, you can launch and stay aloft for as long as you wish. Long out and back cross country flights may be possible. You are not constrained by the need for fuel. The natural lift from the wind represents freedom. In other cases where you must depend on thermal lift, soaring is a lot of work as well as tension as you seek lifting winds that continually move from place to place. I one time took a glider ride with an instructor on the Island of Oahu where there was a very dependable wind. After our launch, we flew about with all the lift we needed from the breeze coming off the ocean along the ridge. We had to land only because that was the end of the time for which I had paid.

Flying dreams. Have you ever had one of those dreams in which you were able to fly simply by putting out your arms and jumping off a building or cliff? I have talked with many people, both pilots and non-pilots, who have had this recurring dream. Of all the topics in my own dreams, these have been the most vivid. When I move my arms downward, the force of acceleration upward is very strong in my inner ears and stomach. Usually, I do my dream flying of this sort at low altitude, from building to building. I don't seem to have a great interest in flying very high. Another common thread in these dreams is that I am the only one in the dream with the capability to fly in this way -- and I am the envy of everyone else because of it! When I tried hang gliding, I found an activity that is frighteningly close to this dream.

A special form of glider flying is called 'hang gliding'. Hang gliders are light enough to hold and launch by running into the wind until becoming airborne. Typically the launch takes place from the top of a hill or mountain where the wind is producing a good upward lift. The glide ratio for hang gliders is not as good as it is for sail planes, so the hang glider pilot must have better lifting forces to stay aloft very long. Nevertheless, hang glider records for time aloft (18 hours) and distance (hundreds of miles) are surprisingly long.

I once took a series of hang glider lessons and, in the end, obtained the 'Hang One' hang glider rating. I found that the hang gliding experience is as close to the dream of self-powered flight as anything I have done in aviation. I took the course at Kitty Hawk Kites in Nags Head, North Carolina a few years ago together with my 14 year old son, Mark. After a few nose down experiences (called lawn darts) into the sand dunes where the Wright Brothers first flew their own gliders, we moved further up the hill. When I had gained enough control to launch from the top, I strapped myself into the hang glider, tested the wind direction by kicking up sand, ran as hard as I could into the wind and felt myself floating aloft. I could control my speed by pushing forward or pulling back on the bar. I could make turns by shifting my weight. For the landing, I could glide close to the ground, push forward on the bar to slow down, and when I did it right, stop in the air before touching down. It was one of the most thrilling experiences of my life!

Judgment factors in soaring

Many judgment factors in soaring parallel those of all aviator's decisions, but several factors take on added emphasis or are unique to the activity. These include glider assembly, launch methods, pilot environment, weather, landing sites and small group norms.

Assembly. Gliders are designed to be easily disassembled for routine transport to and from soaring sites as well as from remote landing sites. Thus inspecting the glider for proper assembly becomes important before every flight. Every year, unconnected and unchecked controls lead to damage and injury. Decisions on how glider assembly is conducted and subsequently checked represent controllable, but often lax pilot decisions.

Launching/towing methods. Towing methods provide an unusual situation in aviation in that two pilots in two separate aircraft tied together by a rope must work together as a team for safe launching. Thus the decisions of one pilot or winch operator significantly influence the decisions that the other must make. Mutually understood procedures and good communications offer the best approach. Launching using powered winches or cars require a different set of procedures but similar communications. Actions take place quickly during glider launches so preplanning what to do in unusual situations and quick decision making is required.

Pilot environment. As flights become longer (3-5 hours) pilot comfort becomes more critical in allowing the pilot to make crucial decisions such as when and how to land at the end of the flight. Pilots experience heat at low altitudes and cold at high altitudes along with temperature fluctuations associated with climbs in lift and descents during glides or in sinking air. The enclosed cockpits with of their large canopies and limited ventilation provide a greenhouse effect, accentuating the heat problem. In high temperatures and at high altitudes, dehydration can be a significant problem that directly affects pilot abilities to make good decisions. Carrying water is easy, but dealing with the elimination of the water creates a more difficult barrier to overcome. Full bladders can rush decisions as well as create a major hazard if ruptured during a rough landing. In the western USA, gliders often operate at altitudes well in excess of 12,000 MSL, which makes oxygen use necessary. Without oxygen, decisions progressively deteriorate with time, often in a very short time. Therefore, managing pilot comfort and physiological status is critical to maintaining pilot decision capabilities.

Weather. While balloon pilots are avoiding thermals and winds over 10 knots, and airplane pilots are climbing above convective turbulence, glider pilots seek out thermals because, within most turbulent phenomena is the lift that they need to stay aloft. Within these thermals is also found sinking air (they are trying to avoid) and possibly turbulence that may exceed the capabilities of the aircraft or pilot. Care must be exercised in choosing which cloud to fly near. The lift in a thermal provides many hours of fun, but lift in a thunderstorm must be avoided. The smooth lift in a mountain wave can provide a ride to great heights seemingly suspended in time and space. However, lurking beneath this wave may be a powerful turbulent rotor capable of exceeding glider load limits. Properly dealing with marginal weather conditions represents a major component of decision making in soaring.

Landing sites. Compared to balloons, glider pilots usually have a much wider choice of landing sites, but the choice can be influenced quickly by local terrain and weather conditions. Should conditions differ from those expected, the glider pilot may be required to land in an unprepared area that they have never seen before. In landing away from the launch site, pilots face rapidly changing unknowns that may be outside of their control, challenging even the best decision makers.

Small group/club atmosphere. Because soaring activities often occur in small groups and clubs, the safety norms of the group have a strong influence on the pilot's decisions. Pressures always to return to the takeoff site on time to allow the next pilot to fly can lead pilots to attempt a glide back to the site when altitude is insufficient to do so. Prohibiting off field landing can lead pilots to attempt to return to a site when conditions make it difficult, dangerous or impossible. Since these small groups may operate for years with little outside influence, efforts to change unsafe norms may not exist or meet strong resistance. Often the need to change a procedure or policy to improve safety may go unnoticed. On the positive side, in the group setting, good decisions by experienced pilots can be observed by novices and reinforced.

Hazards of soaring

Soaring is also considered a relatively safe sport as judged by its participants, but accidents and fatalities occur. While mid air collisions are relatively high on the list, the highest rate of fatal accidents occur during launching and landing with the stall/spin accident being the most common. Equipment failures are rare as an accident cause but accidents due to improper assembly of the glider occur and should be attributed to pilot error.

Accident statistics for soaring activities are difficult to compile because many accidents are not reported and those statistics that are reported tend to be scattered among the various national organizations and their insurance companies. Because there are fewer mechanical elements in soaring, the percentage of accidents cause by mechanical failure are lower than for general aviation (Hammond, 1992). Table 13.3 shows that most accidents are related to pilot error. The high rate of stall/spin accidents reflects the tight maneuvers that sailplane pilots frequently make to get into position for a landing with no power. A look at the other causes seem to show that a higher percentage of soaring accidents are caused by perceptual judgment than cognitive judgment (see Chapter 1). Nevertheless, cognitive judgment is important as shown in Table 13.4.

Table 13.3
Causes of sailplane accidents and fatalities during 1981-1991 by percentage of total (from Hammond, 1992)

	Accident %	Fatalities %
Stall/spin	16	39
Premature tow termination	14	12
Landing undershoot	12	10
Landing overshoot	4	0
Hit object while in air	12	8
Hit object while on ground	8	3
Weather/wind shear	6	3
Mid-air collision	5	9
Ground loop	5	0
Hard landing	3	3
Structural	2	5
Mechanical	4	1
Medical	2	3
Misc.	1	1
Unknown	4	2

Decisions are the challenge that keep sailplane pilots in the sport for years. Competition pilot Ed Byars (Byars and Holbrook, 1974, p. 49) states it plainly, 'the crux of cross country (soaring) is decisions'. This heightened rate of decision making increases the challenge inherent in the desire to remain aloft by searching for lift and avoiding sink. The challenge is also present in attempts to maintain alternate landing site awareness and in circling closely with other gliders in thermals. Unfortunately, these risky decisions often lead to accidents and incidents in soaring. Table 13.4 presents some of the typical faulty decisions that have resulted in accidents and incidents.

Table 13.4
Some typical faulty pilot decisions in soaring for various activities

Activity	Faulty decisions
Preflight	Assuring maintenance work properly completed Improper preflight inspection Failing to check for proper glider assembly Incomplete check for freedom of control movement Failing to secure loose objects
Weight and balance	Poor estimation of passenger weights Knowingly exceeding allowable gross weight or cg limits Failing to remove or install ballast Regulations/procedures Failing to keep current on regulations and/or procedures Failing to use current navigational charts
Cross country flights	Careless or missing weather briefings Lack of familiarity with aircraft Lack of soaring skill Poor planning of route, support crew, trailer, etc. Lack of position awareness Continuing into unsuitable weather or terrain Failing to keep a landable area within gliding range Delayed decision to land Incomplete or lacking pattern for off field landing
Local flights	Failing to review procedures when local situation changes Lack of proficiency in basic flight skills Unwillingness to land off field if unable to return to the takeoff site (glider version of get-homeitis)
Contest flights	Over-extending skills to win or score higher Unfamiliarity with emergency egress and parachute use Lack of skills and procedural knowledge to fly in close proximity with other gliders Insufficient cross country skills
Physiological	Violating the 'I AM SAFE' checklist Developing dehydration Flying into conditions that promote vertigo Misunderstanding disorientation in poor visibility

Glider accident investigation

Soaring related accidents, even fatal ones, tend not to be investigated and reported in any depth because no one has clear responsibility to do so. Investigators often have a difficult time making an analysis due to the lack of evidence. National bodies would like to investigate further, but lack the financial and personnel resources. Glider organizations have no legal authority to investigate accidents. As a result, discussions and learning opportunities of most value to the sport's participants are provided by those survivors of accidents that have the courage to conduct their own analyses and to write about them. While these anecdotal accounts provide little scientific data, they do support the concept that accidents can happen to any pilot, the novice or the expert alike.

> *Too dry.* An experienced pilot with 6000 hours in jet fighters including combat in Vietnam, 1,000 hours in other powered aircraft and 3000 hours in gliders hit the ground in a steep nose-down attitude from about 75 feet after stalling his sailplane during an attempt to land off field near Reno, Nevada. The accident took place on the sixth day of a contest in which the pilot had flown for over 23 hours on the first five days and had been airborne for nearly six hours this August day. During the assigned task, a maximum altitude of 14,300 MSL had been reached in what was characterized as 'squirrely', widely spaced thermals. After leaving the last turnpoint, the pilot struggled at low altitude in an attempt to return to the takeoff airport -- a decision to leave a safe landing site in the Red Rock Valley to make it 'home'. While gliding over a dirt road, the pilot failed to realize that he was only 200 feet AGL when he started to land. Only when he noticed that his shadow was very big did he realize how low his 180 degree turn was. At this point the glider stalled.
>
> The pilot was severely injured but conscious. He took a big drink of the ice water still beside him. His post-crash decision would be critical as he sat in the crumpled glider 10 miles from civilization with a broken radio. With his leg bleeding badly he recognized the need to apply a tourniquet. He started to shiver in the 95 degree heat, a sign of shock. He wrapped himself in his parachute, elevated his legs and waited. An hour later, a passing motorist stopped to help. His recovery was long.
>
> Contributing factors in this accident were: (1) Fatigue -- a six day contest with thirty hours of flying for an older pilot. (2) Dehydration -- no water was drunk during the five-hour-plus flight. (3) Low altitude stall -- different perspective of low turn and possible wind gradient effects and (4) Get-homeitis -- leaving a safe landing place at low altitude to get back despite deteriorating lift (Thelen, 1992).

By the nature of soaring activity, the number of individuals involved in any particular accident tend to be one, occasionally two and only rarely three. Seldom do these situations make the national news unless they are unusual or involve a well known individual. A local glider instructor related several accidents that occurred over the last few years (Kirkbride, 1994).

> *The cactus.* The first one recalled by local glider pilots involved an unusual occurrence in Arizona. While rolling out after making a landing, the glider struck a large cactus. The cactus weighing several hundred pounds fell onto the person seated in the front seat crushing him to death.
>
> *Spin.* Another occurred several years ago at one of the Ohio clubs. In this accident, an experienced instructor and a rated airplane pilot were approaching to land in a large three-placed glider. While the observers were unable to determine who was actually flying, the sailplane spun into the ground on final killing all three.
>
> *Hidden powerline.* In the third situation, an experienced cross country pilot was attempting to land in a mountain clearing in a neighboring state. At approximately twenty feet above the ground the glider struck a powerline. The poles were hidden by nearby trees. The powerline caught the pilot under the chin and fatally injured him.
>
> *Tow plane.* In the fourth situation, a tow pilot who had recently regained his medical was towing a solo student glider pilot in a training aircraft. The towplane nosed over, hit the ground after attaining only a couple hundred feet on the climb and the towpilot was found dead. The student glider pilot fortunately released and landed safely.

No mechanical failures were discovered in any of these accidents. Therefore, the investigations centered on the proficiency of each pilot and their decisions. In the second case, investigators raised the question, 'Was an experienced instructor allowing a non-glider rated pilot fly a glider noted for unforgiving stall/spin characteristics without adequate supervision'? In the third case, they asked, 'Was there any way the pilot landing in a mountain clearing could have avoided the need for an outlanding or could have located the hidden powerlines'? In the fourth case, the question was, 'Could the tow pilot's medical situation have been anticipated'? As is often the case in glider accidents, the answer to these questions will never be known because of limited data available.

To sailplane pilots, probably the most shocking accident in recent years, was the 1992 death of three time world champion and German soaring coach Dr Helmut Reichmann while training members of the German Junior National Soaring Team in the French Alps. While thermaling on the side of the mountain with a team member, a mid air collision occurred. Reichman was unable successfully to parachute from his damaged sailplane after jettisoning its canopy. Speculation suggested that the contact had occurred too close to the ground to permit time to egress from the sailplane.

Reichmann's credentials were significant as an experienced pilot, instructor and coach. He was a proponent of safety, had lectured worldwide on sailplane safety and had studied the sport for over thirty years. Did he make a faulty decision? Did he allow his student to get too close to the terrain or into a blind spot? Were they thermaling too low? Again, we will never know, but if it can happen to the best, it can happen to anyone. We can only reinforce in our students how hard all pilots must work to avoid similar situations.

Incident reports from soaring pilots

Incident reporting, which has become a major factor in airplane risk management (see Chapter 11), is non-existent in any formal way in the soaring community. Informally, incidents are shared between members of soaring clubs and to some extent through soaring magazines. Thomas Knauff and Doris Grove (1992) have prepared a useful judgment guide for glider accident prevention that contains a number incident reports. This guide was prepared based on the aeronautical decision making manuals provided by the FAA (Buch, Lawton and Livack, 1987; Diehl, Hwoschinsky, Livack and Lawton, 1987; Jensen, Adrion and Lawton, 1987; Jensen and Adrion, 1988; Jensen, 1989). Two incidents are given below which illustrate the kinds of problems sailplane pilots sometimes face.

Disconnected. Despite twenty-five years of experience a glider pilot forgot to connect all the controls and to check them independently prior to launch. He relates the story as follows:

> I had a polyethylene tube with a red flag on it and the statement, 'five pins' written on the side to place over my elevator push rod as a reminder (of the assembly procedure). I have had problems in the past with hookups by virtue of being interrupted in my procedure of putting the glider together. On this occasion, early one spring, my son was with me to help assemble the sailplane. I did indeed notice the reminder and thought to myself, 'Gee, Tom hasn't seen this neat little memory tool, I think I will show it to him'. In doing so I interrupted my procedure and right, you guessed it, I failed to hook up the elevator on my Standard Class (no flaps) Pegasus (Thelen, 1993, p.46).

On takeoff, the glider was uncontrollable in pitch, but the pilot was quick enough to release at a low altitude and 'land (crash?)' in an unobstructed area with no damage. Contributing factors were, (1) failure to use a checklist for assembly, (2) failure to do a positive control check prior to flight, (3) allowing distractions during assembly.

Multiple decisions, multiple lessons. A relatively new pilot who classified himself as 'your basic beginning, weekend, chicken pilot' made a series of good and bad decisions that almost got him trapped in a wave at altitude on top of an ever increasing cloud deck. While trying to qualify for the five hour requirement for a soaring badge, the pilot reached 14,000 MSL without oxygen, was unsure of his position above the clouds. He ended up being escorted to an alternate landing site by towplanes who conducted an airborne search for him. His story follows:

> With just over 200 hours, one cross country and landings at only three airports, the lure of a mountain wave and the possibility of a five hour flight lured the pilot into quickly preparing his glider and taking off. During the early morning the pilot did not check the weather but expected the wave to last throughout the day. As he rushed to prepare for his flight, the bits and pieces of weather conversation left him with 'an erroneous idea that it might get cloudy, but that it would pass'.
>
> As he hurried to prepare, he constantly reminded himself to slow down and follow his checklist. Others helped him prepare. However he constantly discovered himself missing items, such as being strapped in the cockpit with the towplane waiting, only to find that the ballast was in the nose of the aircraft when his weight dictated that none be present. A final item missed was that the radio was not operating because of a speaker wire being disconnected.
>
> After entering the wave and reaching 7000 MSL he discovered the radio problem but continued the flight. He was not sure that this opportunity to complete the five hour task and possibly an altitude gain of 3000 meters (FAI gold altitude gain) in the rental ship would ever present itself again. After reaching 12,500 feet in lift that continued, he decided to go higher. Without oxygen, the regulations required that he only be above this altitude for less than 30 minutes and not exceed 14,000 MSL. He went on to 14,000 and the lift continued higher. At this point he decided to open the spoilers and descend to 12,500. However the climb and the decent probably took longer than planned.
>
> With the success of achieving the gold altitude gain fresh in his mind he made a fateful decision. He would try for the five hour duration also. The takeoff airport was just beginning to be obscured by cloud as the front moved in. Would it be possible to descend under the cloud deck and still make the airport. Without a radio, no help was available and the pilot's experience left him without the knowledge to judge. But the pilot recalls 'in the back of my mind was the mistaken perception that I had been told that this was to be expected'. Mistakenly, the pilot expected the clouds to dissipate later in the day. Besides an alternate airport was visible in the valley below and there were hundreds of landing places elsewhere in the valley. The pilot considered this decision to be 'the big mistake of the flight'. Had he landed at the alternate airport there would be no story.

> Over the next three hours, the clouds obscured the alternate airport. Waiting for the clouds to clear over the takeoff airport he decided to stay high but under the magical 12,500 feet that the regulations require the unavailable oxygen. He monitored himself for hypoxia symptoms but recognized none. The pilot relates little about these three hours except his cold feet at altitude and the shot of excitement as a large military transport climbed through the slowly increasing cloud deck to pass within a thousand feet or so.
>
> At the end of the three hours, the clouds had crept slowly over the area obscuring the major landmarks familiar to the pilot. As he decided on which other visible landing sites were available, a new sense of urgency presented itself in his decision making. He noticed a vertical cloud formation coming from the west toward him. This meant rapid weather deterioration and the requirement to take action now. He decided to descend, NOW! He also radioed his intentions and asked for a towplane escort, not knowing if he was being received. As expected, an unfamiliar airplane appeared above the cloud deck, as did a recognizable towplane, eventually to escort him to a safe landing at an airport.
>
> The descent and landing were not uneventful. The holes in the clouds were quickly closing. So, a fast descent was required. The wind had increased as had the turbulence, especially after being in the smooth air of a wave. Being unfamiliar with the airport, the pilot landed before the displaced threshold just missing the lights. As a final good decision the pilot decided to forgo a tow back to the airport, so the trailer was summoned. As this occurred, the front moved in quickly.

In this instance, the pilot claimed several lessons learned including, (1) 'When you take chances you don't always risk just your own hide' (The airplane pilots were also at risk here.), (2) land at the best alternate and do it early, (3) it takes time to descend from altitude, (4) large transports move quickly and may never notice something as difficult to spot as a glider and (5) 'don't land beside the runway at an unfamiliar airport'. You never know what you have to contend with. The white line in the middle of the runway is much safer (Dunkel and Thelen, 1992). The decisional factors in this instance suggest the following rules for glider pilots:

1 Hurrying to launch invites missing checklist items. At least the pilot attempted to use a checklist.
2 The lure for personal goals and those recognized by peers offer a strong attraction to continue.
3 Even though the regulations allow flight to 14,000 feet, what effect did this ascent and the subsequent time at 12,500 have on this pilot's decisions? The effects can be subtle.

4 Some of the decisions were based on secondhand and misinterpreted weather information. Monitoring the conditions becomes important to making the proper decision to continue or abort.

5 Once everyone is safe on the ground, the cheapest solution to having aircraft away from home base may not be flying them home but leaving them there and returning another day.

References

ASRS (1989). 'On the horns of a dilemma'. *Callback* (No. 125), November.

Boyne, W.J. (1987). *The Smithsonian Book of Flight*. Washington, DC: Smithsonian Books, 35-52.

Buch, G.D., Lawton, R.S. and Livack, G.S. (eds.) (1987). *Aeronautical Decision Making for Student and Private Pilots*. Springfield, VA: NTIS, DOT/FAA/PM-86/44.

Byars, E. and Holbrook, B. (1974). *Soaring Cross Country*. Cumberland, Maryland: Soaring Symposia, p 49.

Desestres, B. (1988). *Silence on Plane*. Chambourcy, France: Seidec.

Diehl, A.E., Hwoschinsky, P.V., Livack, G.S. and Lawton, R.S. (eds.) (1987). *Aeronautical Decision Making for Student and Private Pilots*. Springfield, VA: NTIS, DOT/FAA/PM-86/41.

Dunkel, D. and Thelen, G. (1992). 'Safety corner'. *Soaring* **56**(1), 43-46.

Hamilton, T. (1989). 'Pilot decision making'. *Balloon Life*, pp 10-14, February.

Hammond, G. (March 1992). *Flight Instructor Refresher and Revalidation Clinic Training Course Outline*. Hobbs, NM: Soaring Safety Foundation.

Jensen, R.S., Adrion, J. and Lawton, R.S. (1987). *Aeronautical Decision Making for Instrument Pilots*. Springfield, VA: NTIS, DOT/FAA/PM-86/43.

Jensen, R.S. and Adrion, J. (1988). *Aeronautical Decision Making for Commercial Pilots*. Springfield, VA: NTIS, DOT/FAA/PM-86/42.

Jensen, R.S. (1989). *Aeronautical Decision Making -- Cockpit Resource Management*. Springfield, VA: NTIS, DOT/FAA/PM-86/46.

Kirkbride, L. (1994). Personal communication.

Knauff, T. and Grove, D. (1992). *Accident Prevention Manual for Glider Pilots*, 2nd Edition. Hobbs, NM: Soaring Safety Foundation.

Schweizer, P.A. (1988). *Wings Like Eagles: The Story of Soaring in the United States*. Washington, D.C.: Smithsonian Institution Press. 1-5.

Thelen, G. (1992). 'Hung out-too dry'. *Soaring* **56**(2), 47-48.

Thelen, G. (1993). 'Disconnected'. *Soaring* **57**(4), 44-46.

World Meteorological Organization (1993). Technical Note No. 158: *Handbook of Meteorological Forecasting for Soaring Flight, Second Edition*. Geneva, Switzerland: WMO. 1-3.

14 The future of aviator judgment

My son, preserve sound judgment and discernment, do not let them out of your sight...then you will go on your way in safety.

Proverbs 3:21,23

Judgment remains the final frontier for understanding the human as an integral part of a man-machine system. It continues to be the best that man has to offer and will continue to be the reason for his superiority over the machine. Although technology will replace man in many functions, man will still have a place in systems where uncertainty remains. For this reason we must continue to learn about human judgment and explore new ways to develop it in aviators.

If I might offer one last personal example, in every check flight I have taken, my goal has been to pass or to be judged 'competent' by the examiner. I didn't really give much thought to being an expert, I just wanted to pass. I remember thinking 'I don't feel that strong in this area, I hope the examiner does not check the areas in which I am not strong'. My first thought was to cover my weaknesses so that the examiner would not see them. The question that comes to mind is 'where is safety in those thoughts'? Wouldn't it be better if I would tell the examiner 'I feel weak in this area so please test me there so that I can be sure I've got it'. Wouldn't it be better if all pilots tried to be more than competent to pass the exams? Suppose all pilots instead had as their goal to be *expert*.

Consider a recent accident that occurred in Columbus Ohio (NTSB, 1994). An Atlantic Coast Airlines Jetstream 4101 stalled and crashed killing five people because the pilots let it get too slow on final approach. Prior to the flight the captain expressed concern over the weather that he would face. The NTSB found that this captain had had trouble passing simulator and flight checks because he could not handle manual instrument flight, especially during approaches. The weather was not all that bad but it would require an instrument approach in possible icing conditions. Clearly, this captain was 'over his head' on this flight. Wouldn't it have been right for this captain to turn down this flight and any future flights until he was confident of his ability to do instrument approaches?

What is the difference between a competent pilot and an expert? After most of a career of study, I have reached the conclusion, in agreement with Dreyfus and Dreyfus (1986), that the difference between a competent pilot and an expert is *judgment*. This chapter discusses a research effort that is probing into the heart of pilot competence and expertise in a effort to discover how to develop expertise in pilots. The chapter concludes with a model of pilot expertise that I believe has the potential for better understanding and will promote the development of improved judgment training strategies for pilots.

This ongoing research program being conducted at the Ohio State University is revisiting human judgment in an effort to develop a better understanding of decision processes in expert aviators. As we have seen, training strategies for the improvement of general aviation pilot judgment based on attitude models of human information processing have been developed and tested with considerable success (Diehl, 1992). However, there is a growing belief that the usefulness of this approach may have reached a plateau and that it is time to consider a more comprehensive model as the basis for new judgment intervention strategies. In this new effort the expertise approach to cognitive task analysis is being used because we believe it has the potential to create new insights into human decision making in the aviation domain. Traditional Instructional Systems Design (ISD) methods do not offer useful guidelines for complex cognitive tasks such as those involved in flying today's complex aircraft (Ryder and Redding, 1993). The expertise approach is being used because its application can reduce duplication of effort in the development of intervention strategies.

The expertise approach

The OSU study of expertise seeks to understand and account for what distinguishes outstanding individuals in a domain from less outstanding individuals in that domain, as well as from people in general. The approach is limited to outstanding behavior that can be attributed to relatively stable, learned characteristics of the relevant individuals. The classical expertise literature says that expertise is acquired over decades of experience and practice (Ericsson and Smith, 1991). The approach includes both a careful analysis of expert performance and of how that performance was attained.

Building cognitive models using the expertise approach involves three steps: (1) identifying representative tasks that capture the essence of superior performance in a specific domain, (2) detailed analysis of the superior performance through several methods including verbal reports during performance of the tasks (or stimulus materials), (3) accounting for the acquisition of the characteristics and cognitive structures found to mediate superior performances of experts.

The identification of tasks that can capture superior performance is often not easy. It may be helpful to identify only a small number of such tasks and test them on a larger population of experts and novices in a particular domain. In our study we used six aviation tasks that will be described below.

The full range of methods of analysis in cognitive psychology can be applied in the examination of phenomena associated with a particular type of expertise including, performance analysis, expert-novice comparisons and extensive studies of single subjects. One cannot, of course, directly observe the mediating steps in information processing but it is possible to specify hypotheses about relations between the internal processing steps and observable behavior. Think-aloud verbalization and retrospective verbal reports are the most common methods of assessing differences in mediating processes as functions of the subject's level of expertise. Subjects are asked to think aloud while carrying out a series of representative tasks.

Although the expertise literature says that aggregation of experience is the single most important factor in the acquisition of expertise (Chase and Simon, 1973), we believe that 'meaningful' experiences are at least important if not essential for learning to occur. For example, some tasks never seem to be learned even though they are performed repeatedly. Several studies have demonstrated that, at best, minor improvement is gained with years of experience in comparison with initial training (Ericsson and Smith, 1991). In aviation, pilots know that flight experience does not equate to expertise -- yet, hours of flight time continues to be the criterion for currency, ratings and certificates. We believe that other important mechanisms are at work including: the teacher, instructional media, motivation and internal adaptation processes.

Compared with statistical procedures, expertise procedures are not likely to yield more reliable results. However, by going through the expertise procedure, we are more likely to encounter and develop a better approach toward interventions than we would using the statistical procedures. This is because much of the expertise approach deals with the acquisition of expertise, not just the explanation of expertise. The statistical approach may retrospectively identify what happened and how often, but it does not suggest the causal factors or remedial actions. The expertise approach actively solicits such information and the results lead more quickly into intervention strategies.

The general aviation domain

The domain selected for this study of human expertise is a subset of general aviation pilots who fly the Cessna P-210, the pressurized Beech Baron and the Piper Malibu. These are considered complex, fairly high performance single- and twin-engined airplanes and are usually flown by a single pilot, often into the most complex airspace in almost all weather conditions. For these reasons, the flying task may be as difficult as any civilian flying task. To add to the complexity of the task, many of the pilots of these aircraft have other professions (doctors, lawyers, etc.) and do not fly very often and, therefore, may not be as proficient as pilots who make flying their profession.

Approach

The first step in our cognitive task analysis was an informal interview survey of pilots to determine a reasonable definition of expert vs novice performance in this domain. We discovered early on that the word 'novice' is not very descriptive of most people in the aviation domain. Instead, 'competent' is a better term to describe the pilots to which we are focusing our attention because they have passed competency checks and have been certified by the FAA. However, I believe that pilots go through the five steps offered by Dreyfus and Dreyfus (1986) given in Chapter 1: novice, advanced beginner, competent, proficient and expert each time they enter a training program to acquire a new skill or rating.

Distinguishing qualities of the expert general aviation pilot. A part of the first step is to define the qualities of the expert in the domain. Our research team settled on the following distinguishing qualities of the expert general aviation pilot presented earlier in Chapter 1:

1 High level of skill and works constantly to improve it,
2 Highly motivated to learn all there is to know about this flight domain,
3 Superior ability to focus (or compartmentalize) his or her attention on the flying task at hand and the mental discipline to change his or her focus of attention when new information suggests that a change is necessary,
4 Keen observer of the flight environment, including location of other aircraft, terrain, navigation features, ATC clearances, weather phenomena,
5 Carefully establishes a baseline for normal instrument indications, aircraft sounds, vibrations and g-forces with respect to control action so that his or her threshold for slight variations is very small,
6 Continuously skeptical about 'normal' aircraft functioning and is constantly making contingency plans for those circumstances when things might go wrong,
7 Superior mental skill and capacity for problem diagnosis, risk assessment and problem resolution,
8 Excellent communication skills and can readily adapt them to the audience and situation and
9 Knows his or her limitations, is motivated to avoid situations that might push his or her skill to those limits, and has the willpower to overcome the pressures of people around him or her to push the limits of his or her skill.

Conversely, a competent pilot in this domain is one who does not possess these superior qualities found in expert. The competent pilot may have as much flight time as the expert, has the stick and rudder skills sufficient to pass all necessary flight tests and has the knowledge necessary to pass the written tests. He or she has demonstrated competency by taking the biennial flight review but that is as far as he or she takes aviation. Because motivation in these pilots is focused on matters outside of the cockpit, the extra skills, knowledge, mental models found in the expert have not been developed in the competent pilot.

In addition to these personal motivational factors, some other reasons why competent pilots often do not become experts include the organization (bureaucracy) for which they work, automatic systems in the aircraft and creativity blocking standard operating procedures for the aircraft they fly. The organization can block a people from becoming experts by turning down their every request for changes that they have found from their experience are needed for safe, efficient operations. Automatic systems can block pilots from becoming experts by forcing them to follow symbols and control actions that are designed into the system without thought for many novel situations offered in aviation. Standard operating procedures and regulations can block people from becoming expert by forcing them to comply with procedures rather than do what they know to be safer and more effective (Fahlgren and Hagdahl, 1990).

An examination of the qualities that differentiate the expert from the competent general aviation pilot reveals that much of the task of the expert involves cognitive activity. 'Stick and rudder' skills play but a small role in differentiating the expert from the competent. If one only considers the tasks necessary to be a competent general aviation pilot, the tasks might be more heavily weighted on the perceptual-motor skills. However, if one wishes to be an expert in this domain, one must excel in cognitive skills or judgment.

Distinguishing domain-specific tasks. The next step involves the design of tasks or scenarios that distinguish the expert from the competent in this general aviation domain. In addition, these tasks must also be such that they can be done in a laboratory simulation. Our research team developed the following six tasks for simulator testing:

TASK 1: Nosewheel shimmy on takeoff
TASK 2: Airframe icing
TASK 3: Subtle cabin pressure loss at altitude
TASK 4: Unexpected hold (including moderate to severe turbulence)
TASK 5: Windshear recovery on approach
TASK 6: Abnormal gear extension

A scenario was developed including all of these tasks for a flight from Pittsburgh to Charlotte. The scenario was designed so that it could be used in various forms of simulation.

Cognitive task analysis. The next step was to conduct an experiment with pilots in structured interviews using the scenario. Audio recordings were made during simulated flight as pilots described their thoughts and decisions as the six tasks were encountered. Verbal data were analyzed through coding into the categories that identify the expert from the competent in this domain. From these data emerged our model of the expert general aviation pilot which we describe below.

Intervention strategies

Because our ultimate purpose in this research is to determine how to change competent pilots into expert pilots, our model focuses on that aspect of pilot expertise. The record shows that it is in the area of cognitive skills that we fail most often. Deliberate teaching of these judgment skills in general aviation (where the foundation for a flying career is established) is almost non-existent. A strong emphasis is being placed on the development of some of these skills in the airline environment but, at that stage, it may be too late for some to change habit and thinking patterns learned in general aviation.

A model for creating the expert general aviation pilot

Figure 14.1 presents a proposed new model for creating an expert general aviation decision maker from a competent pilot. There are four steps in this process: aviation experiences, risk management, dynamic problem solving and attentional control. This model is the result of a series of studies conducted over the past year (Jensen, Kochan and Chubb, 1994).

Aviation experiences

Proponents of the expertise approach to studying human behavior have indicated that expertise is gained through years of experience. They often say that 10 years of experience dedicated to one field makes an expert. Some have used hours of experience in flight as an indication of expertise. We believe that it takes more than hours of flying experience to make an expert pilot. In this step we believe that there are five aspects of experience that are necessary: number, variety, meaningfulness, relevance and recency. Although many hours of flight are good, if they are all done in Ohio they do not make an expert pilot in mountains -- a variety of experiences is need. Experiences that have no meaning (e.g., boring lectures) do not change behavior -- the experiences need to occur in ways that are meaningful to the learner (e.g., experiences with sound, visual effects, motion -- simulation). The experiences must be relevant to the kind of flying that is anticipated of the expert (e.g., in teams if for a multi-person crew aircraft). Finally, the experiences must be recent. Although some experiences remain in one's mind for life, most need to be reinforced periodically to be available for expert decision making.

Risk management

The second step in creating the expert pilot decision maker is to develop risk management techniques. This step requires the establishment of a proper set of values consistent with societal norms (i.e, the pilot must know the safety expectations of his passengers and company). It also requires that the pilot studies carefully all of the possible hazards and the probability that these hazards could affect his flight practice. Pilots must establish their own personal minimums applicable to all flight situations as their personal risk management tool. Finally, pilots must be aware of and make attempts to influence the risk management procedures used in the companies for which they work.

Dynamic problem solving

The third step is to develop a technique for solving dynamic problems known as satisficing. This is an iterative process of deductive and inductive reasoning that asks questions uses what principles are known and is open to new data and approaches. This technique for solving ill-defined problems, which has been identified in experts, consists of beginning with a clear understanding of the situation and making decisions that have a good chance of leading to the optimum solution, always keeping the safest options open. Feedback from the action is used to gain further information leading to additional decisions. Finally, the optimum choice is made. *The key is to make decisions that avoid closing safe options.*

Attentional control

The final step is to develop an ability to control your attention so that you can focus on the task at hand leaving all other matters out of your mind. In conjunction, the expert pilot must also be able to perceive the smallest indication that something else in the cockpit deserves his attention and switch to that matter quickly and deliberately. It means being able to put matters outside of the cockpit out of your mind. Finally, it means not allowing pressures to make decisions based on non-aviation concerns to influence you away from the safe choices for which you are committed.

We believe that this model will be useful in identifying the expert pilot and in showing the way toward the develpment of the expert pilot. It was developed with an understanding of all of the material presented in this book as well as the series of interviews and experiments with expert pilots briefly described. I would invite you to consider the use of this model as you develop your own expertise in aviation.

References

Chase, W.G. and Simon, H.A. (1973). 'Perception in chess'. *Cognitive Psychology,* **4**, 55-81.

Diehl, A. (1992). Does cockpit management training reduce aircrew error? *ISASI Forum,* **24**(4), January.

Dreyfus, H.L. and Dreyfus, S.E. (1986). *Mind over machine: The power of human intuition and expertise in the era of the computer.* New York: The Free Press.

Ericsson, K.A. and Smith, J. (eds.) (1991). *Toward a General Theory of Expertise: Prospects and Limits.* Cambridge: Cambridge University Press.

Fahlgren, G. and Hagdahl, R. (1990). Complacency. *Proceedings of the 43rd Annual International Air Safety Seminar,* Rome, Italy: Flight Safety Foundation.

Jensen, R.S., Kochan, J. and Chubb, G.P. (1994). *A New Approach to Aeronautical Decision Making: The Expertise Method.* Columbus, OH: OSU Department of Aviation.

NTSB (1994). *Stall and Loss of Control on Final Approach, Atlantic Coast Airlines, Inc./United Express Flight 6291.* Washington, DC: NTSB Accident Report No. PB94-910409. NTSB/AAR-94/07.

Ryder, J.M. and Redding, R.E. (1993). 'Integrating cognitive task analysis into instructional systems development'. *Educational Technology Research and Development, 41*(2), 75-96.

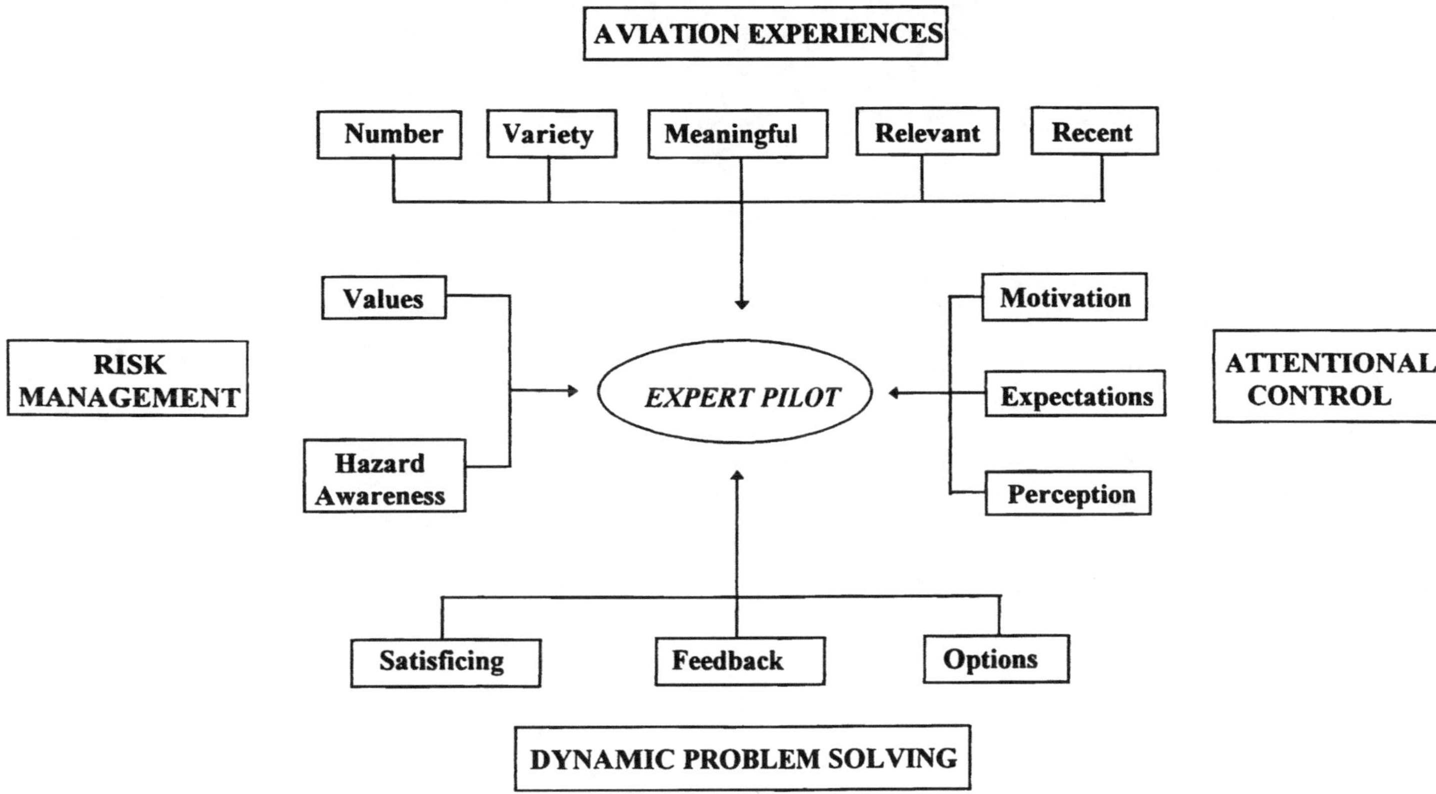

Figure 14.1 Model for creating the expert pilot decision maker

Index

www.ingramcontent.com/pod-product-compliance
Lightning Source LLC
LaVergne TN
LVHW011109300926
849160LV00008B/324